P9-BHX-874

About Island Press

Island Press is the only nonprofit organization in the United States whose principal purpose is the publication of books on environmental issues and natural resource management. We provide solutions-oriented information to professionals, public officials, business and community leaders, and concerned citizens who are shaping responses to environmental problems.

In 1998, Island Press celebrates its fourteenth anniversary as the leading provider of timely and practical books that take a multidisciplinary approach to critical environmental concerns. Our growing list of titles reflects our commitment to bringing the best of an expanding body of literature to the environmental community throughout North America and the world.

Support for Island Press is provided by The Jenifer Altman Foundation, The Bullitt Foundation, The Mary Flagler Cary Charitable Trust, The Nathan Cummings Foundation, The Geraldine R. Dodge Foundation, The Charles Engelhard Foundation, The Ford Foundation, The Vira I. Heinz Endowment, The W. Alton Jones Foundation, The John D. and Catherine T. MacArthur Foundation, The Andrew W. Mellon Foundation, The Charles Stewart Mott Foundation, The Curtis and Edith Munson Foundation, The National Fish and Wildlife Foundation, The National Science Foundation, The New-Land Foundation, The David and Lucile Packard Foundation, The Surdna Foundation, The Winslow Foundation, The Pew Charitable Trusts, and individual donors.

About The Nature Conservancy

The mission of The Nature Conservancy is to preserve plants, animals, and natural communities that represent the diversity of life on Earth by protecting the lands and waters they need to survive.

Since 1951, the Conservancy, a nonprofit membership organization headquartered in Arlington, Virginia, has worked with local conservationists both in the United States and internationally to identify and protect critical habitat. To date, the Conservancy and its members, who now number more than 820,000, have been responsible for the protection of more than 8 million acres in the United States and Canada, and have helped like-minded organizations in Latin America, the Caribbean, Asia, and the Pacific safeguard more than 53 million acres. The Conservancy is well known for its expertise in land acquisition for conservation purposes; in the United States, the organization owns more than 1,300 preserves—the largest private system of nature sanctuaries in the world.

The Conservancy recognizes that local organizations have the best insight into local conservation issues. It therefore works to support partners, ensuring that they have the resources and skills to make decisions that will guarantee a rich natural legacy.

For more information about The Nature Conservancy and its work in Latin America and the Caribbean, visit its website at www.tnc.org.

Parks in Peril

Parks in Peril

People, Politics, and Protected Areas

Edited by
Katrina Brandon
Kent H. Redford
Steven E. Sanderson

The Nature Conservancy

Island Press
Washington, D.C. • Covelo, California

Library of Congress Cataloging-in-Publication Data
Parks in peril : people, politics, and protected areas / edited by
Katrina Brandon, Kent H. Redford, Steven E. Sanderson.
p. cm.
Includes bibliographical references (p.) and index.
ISBN 1-55963-607-6 (cloth). — ISBN 1-55963-608-4 (pbk.)
1. National parks and reserves—Latin America—Management—Case studies. 2. Biological diversity conservation—Latin America—Case studies. I. Brandon, Katrina, 1957– . II. Redford, Kent Hubbard. III. Sanderson, Steven E.
SB484.L38P37 1998
333.78'315'098—DC21 98-23680
CIP

Printed on recycled, acid-free paper
Manufactured in the United States of America
10 9 8 7 6 5 4 3 2 1

We dedicate this volume to all of the protected-area professionals in the Parks in Peril program, who are laboring under trying circumstances, and frequently with significant sacrifice, to save the beauties of nature.

Contents

Foreword

In 1989, The Nature Conservancy's Latin America and Caribbean Division evaluated the experience it had gained during its first five years of protected-area conservation. We were humbled by the enormity of the task but emboldened by some signs of promising approaches. From this analysis, we developed two fundamental strategic approaches to drive our work. First, we would focus on what the Conservancy knows best from our forty years of experience in the United States: site-based conservation. Second, we would work with partners and strengthen their capacity to overcome the challenges faced in protecting the biodiversity in parks. Parks in Peril (PiP), launched in 1990, became the flagship program to implement this strategy.

With the help of many talented and dedicated partners and supporters, Parks in Peril has become the largest single program supporting parks in the Western Hemisphere. Parks in Peril encompasses sixty parks in eighteen countries and covers an area of over 30 million hectares. Over fifty nongovernment and government organizations work together to implement the objectives of the Parks in Peril program. Three key goals move these protected areas toward minimum critical management:

1. Establish on-site protection.

2. Integrate these protected areas into the economic and cultural life of local communities.

3. Create long-term funding mechanisms to sustain the local management of these areas.

In 1995, in an evaluation of our results and progress, we introduced an important fourth goal:

4. Use the experiences of PiP site-based activities to influence conservation in other sites in the region's most imperiled ecosystems.

The fourth goal challenges us to take a critical look across the program's site-based activities, to harvest experiences and lessons, and to disseminate this knowledge to others. In the Parks in Peril program we strive to learn from our work, which occurs at a small number of sites, and to influence the work in many arenas of biodiversity conservation. This goal of disseminating our experiences also forces us to undertake the difficult and unpopular activity of critical analysis and documentation. However, we recognize that only through self-scrutiny can we develop positive programmatic change. Only through sharing our knowledge can we maximize the impact of our efforts.

You now have in your hands one of the products of the pursuit of this learning and leverage objective. The nine case studies and the accompanying analysis have resulted from a collaboration of Conservancy staff, protected-areas professionals working at PiP sites, and outside experts. Their good work provides valuable information and much to ponder regarding the dynamic environment in which protected-area work exists. We are sobered by the difficulties in achieving goals so easily stated. Yet the parks and the nongovernment organizations (NGOs) that have emerged to support them have made significant strides in converting these crucial areas to resilient parks. The parks have been in existence for fewer than twenty-five years, and the NGOs have emerged only in the last twelve years.

Evidence for a big premise of this book emerges clearly—protected areas cannot carry the entire burden for biodiversity conservation. The destructive forces at work range far beyond the boundaries of these parks. Yet protected areas provide the laboratories for understanding these forces and documenting their impact. They provide the foundation for influencing change on a larger scale. Indeed, the analysis and conclusions in this book have contributed to a consensus in the Conservancy that if we are to effectively conserve biodiversity, we must use our experiences and focus on site-based conservation to plan and work at a larger geographic scale. In 1996 we adopted a set of strategies and guidelines committing us to an ecoregion-based approach to our conservation work that will challenge us to think and act across larger landscapes and work in new ways to effect biodiversity conservation.

On behalf of the Conservancy and our partner organizations, I want to acknowledge and express our appreciation for the extraordinary contribution of USAID in making this book possible. With its continued guidance and support we hope to distribute additional similar documents devoted to learning and sharing information. We thank all of our partners who have contributed so much to our understanding of what it takes to achieve enduring conservation. And to all the people out there who have never heard of PiP but are fighting the good fight along with us, it is our greatest hope that you will find this book of interest and use.

Bradford C. Northrup
Director, Parks in Peril Program
Latin America and Caribbean Division
The Nature Conservancy

Acknowledgments

The editors would like to thank all of the many people in The Nature Conservancy who have worked to make this book possible. We would like to single out Brian Houseal, Joe Keenan, Susan Anderson, Kathy Moser, Domingo Marte, John Tschirky, Randy Curiñtis, Gina Green, Hugo Arnal, Greg Miller, Dan Quinn, Mónica Ostria, Christine Campbell, and Brad Northrup. The book would never have been possible without the hard work and good spirits of Karin Wall, Eva Vilarrubí, and Jane Whitten.

The executive directors of the partner organizations agreed to allow us to use their hard and unstinting work at the nine case study sites as a basis for learning by others. To them we offer our many thanks: Susana Rojas, Oscar Núñez, Vera Varela, René Ledesma, Joy Grant, Teodoro Bustamante, Fausto López, Hermes Justiniano, and Gustavo Suárez de Freitas.

Partial support for the preparation of the manuscript for this book was provided by the Office LAC/RSD/EHR, Bureau for Latin America and the Caribbean, U.S. Agency for International Development, under terms of Grant No. LAG-0782-A-00-5026-00. The opinions expressed herein are those of the authors and do not necessarily reflect the views of the U.S. Agency for International Development. We wish in particular to thank Jeff Brokaw, Eric Fajer, and Cynthia Gill of USAID.

Many thanks also to spouses, children, and childcare providers for patience during extended travel.

Introduction

Katrina Brandon, Kent H. Redford, and
Steven E. Sanderson

While there is a general, global consensus that biodiversity needs to be conserved, there is little agreement on exactly what biodiversity means or how to preserve it (Sanderson and Redford 1997). In the past two decades there has been a shift among conservationists on how best to protect rapidly disappearing wildlands and wildlife. Traditionally, protected areas were viewed as the cornerstone of biodiversity conservation. From the mid 1980s to mid 1990s, it became more fashionable to downplay the importance of protected areas and focus instead on approaches outside of these areas. Optimism about conservation outside protected areas increased with the popularity of sustainable development. In the conservation arena, this was translated into the concept of sustainable use (WCED 1987; Redclift 1989). Sustainable-use proponents assume that conservation can best be accomplished through people's use of resources (Robinson 1993). From this perspective, parks are threatened because their resources are not being used and therefore valued. This reflects the conviction that conservation can best be accomplished if people wisely use and exploit resources.

This view was echoed at the 1992 World Parks Congress in Caracas, where there was a call to support "win-win" combinations—especially of systems that encouraged human use of wildlands and species as a way of preserving them. Advocates of this position have even gone so far as to suggest that there shouldn't be protected areas and that all areas should be open to some kind of use, or that we give up on strictly protected areas altogether (Pimbert 1993; Wood 1995; Ghimire and Pimbert 1997).

Since then, a number of catchy phrases have become the guiding principles for conservation: grassroots community-based conservation, parks for people, sustainable development and use, and conservation for development (Stevens 1997; Western and Wright 1994). These slogans imply that such win-win combinations are possible and that conflicts over resources can be resolved with relative ease. Yet there has been little rigorous analysis of what happens, in ecological terms, when such phrases are used to guide actions.

This book challenges these catchy phrases and guiding principles for conservation from the three disciplinary perspectives of the editors: biological, social, and political. It uses analysis grounded in case studies of parks in Latin America and the Caribbean to identify the inherent conflicts in translating these slogans into actions, and the impacts of those actions on biodiversity conservation. The major premise of this book is that *protected areas are extremely important for the protection of biodiversity, yet requiring them to carry the entire burden for biodiversity conservation is a recipe for ecological and social failure.*

This major premise has four cutting edges: political, biological, social, and conceptual.

- *Biodiversity conservation is inherently political.* Many of the most significant challenges in protecting biodiversity are not biological challenges but challenges that will be played out in social and political arenas. A struggle is underway to define ownership of the meaning of biodiversity—who wins will largely determine how and whether biodiversity is conserved. This struggle, which is inherently political, encompasses the quest for strong, resilient, dynamic parks.

- *There are limitations on sustainable use as a primary tool for biodiversity conservation.* Serious questions as to whether sustainable use is axiomatically compatible with biodiversity conservation have been raised (Redford and Richter 1998; Robinson 1993; Sanderson and Redford 1997). The trend to promote sustainable use of resources as a means to protect these resources, while politically expedient and intellectually appealing, is not well grounded in biological and ecological knowledge. Not all things can be preserved through use. Not all places should be open to use. Without an understanding of broader ecosystem dynamics at specific sites, strategies promoting sustainable use will lead to substantial losses of biodiversity.

- *Effective park protection requires understanding the social context at varying scales of analysis.* Parks, and the management they require, are influenced by the historical forces and social actors that shape the local context at each site. These local forces are, in turn, affected by larger national and international social, economic, and political forces. Understanding these contexts is essential to developing lasting management approaches.

- *Actions to protect parks need a level of conceptual rigor that moves beyond slogans and stereotypes.* Catchy phrases, slogans, assumptions, and stereotypes have shaped conservation policy, to the detriment of both people and wildlife (Brandon 1997). Implementing these slogans and basing actions on stereotypes have not led to progress in conserving biodiversity either outside or inside of parks. Furthermore, such behaviors have constrained creative thinking on park protection and ways to solve the biodiversity crisis outside of parks. Finding and implementing solutions requires moving beyond assumptions and stereotypes to a new era of conceptual clarity, analysis, and action.

In order to better analyze the propositions above, this book uses the experience of The Nature Conservancy (TNC), which launched the Parks in Peril Program (PiP) in 1990. The Conservancy is a large, conservation, nongovernmental organization in the United States with programs in every U.S. state, in several countries in the Asian Pacific region, and in eighteen countries in Latin America and the Caribbean (LAC). PiP is the largest single program to support parks in Latin America, and perhaps in the entire tropical world. The PiP program was initiated to respond to the rapid creation of parks with no effective management and the high degree of threat faced by most. It includes sixty parks in eighteen countries, covering an area of over 30 million hectares (75 million acres). The program is administered by the Conservancy through its partner organizations in each of the eighteen countries. The United States Agency for International Development (USAID) has been a key partner in these efforts, as part of the agency's mission to promote biodiversity conservation in developing countries.

The PiP program is based on building a collaborative partnership among national, international, public, and private organizations. The Conservancy works directly with its partner organizations, sharing experience and technical resources, and working to help partners build their capacity as independent, self-sustaining conservation organizations. Through the PiP program, the Conservancy has provided direct grants to partners to assist government organizations in protected-area management. At many sites, the partners directly manage conservation activities for governments or work closely to support and assist government efforts.

While there has been an explosion of literature on how to undertake conservation planning and ways to stem biodiversity loss and tropical deforestation, surprisingly little attention has been paid to what the real challenges are in saving parks. To date, there is little knowledge about the interaction of ecosystems, local peoples, and large-scale management schemes that is applicable beyond the boundaries of a single site (Sanderson and Redford 1997, p. 129).

Ensuring the long-term protection of sites means that the conservation community must be able to adequately analyze the impact of historical, social, and political trends across sites and use such analysis to develop long-lasting, effective solutions. What has been written about parks and protected areas has largely been oriented "inward" toward park management and conservation biology. Faced with the task of understanding and responding to the "outward" aspects of park protection, The Nature Conservancy, jointly with its partner organizations in LAC, decided to undertake this broader analysis of the ecological, social, and political issues faced by the parks in the PiP program.

Nine case studies of Neotropical parks, written jointly with representatives from the organizations that must respond to these issues daily, here serve as the basis for analyzing the social and policy context for park management. This is the only book that presents an analysis by conservationists in Latin America and the Caribbean of the parks they are trying to protect and the challenges they face daily in carrying out this mission. This book is the first to analyze the impact of popular slogans and stereotypes on actual conservation policy, and to interpret likely impacts on biodiversity conservation from three disciplinary perspectives. While the focus is on Neotropical parks, we believe that the key themes, convictions upon which this book is based, and findings have implications for worldwide biodiversity conservation.

Protected Areas in the Larger Landscape

Worldwide, over 13,000 sites protect about 8.9 percent of the earth's surface, if marine sites are included (IUCN 1998). These protected areas—also referred to as parks throughout this book—represent the most important method of conserving biodiversity in situ. There has been a dramatic expansion of parks during recent decades, with more parks and reserves having been established since 1970 than in all previous periods (WCMC 1992). In the face of unprecedented land conversion, parks were viewed as the cornerstone in the national strategy to protect species, habitats, and ecosystems.

It is vital to realize that biodiversity conservation is often not an integral management objective even for lands that are officially protected. The World Conservation Union (IUCN) lists six categories of protected areas; one explicitly describes management for biodiversity conservation as an objective, and three others implicitly do so. Within these four categories of protection, an array of human activities is allowed in all but Category 1, the Strict Nature Reserve/Wilderness Area. Human uses in the other categories include "sustainable production, subsistence resource use, or traditional use and visitor use for inspirational, educational, cultural and recreational purposes" (WCMC 1996).

In many countries, national designations such as forest reserves make up a significant share of the protected land. For example, 1992 data for Costa Rica indicate that while 27 percent of the area of the country was officially designated as protected, 15 percent of those lands were open to a variety of human uses. Within Latin America and the Caribbean, protected areas in the least restrictive uses occupy twice the land area of more restricted categories. Whether any of these protected areas contribute to biodiversity conservation, and, to what component of biodiversity (genetic, species, ecosystem), is largely dependent on their ownership and management (Redford and Richter 1998).

Further complicating notions of protection is the fact that much of what is called protected isn't really protected at all. Areas that are declared protected on paper but have no active management are known as paper parks and reserves. The intent of USAID in supporting the work of the Conservancy and its partner organizations was to transform these paper parks into effective conservation management units. Yet in Latin America alone, scores of sites lack the financial and technical support to insure that they will effectively conserve biodiversity. As the case studies themselves demonstrate, such a transformation can be difficult, even when there is adequate financial and technical support.

This brief overview indicates that only a small amount of land is truly managed and protected *primarily* for biodiversity conservation. From an ecological perspective, expecting those few areas to effectively protect biodiversity is dangerous.

Biodiversity Conservation As a Political Force

Definitions of *biodiversity* have been dynamic and have shifted dramatically in the past two decades. Understanding the history of the term helps define the international politics and players involved in shaping park policies today. The Conservancy used the term *natural diversity,* which focused on the richness of species, in the 1970s. Biological diversity was defined using concepts of genetic and species diversity by 1980 in the U.S. Council on Environmental Quality's annual report. The passage of the International Environmental Protection Act in 1983 by the U.S. Congress directed federal agencies to help developing countries conserve biological diversity, without providing any firm definitions. A series of publications in the mid 1980s began filling the definitional void, and biological diversity was defined to include "genetic diversity and ecological diversity, thus encompassing the full range of species and ecosystem dynamics (interactions between species, among species, and between species and their habitats)" (Burley 1984, 2).

Biodiversity as a term was introduced and popularized by E.O. Wilson in 1988, though *biological diversity,* which has its roots in the 1950s work of Hutchinson and MacArthur, was used by Thomas Lovejoy in the *Global 2000* report. But where earlier definitions had focused on genetic, species, and community or ecosystem elements of biodiversity, Wilson and others began to use the term to mean species richness (Wilson 1988; Ehrlich and Wilson 1991). Within the community of biologists and ecologists, a debate on the meaning of the term ensued.

As the term increased in popularity and political importance, other groups expediently began tacking their agendas onto it. *Biodiversity* united the following groups and their focus: plant breeders and crop germplasm; ethnobotanists and the diversity of plants and their uses; pharmaceutical companies concerned with potential new drugs; and anthropologists noting the links between cultural and biological diversity. Indigenous and traditional peoples saw the potential to gain power and prestige at international fora by their claims of defending biodiversity. The death of rubber tapper Chico Mendes focused attention on the burning and conversion of Amazonian rainforests, drawing in human rights groups, religious organizations, and rock and movie stars, among others. Government officials saw a chance to gain international recognition by supporting wildlife and habitat protection—with the twist of emphasizing the importance of national sovereignty to stay in the spotlight back home.

All of these groups, and others, jockeyed for position to make sure that their view of biodiversity was included in the Biodiversity Convention. Therefore the 1991 treaty draft included human cultural diversity in its definition of biodiversity: "Human cultural diversity is manifested by diversity in languages, religious belief, land management practices, art, music, social structure, crop selection, diet, and any number of other attributes of human society" (WRI et al. 1992, 2).

Such broad definitions are politically expedient in getting agreement from a broad coalition of players at the international level. However, implicit in this definition is a need to include management of human cultural diversity alongside biological and ecological aims. This introduces further conflict into meeting conservation objectives. One reason for this is that *biodiversity* is scale dependent: its meaning varies according to the focus on structure, function, or processes. As a result, the management mechanisms brought to bear on conservation are fundamental to the kind of biodiversity conserved (Sanderson and Redford 1997). For example, the governance or management of biodiversity conservation is based at national levels. At the 1992 U.N. Conference on Environment and Development (UNCED), opening paragraphs noted the sovereign rights of nations over "their" biodiversity. Yet the international pressures for *use*, or sustainable use, which have become intertwined with biodiversity conservation have created a conflicting set of values and of policies. These conflicts are evident throughout the case studies, and they highlight the need to better understand the political underpinnings of land and resource ownership and use, including parks, at different scales. They also highlight the need for agreement on what biodiversity is and means.

Limitations on Sustainable Use

While many in the conservation community view sustainable use as a necessary component of biodiversity conservation, it has become increasingly apparent that not everyone or everything wins in the win-win combinations. In the first place, what resources are used, and who gets to use them, are decisions usually based on political expediency rather than biological knowledge. In protected-area systems, who has access to resources, how they can harvest the resources, and what they can do with the resources once they are harvested are decisions based on the social and political context. On the surface, it sounds socially appropriate and equitably justifiable, for example, to say that indigenous peoples can harvest resources using traditional techniques. Yet the external imposition on traditional peoples of restrictions that aim to keep them "traditional" represents a naive view that this is the best way to conserve biodiversity (Redford 1990).

While decisions to allow certain uses are made precisely because they appear to be more sensitive to the local social context and are politically palatable, they ignore a range of biological and ecological concerns that, over the long term, are likely to deplete species and lead to local livelihood crises. There has long been a casual assumption that it is possible to both use and conserve biodiversity. However, biodiversity, per se, doesn't exist to be used. In order to assess whether or not a given use is sustainable or not it is vital to divide "biodiversity" into its components, most usually defined at three levels: genetic, populations and species, and communities and ecosystems. Each of these components in turn has attributes of composition, structure, and function (Redford and Richter 1998).

Any significant use of biodiversity has clear impacts on the component targeted for use. From trophy hunting and highly selective logging, which can alter the genetic composition of the species targeted, to row agriculture and dams, which affect biodiversity at its three levels (genetic, population-species, and community-ecosystem), human use has always impacted biodiversity (Redford and Richter 1998). The win-win solutions promulgated by those desperate to advocate cost-free development do not exist.

Social Forces Shaping Park Protection

The recent creation of many parks in the Neotropics is a result of at least three factors: the heightened awareness at international levels of the need to conserve biodiversity; the rapid conversion and loss of significant areas, particularly tropical forests, where levels of endemism are high; and the importance of parks as a way to halt this process. The areas declared as parks or protected areas were often declared as such for three reasons: they were the remaining areas that had not been transformed, they were at high risk of being transformed to other uses, or dedicated individual actors and unique local circumstances led to their creation. Yet the process of creating a park rarely prevents such transformations.

It is evident that just as biodiversity is scale dependent, so too are the social forces that shape how the biodiversity is used, managed, or destroyed. At the most local levels, one finds a variety of issues that affect conservation and resource use: farming and grazing techniques and changes in those techniques, infant mortality and local rates of population increase, local settlement patterns, local tenure security, access to markets, technical changes, and changes in consumption or standard of living are all relevant. These local forces jointly affect the local rates of conversion or use of wildlands and wildlife. Often these local forces can lead to significant changes in resource use (Robinson and Redford 1991; Robinson and Bennett, in press).

Regional-level changes are usually generated by forces external to that region: the government decides to put in roads; this leads to access to markets, causing changes in agricultural production patterns; migrants arrive to claim lands and colonize; the value of beachfront land rapidly changes when tourism and vacation homes become a possibility. These regional forces will have their own dynamic yet will influence the decisions and resource management activities at local levels as well, and may cause profound changes in local-level systems. Finally, there are national- and international-level forces that reach down into the regional and local arenas. These include development policies and structural adjustment; unpredictable and changing market forces, such as demand for particular resources (e.g., the increased global demand for shrimp or North American demands for fresh fruits and vegetables in winter); and even war.

Looking at the conversion of once forested lands in the past three decades leads to the overwhelming conclusion that some combination of forces has rapidly altered once stable patterns of resource use. The rapid globalization of the world economy and information has had a profound effect, leading to rapid changes at national levels, which in turn affect regional patterns, which then affect local levels. Occasionally, actions happen in the opposite direction, escalating outward from local levels. It should not be surprising then that parks and their surrounding areas are affected by changes and forces at national, regional, and local levels. Once a park is created, these forces crash into the park's invisible boundary, frequently leading to conflict at a variety of levels. As noted by Jeff McNeely,

"Protected areas are on the front lines in the battle to conserve biodiversity, and since these areas are often selected and managed specifically to protect species and ecosystems of outstanding value from human degradation, they are also sites of highly contentious debates between local and national interests and conflicts often result . . ." (McNeely 1995, vii). From within a park these external forces are appropriately looked upon as threats, for in most cases they do threaten the long-term integrity of park systems.

While rapid social change has led to devastation in many areas and within many social groups, it also offers opportunities if resource stewards are able to understand these trends and use them to their benefit. Understanding the social context is an essential base for effective action. But as described below, old assumptions and stereotypes that are prevalent in the literature on people and parks must be discarded to allow for the development of creative and lasting biodiversity conservation.

Toward Creative Conservation

One of the themes that is addressed throughout this book is the broad set of questionable assumptions and stereotypes that have entered into conservation (Brandon 1997). One can point in several directions in trying to figure out how untested assumptions are so readily adopted as gospel. At the root of most assumptions and stereotypes is the tangled and multilayered definition of biodiversity that means all things to all groups. The different stakeholders in the politics of biodiversity all have their own perspectives, biases, and agendas. The definition of biodiversity has implicitly come to include the belief that we can have it all: biodiversity conservation and development. This belief represents a grave risk for protected areas and can lead to management strategies that try so hard to be and do everything to please all groups that they do nothing effectively.

In practical terms, conservationists would like to think they are knowledgeable about how to plan and execute conservation activities. But as increasing numbers of evaluations of these activities suggest, many of today's field-based initiatives are not living up to their proclaimed potential. (For more than sixty case studies, see West and Brechin 1991; Wells and Brandon 1992; Western and Wright 1994.) Many of the shortcomings in today's conservation projects are due to a belief among conservationists that what they are doing is conservation when, in fact, they are really doing large-scale social interventions in complicated macropolitical settings.

The case studies in this book suggest that there needs to be substantial rethinking about parks and about what can realistically be expected of efforts to manage and protect them. This will require a tremendous amount of focus and conceptual clarity. As we enter a new phase pushing for the consolidation of parks, we must learn from what is happening in the field and integrate such knowledge with research in social and political sciences. We must enter a new era if long-term solutions to the biodiversity crisis are to be found. Finding and implementing such solutions require moving beyond assumptions and stereotypes.

The Case Study Analysis for This Book

In 1995, The Nature Conservancy, jointly with its partner organizations in Latin America and the Caribbean, decided to undertake a broad analysis of the ecological, social, and political issues faced by the parks in the PiP portfolio. Such an analysis was viewed as essential to design better programmatic responses to the threats faced by the parks. The PiP program had begun collecting preliminary information on many of the park sites themselves: work plans and evaluations existed for all of the sites. But there was no formalized process to broadly look across the portfolio at the context in which the PiP program was operating.

Each of the regional programs within the Conservancy's LAC program nominated sites that they believed were representative of the diversity of issues challenging effective park management and protection within their region. The sites in the analysis were not selected through a rigorous sampling process, but we believe that they are representative of both the day-to-day and the long-term issues that arise in park management.

For each of the sites, the Conservancy identified case study collaborators who were external to the Conservancy and the PiP program but had expertise in protected-area issues and were fluent in Spanish. As a result, an external analytical perspective was brought into the process of information collection, site visits, and writing up the case studies. At most sites, case study collaborators traveled to the field with one or more individuals from the partner organization. The case study collaborators and the partner representative, coauthors on each of the case studies, are generally referred to as teams throughout the book.

Site visits took place between June and August of 1995 and ranged from six to ten days, although most of the partners spend extensive amounts of time at the parks each year. The site visits were intentionally short and were not intended to produce original research. Instead, they represented an effort by the Conservancy collaborator and the partner to pull together, synthesize, and analyze existing information about particular parks and the surrounding areas. Project plans, progress reports, and evaluations were reviewed. When possible, discussions were held with: (1) principal staff concerned with the park at the Conservancy's partner organization; (2) protected-area managers and their staff; (3) representatives of national agencies charged with protected-area administration; (4) staff of other NGOs participating in the project; and (5) key informants at the sites, or in the particular countries, with relevant knowledge and experience. Discussions took place in both formal meetings and informal settings. Visits to the parks and the communities around them were useful to verify written information, meet with key informants, clarify issues, and "ground-truth" impressions developed prior to travel.

All of the case studies have been reviewed by the three editors of this book, relevant Conservancy staff, and the partner organizations. All participants and Conservancy staff met in September 1995 to complete case study write up and review. These cases have not been revised since that time, although some citations have been updated. Instead, postscripts have been included for each of the cases to describe key information that has changed between the time the studies were written and March 1998. Also, glossaries with acronyms and relevant translations follow each of the case study chapters.

Key Themes

The use of a common set of themes at each site allowed for a comparative analysis of the relevance of those topics across all sites. Eight topics were defined that reflect the social and political context of threats to parks. These eight themes were reviewed at all of the sites. Three additional themes—traditional peoples and social change, transboundary issues, and resettlement—were reviewed when relevant at a given site. All of these themes are briefly described below and analyzed in greater detail in subsequent chapters.

These themes provide a useful way to gain insights into the challenges in trying to undertake conservation at nine protected areas. These nine sites are representative of some of the most biologically significant yet threatened areas in Latin America and the Caribbean. The case studies that follow (chapters 4–12) describe the context of implementing field-based conservation and park management from the perspective of those charged with those actions on a daily basis.

Park Establishment

More parks and reserves have been established since 1970 than previously existed. The increased establishment of such protected areas was spurred by recognition of the rapid destruction of many tropical areas and species and the view that parks represent the most important method of conserving biological diversity in situ. In Central America "officially gazetted protected areas increased from only 30 in 1970 to more than 230 by 1990" (Cornelius 1991). Many of the sites were established by decision makers in capital cities with little or no prior analysis of the ecological or social reality at the site. They lacked active management, and they remained little more than paper parks.

Teams were asked to analyze how parks were created and established and how and why the boundaries were drawn where they were drawn, as well as to provide their view on what, if any, changes are needed to insure the ecological viability of the park. Such changes may be necessary when significant ecosystems were excluded from a park, when a park is too small to accommodate the range needed by species the park is trying to protect, or when edge effects pose particular problems. Teams were also asked to clarify whether multiple agencies are responsible for biodiversity conservation activities and if jurisdictional responsibility for parks is clearly mandated.

Land and Resource Tenure

Tenure is the form of rights or title under which property is held and that determines how an individual or group may use, share, sell, lease and inherit, or otherwise control property and resources. Worldwide, there are many different forms of land and resource tenure. But in all cases, tenure provides the basic framework for who can use resources and how they can be used. Therefore, rules of tenure have a direct impact on how lands and resources are used and managed. Understanding what is traditional, what is legal, and what are actual claims on both land and resources is an often overlooked element of the social complexity surrounding parks.

We asked teams to identify the current land and resource tenure situation in and around the parks as best as possible. Because the set of issues surrounding land and resource tenure is so complex and could easily become a major area of research for each of the parks, we asked the teams to pay particular attention to tenurial issues that thwart the development of effective park protection and positive relationships with local communities. We also asked them to identify national policies and local practices that influence tenure and affect park security.

Resource Use

The debate over what kinds of use are appropriate in protected areas, and whether sustainable use is possible, has been described as "the most volatile and divisive conservation issue of the decade" (Satchell 1996). The basis of debate is as follows. Although many parks have been subject to human use for thousands of years, in some, biological integrity has remained sufficiently high—meaning that the ecological processes are still intact—for these areas to be of high importance for biodiversity conservation. Human use in these areas has traditionally been at low levels; however, the social forces that maintained such low use levels may be changing and use levels could rapidly increase, leading to diminished value of these areas for biodiversity conservation. At the same time, biodiversity outside of many protected areas is seriously threatened through changes in land and resource uses. Therefore, maintaining large protected areas with as little human use as possible is viewed as the best way to insure long-term effective biodiversity conservation (see Kramer, von Schaik, and Johnson 1997). Use advocates disagree with this position and believe that all areas should be open to some kind of human use; some even believe that strictly protected areas shouldn't exist (Pimbert 1993; Janzen 1994; Wood 1995).

The more moderate position recognizes that what is appropriate at each site will depend on a variety of social and ecological factors. This position is implicit in the six categories of parks and protected areas set by the World Conservation Union, which recognizes different uses, ranging from strict protection to multiple-use zones in parks.

Use within or adjacent to a protected area can therefore be seen as either a substantial benefit or a serious threat. Teams were asked to examine the types of resource use in and near parks. Particular attention was to be given to who was using the resources and to what end—whether use was for local consumption or commercial gain. Teams were also asked to identify whether uses were legal or illegal, the levels of use if known, and the existing or potential impacts of exploitation on the resources themselves and on the protected area.

Organizational Roles

Often an amazing array of organizations are involved in conservation activities at a particular site, and each of these organizations may have a different way of implementing their objectives. For example, the Conservancy's partner organizations vary from national-level research and policy organizations, to national-level advocacy groups, to local-level NGOs with expertise in site management. The roles that partner organizations assume are closely linked to the role and capacity of government agencies charged with conservation and the other kinds of NGOs and agencies working near each park. In the past decade there has been an explosion of new conservation organizations in Latin America and the Caribbean. In some places, this has led to specialization, with different groups focusing on different things, such as environmental education, rural development, ecotourism, and site protection. In other places, it has led to groups undertaking a variety of functions at a given geographic locale.

Teams were asked to identify the organizational roles undertaken by partner organizations and other NGOs active in the project. If other government or research agencies, or the private sector, had important roles in protected-area management, they were asked to describe such activities. Finally, they were asked to identify activities (such as litigation, advocacy, or other measures) that deal with special interests or problems.

Linkages between Park and Buffer Areas

In the past decade, a range of approaches have attempted to link protected-area management with the needs of nearby communities through local social and economic development. These approaches are collectively described as *integrated conservation development projects* (ICDPs Wells and Brandon 1992). ICDPs are based on the premise that protected-area management must reach beyond traditional conservation activities *inside* park and reserve boundaries to address the needs of local communities *outside*. One of the most important elements in this strategy is linking conservation objectives to development activities in a way that will act as a long-term incentive for conservation. A wide array of activities are promoted as part of this approach, including: improved natural resource management, use, production, and marketing; ecotourism; conservation education; social services (schools, health clinics); and income-generating projects.

Team members were asked to briefly review the variety and intensity of ICDP activities underway and determine whether those activities were bringing about change that reduced pressure on parks and their resources.

Conflict Management and Resolution

Conflict exists at most protected-area sites at varying levels and degrees. Conflict is found between communities and park authorities; among communities over resources and their use; between established communities and migrants; between different social and class groups; and within communities over different visions of resources, their use, and the future. Conflicts between men and women, between clans, and between neighbors may be particularly prevalent in resource-dependent communities near protected areas (Redford and Mansour 1996). On a wider scale, competing interests arise over land uses. In parts of the Amazon, for example, rubber tappers, indigenous groups, cattle ranchers, and small farmers often come into conflict. Superimposed on local-level conflicts are the conflicting interests of different groups and agencies. Numerous government agencies may have conflicting agendas, as do national and international NGOs, tourism lobbyists and operators, and private-sector or parastatal companies, particularly those involved with logging or mining. Each of these different actors has some vested interest in the protected area and its management. Conflict may take many forms, from friendly disagreements and debate to violence and bloodshed.

Teams were asked to identify what levels of conflict exist, why the conflict exists, and among which groups. They were asked to identify any strategies that have minimized or avoided conflict. In most cases, only those conflicts that were long standing or obvious and dramatic were identified, since a thorough analysis would have required substantial original research. Nonetheless, identifying even the obvious conflicts and what, if anything, has been done to mitigate them provides a glimpse into the daily challenges of park management.

Large-Scale Threats

Timber and mining concessions within or adjacent to protected areas, large-scale development activities, road construction, and weak government institutions are all examples of large-scale threats. These threats often have a much greater impact on parks than do locally based resource uses. Sometimes, these large-scale threats are clearly illegal—such as loggers coming into a protected area and clear-cutting a section. But more often large-scale threats are the result of government policies that unknowingly conflict with or undermine park management objectives. The themes of park establishment and national policy are closely linked to large-scale threats. For example, in many cases, logging and mining concessions have been granted prior to declaration of a park. When the park was established, there may have been no ruling or legislation for handling these concessions: are they voided, are lands granted elsewhere as compensation, or are they still legitimate? Clarifying overlapping claims may take years; resolution has sometimes unfortunately depended on the strength of the competing government agencies in front of appropriate decision makers or courts. In other cases, development priorities conflict with park legislation; roads or rights-of-way for electricity, water, or other services may be built through or near national parks. Tourism poses another large-scale threat at some areas. While some level of tourism may be important in providing financial support and justification for maintaining a protected area, too much tourism can cause serious disruption to wildlands and wildlife and can also attract development (Brandon 1996).

Teams identified the primary source of external threats to the protected area. To the extent possible given the time constraints on site visits, they were asked to review patterns of overlapping or unclear mandates, such as past conferral of mining or timber concessions that posed a threat to the park. Teams were also asked to look for mitigating factors to deal with these threats, especially creative solutions such as buyouts, conservation easements, or tax breaks in exchange for termination of agreements.

National Policy Framework

A wide variety of policies affect biodiversity conservation. Pricing policies for agricultural commodities, fuels, and wood; transportation and national integration policies such as road building; subsidies for cattle ranching and other land uses that may not be appropriate; and land-tenure policies that encourage colonists to settle frontier areas while allowing productive lands to remain underutilized are all examples of national policies that can undermine conservation (Ascher and Healy 1990).

The absence of political commitment for biodiversity conservation and park protection can have serious consequences. Large-scale threats are likely to be serious, and legislative, financial, and institutional support for park management is likely to be low. The national framework can include policies that affect conservation and park protection either directly or indirectly, positively or negatively. Legislative issues, national fiscal policies and structural adjustment programs, and civil service programs are among the variety of factors that can directly affect the issue central to park management. Less direct are national development plans that promote road building to create markets and link rural areas to the broader economy or to emphasize national borders. The case studies identify the overall policy environment, giving particular attention to activities or policies (e.g., forestry or infrastructure) or other actions by the state that affect conservation and the long-term capacity for parks to be effectively managed.

Indigenous Peoples and Social Change

Many of the lands that are critical for biodiversity conservation in Latin America and the Caribbean are inhabited by traditional peoples (see Redford and Mansour 1996). The correspondence between indigenous land and forest cover is strong enough that one geographer has dubbed it the "rule of indigenous environments." This rule states that "where there are indigenous peoples with a homeland there are still biologically rich environments" (Nietschmann 1992). In most protected areas, traditional peoples are allowed to practice "traditional" kinds of resource use. Yet a host of factors are leading to changes in the patterns and levels of traditional resource use. Population increases; a reduced area for exploitation; changes in the value of commodities and integration into market economies; cultural changes; and changes in technologies all affect the patterns of resource use and extraction in and near parks. When relevant, teams were asked to identify traditional groups with a probable influence on parks and assess the social changes in traditional communities and the potential impact on the park.

Transboundary Issues

Transboundary issues have emerged as critical to biodiversity conservation in many parts of the world, since ecosystems don't recognize national boundaries, yet effective management of parks that straddle state or national boundaries depends on compatible uses by the neighboring country. Examples exist of both compatible and incompatible uses across boundaries. For example, one country creates a park with boundaries that overlap national boundaries, while a neighboring country decides to construct a road along its boundary in the interest of national security. Such actions are clearly incompatible and may even lead to conflict. Migrants are likely to arrive on the road, and illegal settlements and resource use are likely to occur inside the park. To circumvent such problems, and to provide the largest possible habitat for wildlife, conservationists have tried to insure collaboration between countries when parks are located along national boundaries. In some cases, adjacent countries have established parks on both sides of a national boundary—these have been known as "peace parks." Teams examined transboundary concerns, looking at factors ranging from how adjacent communities in different countries get along and develop different patterns of resource use, to national-level development and military plans that might affect parks.

Resettlement

Resettlement has been a controversial component of park establishment in many parts of the world. Socially, resettlement is almost always likely to be controversial depending on how the process is carried out: whether people were consulted or compensated, whose lands were expropriated, who was compensated, at what value, within what time frame. There are examples in which people have been willing to be resettled from parks if compensation and participation in the resettlement process were adequate. In other cases, resettlement has been involuntary and has led to conflict, especially when people have had strong ties to particular areas. When there were examples of voluntary or involuntary resettlement, either at the time of park creation or later, teams were asked to review the social and ecological objectives of resettlement and the consequences of resettlement on the park and the resettled populations. In cases of voluntary resettlement, team members were asked to review the design of appropriate incentives and the selection of relocation sites.

Preview

This book analyzes parks from ecological, political, and social perspectives, along with management responses from their creation to the current threats they face. It is structured in three sections, described in greater detail by chapter below. Part I consists of three chapters that frame the issues and describe the Parks in Peril program and its ecological and social context, with each successive chapter focusing more pointedly toward part II. Part II consists of nine chapters, each a case study analyzing a park in Latin American or the Caribbean to determine the social, political, and ecological realities at that site. The four chapters in part III build from the case studies and specific park-level concerns to the political context and lessons. The first summarizes comparative information across the case studies and focuses on the parks themselves. The second and third are synthesis chapters that move beyond the themes to extract the key findings. The concluding chapter offers lessons on what issues emerge from the case studies and the synthetic chapters as most critical to insuring effective park management.

Part I: Neotropical Parks: Challenges and Context

In 1990, The Nature Conservancy launched the Parks in Peril (PiP) program, the largest single program to support parks in Latin America and perhaps the entire tropical world. In chapter 1, "Parks in Peril: A Conservation Partnership for the Americas," Houseal, Ostria, and Touval describe the Conservancy as an organization, and discuss its mission, the partnership approach, and the genesis and development of the PiP program. In chapter 2, "The Parks in Peril Network: An Ecogeographic Perspective," Sayre et al. provide an introduction to the status of Neotropical parks. The system of existing Neotropical parks and how well those parks represent the diversity of ecosystems found in Latin America and the Caribbean are reviewed. Chapter 3, "Analyzing the Social Context at PiP Sites," describes the social context of the PiP sites, with particular emphasis on the local communities and resource uses surrounding the sites in this chapter by Dugelby and Libby.

Part II: Nine Neotropical Parks

"Mexico: Ría Celestún and Ría Lagartos Special Biosphere Reserves" (chapter 4) describes two estuarine wetlands located in the Yucatán Peninsula that are home to a rich variety of migratory waterfowl and other wildlife, especially the greater flamingo. The sites must cope with a variety of small- and large-scale threats, including resource-damaging fishing practices, salt extraction, cattle grazing, and vacation home construction. The potential for ecotourism and the need to provide new livelihood alternatives and limit ecological impacts are described in this case.

The case study in "Guatemala: Sierra de las Minas Biosphere Reserve" (chapter 5) offers an example of consultation in reserve establishment, using research as the basis for management decisions, and of delegation of management authority. The Guatemalan government has delegated management authority for this reserve to an NGO—a trend in park management first tried in Guatemala. The reserve is renowned internationally as the largest remaining habitat of the resplendent quetzal and four hundred other birds; within Guatemala it is regarded as a key watershed and source of sixty-three rivers. The social complexity surrounding the reserve is high; different Indian groups and mestizo colonists reside in and around it, complicating community outreach.

Costa Rica's Corcovado National Park, described in chapter 6, is often proclaimed as one of the most important sites for conservation in Central America. Yet the park is largely dependent on the adequate protection of the forest reserves that surround it—reserves that are being rapidly cleared. Rich historical information in this case details the site-based issues and politics of park establishment. It provides insight into the difficulties of implementing official policies of sustainable development in the face of governmental gridlock and intense local pressure.

Del Este National Park in the Dominican Republic (chapter 7) is representative of the numerous coastal parks where protection of the marine environment must be linked with land-based protection. Del Este includes terrestrial, mangrove, and marine communities; human users include small-holder agriculturalists and fishermen who rely on the resources. Tourism from nearby resort communities offers both potential benefits and management challenges.

Chapter 8 discusses the case of the Rio Bravo Conservation and Management Area in Belize, a unique protected area in the hemisphere—it is remarkably free of threats. It is a private conservation area, managed by an NGO called Programme for Belize. The site is largely untroubled by many of the problems that face other Neotropical sites. The threats to the park are few, and the park has the potential to become financially self-sufficient through tourism. The historical and national policies that have directly affected site conditions are analyzed in this case.

Machalilla National Park in Ecuador (chapter 9) includes a variety of ecosystem types, from fog and dry forests to beaches and marine areas, and Ecuador's only continental coral reefs. Threats to the park include a government policy that allows the park to serve as a corridor for power and oil lines, and the more than one thousand people who reside in the park, graze cattle and goats there, and engage in commercial charcoal production and whose well-being is a concern.

Also in Ecuador, Podocarpus National Park (chapter 10) protects the largest Andean coniferous forests of the genus *Podocarpus*, and it may also be one of the richest spots in the world for birds. The social complexity includes overlapping claims by mining company concessions, national agrarian reform agencies, colonists, and Shuar Indians. Recent supreme court decisions have favored the park; however, transboundary threats from Peru and decreasing land availability complicate park management.

Amboró National Park in Bolivia (chapter 11) is one of the most biologically rich parks in Bolivia, perhaps in the Neotropics. When the park's boundaries were established, nearly one-third of the area included in the park was already claimed. This case describes the intensive process of redefining park boundaries to recognize local land claims, as well as the position of various groups, including labor unions, to open the park to sustainable resource utilization. Management responsibility has undergone a variety of changes in recent years as a result of national-level policy changes, including significant changes in 1996.

Yanachaga-Chemillen National Park in Peru (chapter 12) was established in 1983 as part of a regional development project with funding from USAID. The project also supported the creation of a nearby protected forest reserve and the Yanesha Indian Communal Reserve. The park is surrounded by over sixty formally recognized indigenous communities; one group, the Yanesha, is involved in the implementation of the park's management plan. Parts of the park are under threat from road construction and colonist migration, which have increased since terrorist activities have declined.

Part III: Reality and Reaction: Saving Neotropical Parks

"Comparing Cases: A Review of Findings," (chapter 13) is a comparison of eleven key themes across the nine case study sites. The chapter provides a quick cross reference of the key themes across sites.

"Perils to Parks: The Social Context of Threats" (chapter 14) highlights ways in which park management is constrained by the social and political context and the limitations of using parks as vehicles for resolving structural, socioeconomic problems facing societies.

"The New Politics of Protected Areas" (chapter 15) is a macropolitical synthesis chapter combining the overall issues of how parks are protected in different and highly variable political contexts, along with observations on how those contexts compare. Parks are central to political processes at the national, regional, and local levels in Latin America. They focus the themes of land and resource tenure, large-scale threats, national policy framework, transboundary issues, indigenous rights, and many other pressing political tests. The synthesis in this chapter raises questions about the possibility of new political coalitions rising from parks with general benefits for the rural poor and for conservation.

In "Holding Ground" (chapter 16) the three editors of the book have synthesized the findings, joining their three disciplinary perspectives to address the most critical steps in transforming these recently created and often contentious sites into areas that can safeguard biodiversity. The synthesis offers lessons on what issues emerge from the case studies as most critical to ensuring effective park management, contrasting those lessons with the set of clichés that has served to truncate the discourse concerning in situ conservation.

PART I

Neotropical Parks

Challenges and Context

1. Parks in Peril: A Conservation Partnership for the Americas

Brian Houseal, Mónica Ostria, and Jerome Touval

A barefooted campesino holding a machete stands on a muddy road that passes through a tropical rainforest. He watches a tractor trailer pass, loaded with trunks of mahogany and Spanish cedar that are five feet in diameter. Although the campesino is a park ranger, the rainforest is a national park, and the logging is illegal, he is powerless to stop the truck. He has no vehicle, no uniform, no radio, and no outside support. Indeed, he does not even know with certainty where the boundaries of the park are located.

In 1990, this scenario was commonplace in many of the national parks and reserves throughout Latin America and the Caribbean. Known as paper parks, these large natural areas did not have any semblance of on-the-ground protection or management. Even though the countries of the region had taken the first step to conserve their globally important biological resources by legally declaring these areas protected, there were only limited national government funds and a few dedicated conservationists who struggled to limit the slash-and-burn agricultural expansion, road construction, and illicit logging and mining that threatened the ecological integrity of this priceless natural heritage.

Concerned about the increasing destruction of biological diversity beyond the U.S. borders, Cliff Messinger, a Nature Conservancy board member, asked a key question: "Where are the most threatened ecosystems in Latin America and the Caribbean, and what must be done to save them?" Under Cliff's leadership, the Conservancy embarked on an emergency campaign to safeguard the hemisphere's most imperiled natural areas.

As the Conservancy designed its initial strategy for the Parks in Peril program, it first assessed the status of conservation efforts in the Latin American and Caribbean region. Many individuals and organizations had already gained valuable experience and provided very good advice. For example, the World Wildlife Fund–U.S. willingly shared its first major evaluation of the USAID-funded Wildlands and Human Needs Project, which focused on local communities and compatible uses of the lands adjacent to protected areas. The regional representatives of the Commission for Parks and Protected Areas of the International Union for the Conservation of Nature also freely gave their data on the status of the Neotropical protected areas, as well as strong encouragement for the Conservancy to focus on the core zones of protected areas.

In Latin America and the Caribbean, the decline in military governments in many countries and the emergence of civil societies gave rise to nonprofit organizations addressing needs in environmental, health, education, and social welfare issues. Looking at this situation, and taking stock of its own organizational strengths developed over more than forty years—building U.S. state chapters and establishing and managing the world's largest network of private nature preserves—the Conservancy decided to develop a strategy based on working with national nongovernment organizations (NGOs) in Latin America. The Latin America and Caribbean Division of The Nature Conservancy had already begun to assist a handful of fledgling conservation NGOs to initiate their work in Costa Rica, Panama, Mexico, Venezuela, Ecuador, Bolivia, Peru, and Paraguay and consulted closely with both the U.S. chapters and these new NGOs on developing the best possible strategy.

In order to begin the Parks in Peril program, the Conservancy had to decide where to work and what should be done there. First, the question of where The Nature Conservancy's efforts and resources should be concentrated needed to be addressed. The Conservancy compiled a list of two hundred officially decreed sites in the most biologically significant and threatened protected areas in the hemisphere. This list reflected the best available information and expert opinion at the time it was compiled and was assembled with valuable assistance from Latin American and Caribbean governments, in-country conservation organizations, scientists, and conservation data centers, and also from the World Wildlife Fund, World Resources Institute, Organization of American States, U.S. National Park Service, Caribbean Natural Resources Institute, and International Union for the Conservation of Nature. The list was considered as preliminary and dynamic and was subject to change as greater ecological information was obtained, and as local, national, and regional conditions changed. The Conservancy realized that without rapid and dramatic conservation actions, these sites would likely face irreversible degradation. Since these were already legally decreed protected areas, they were believed to offer the best opportunity for success in protecting a significant portion of the hemisphere's diversity.

The selection of Parks in Peril sites was based on the following criteria:

- *Biological significance*—Based on coverage of important biogeographic regions of the Western Hemisphere, including a variety of the hemisphere's ecosystems. Size, ecological integrity, and proximity to contiguous wildlands areas were considered important factors for long-term viability of ecosystems and wildlife populations.

- *Socioeconomic and cultural value*—Assessment of local land tenure and resource use patterns, regional economic infrastructure, access, and economic resources, including but not limited to the presence of marine systems, watersheds, wildlife, fisheries, forest products, and tourism resources.

- *Endangerment*—Examination of specific threats to sites, including but not limited to road construction, poaching, logging, overfishing, mining, inappropriate agriculture and grazing, along with assessments of the rate of change in the area. Among the factors considered were national development projects and commercial ventures.

- *Management capacity and opportunity*—Analysis of current levels of management (including personnel, assigned budgets, on-site infrastructure, etc.), existing legislation and interinstitutional coordination, opportunities for local community participation, and local organizational support.

Based on the preliminary list of two hundred sites, and with the strong endorsement of over thirty-five governmental and nongovernmental partner organizations in Latin America and the Caribbean, The Nature Conservancy entered into a cooperative agreement with the U.S. Agency for International Development. Recognizing that to achieve its mandate for sustainable development USAID must involve itself in support for sustainable conservation, USAID became a full partner of The Nature Conservancy in the planning, launching, and implementation of the Parks in Peril program. To date, this extraordinary program has raised over $35 million in public and private funds.

Since the Conservancy did not have the institutional capacity to work in all two hundred sites at once, the Parks in Peril program, officially launched in 1990, began work at twenty-two sites. A subset of ten of those sites was supported with funding provided by USAID. These initial ten USAID-supported sites were selected in consultation with non-governmental organization partners and national government natural resource management agencies, as well as with USAID personnel.

Having decided on the sites where Parks in Peril would work, the Conservancy also had to define exactly how the program would approach the challenges of conserving these imperiled areas. Several options were available. With a long history of land acquisition and preserve management by Conservancy staff and partners, this could have been the path followed by the Parks in Peril program. After careful consideration however, the Conservancy decided that the best investment of its resources would be in building local capacity to manage these parks and protected areas in a self-sustaining manner. If the job were done right, eventually there would be no need for The Nature Conservancy to support these sites. For that reason, the Parks in Peril program was designed and implemented in partnership with in-country conservation organizations that share The Nature Conservancy's mission of "preserving plants, animals and natural communities that represent the diversity of life on Earth by protecting the lands and waters they need to survive."

With its partners, the Conservancy recognized that because the governments in the regions were faced with pressing social and economic problems as well as heavy international debt burdens, they could not dedicate the necessary funds to finance the operations of national parks and protected areas. The Conservancy and its partners decided that a first vital step was to strengthen countries' national parks agencies by focusing the Parks in Peril initiative on decreed protected areas and working to build on-site infrastructure. It did not make sense to pursue the declaration of new protected areas until a country had the capacity to manage already decreed areas. The Conservancy, its NGO partners, and the government park agencies also agreed to focus on the need to create a permanent on-site presence in the protected areas. This singular focus became a major undertaking for each protected area, in most cases entailing the acquisition of four-wheel-drive vehicles or boats and horses; obtaining field equipment such as backpacks, boots, jungle hammocks, compasses, and machetes; recruiting and training rangers from the local communities who would be capable of working alone in remote areas for extended periods of time; contacting local residents and obtaining their support for the areas; surveying protected-areas boundaries, opening up patrol trails, building cattle-control fences, and placing signs at access points; building the first ranger stations, and furnishing them with first aid, fire-fighting, and rescue equipment and battery- or solar-powered radios. These activities would build the foundations for world-class national parks and protected areas.

The stories that drifted back to the city offices of the Conservancy and its partners in LAC over the early years have become almost legendary. How a ranger saved the life of a snake-bitten campesino who walked into the station. How a partner who was a pilot jump-started his plane on a remote protected-area airstrip by using a horse and a lariat. How a jeep was floated across a rain-swollen river on two tiny dugout canoes. How a ranger and his son were shotgunned and almost killed by an illegal logger. How a botanist collected new species because she could work safely inside the reserve. How quiet and beautiful and pristine it really is at the end of the road, inside the forest on top of the mountain. How good it was to finally see on-the-ground results.

Recalling the barefoot ranger at the side of the logging road, the Parks in Peril program is a monumental effort to build a logistical connection to each protected area so as to make it capable of sustaining its conservation promise into the future. In order to make the selected protected areas functional and sustainable, The Nature Conservancy defined the three essential goals of the Parks in Peril program as being:

1. *To establish on-site protection.* This is the fundamental component of the program and includes basic protection, infrastructure construction and maintenance, and applied conservation science activities.

2. *To integrate these protected areas into the economic and cultural lives of local communities.* Community outreach activities such as environmental education and pilot projects in sustainable resource use are included in this component.

3. *To create long-term funding mechanisms to sustain local management into the future.* This component of the program supports technical assistance in long-term protected-area management and financial planning, and assists conservation organizations to promote local, national, and international policies that advance conservation.

In 1995, a fourth goal was added to the program:

4. *To use Parks in Peril site-based activities to influence conservation in other sites in the region's most imperiled ecosystems.* This component seeks to leverage the experiences gained through the program in the areas of conservation science, community conservation, conservation finance and policy, institution strengthening, and interdisciplinary analysis. This will be accomplished through a series of outreach activities that include program-related publications and training.

Of course, these goals are easily stated but difficult to achieve. As described previously, Parks in Peril began as an emergency measure to safeguard the hemisphere's most imperiled ecosystems. However, it became evident early in the program that it would be necessary to define the point at which emergency assistance was no longer needed, to identify when the program could safely consider that its mission had been accomplished at a given protected area. In order to define the conditions that were to be achieved with Parks in Peril's assistance to a protected area, Conservancy and USAID staff developed the concept of "site consolidation." The term was meant to denote a site that had reached a predefined degree of functionality, a site that could be considered no longer in need of emergency assistance.

As one might imagine, a lively debate ensued as the Conservancy, USAID, and Latin American partner organizations tried to develop a definition for when a site had reached consolidation status. After much discussion, it was decided that this end-of-project status would be defined by sixteen parameters, known as the PiP Site Scorecard Criteria (see appendix at the end of the book). These site consolidation parameters included:

- the capacity of the local management authority to provide on-site protection and management (including the number and training of on-site personnel and the presence of adequate equipment and infrastructure);
- the state of information available to protected-area managers regarding land tenure in and around the protected area;
- the capacity of the local management authority to provide long-term management of the area (including development of a long-term management plan, usage zoning within the protected area, availability of scientific information to inform management decisions, and development of a threats-related monitoring plan);

- the development of a long-term financing plan for the protected area;
- the presence of a local constituency that participates in management decisions and supports the presence of the protected area.

By 1997, on-site protection and management were progressing well for most of the Parks in Peril sites. During the first seven years of its existence, the program has become the largest in situ biodiversity conservation project in the tropical world. It includes a total of sixty sites, of which twenty-eight are funded by USAID. In those twenty-eight protected areas, located in twelve countries, the program has constructed or renovated more than one hundred headquarters buildings, visitor centers, and other protection facilities and has trained approximately four hundred rangers and extensionists. On 18 million acres, the program has worked to protect cloud forests, tropical forests, savannas, and *páramo* (tropical alpine grasslands). At each of the sites, local communities have gradually become involved in management decisions, fostering a deep-rooted pride in, and support for, the conservation of these areas. Parks in Peril has been responsible for catalyzing innovative methods for ensuring the long-term financial stability of these sites.

Park directors and rangers have been recruited, trained, and equipped to spend long periods of time in these remote areas. The protected areas' boundaries have been surveyed and posted, ranger stations have been built, and local community education and outreach programs are in place. Inappropriate logging, mining, and settlement activities are declining, and ecological monitoring programs are underway. National governments are also increasing protected-areas budget allocations and enforcing the laws. In short, the ranger on a Parks in Peril site is no longer barefoot and has become one of the first generation of professional rangers in that country with the training, equipment, and infrastructure necessary to manage the country's most important natural areas.

Parks in Peril investment has substantially leveraged other bi- and multilateral investments in the conservation of these sites. For example, the Japanese government (SSGA) has invested in equipment purchase, and the program has become part of the common agenda between the U.S. and Japanese governments. Likewise, Parks in Peril sites have been the focus of Global Environmental Facility (GEF) investments in strengthening local capacity for long-term protected-area conservation. Other bilateral and multilateral investors at these sites include the German Technical Cooperation Agency (GTZ), European Economic Community

(EEC), Spanish Agency for International Cooperation (AECI), Swiss Technical Cooperation (COTESU), Inter-American Development Bank (IDB), and the United Nations Development Program (UNDP). Most recent investments include that of U.S.-based utilities companies that are investing in carbon sequestration projects at two Parks in Peril sites, namely Rio Bravo in Belize and Noel Kempff Mercado in Bolivia. Despite the investments from external sources, however, the most significant indication of the success of the Parks in Peril program in achieving the sustainability of these sites has been the dramatic increase in the funds provided for protected-area conservation by national governments in twelve Latin American and Caribbean nations.

Notwithstanding this significant progress, we have learned that sustainable, long-term conservation of the Parks in Peril sites remains elusive. A number of key obstacles remain to be overcome, particularly the integration of these areas with the economic and cultural lives of the local communities. For example, protected-areas policies concerning compatible land and resource uses are often in conflict with communal landholdings, traditional practices, or individual property rights. In the absence of creative mechanisms that promote and enforce appropriate land and resource uses, many terrestrial and marine protected areas are still treated as open-access resource areas with few possibilities for sustainable uses that also ensure longer-term economic benefits. In general, the protected-areas directors throughout the region have made only limited progress in building the local constituencies they need to devise innovative solutions to protect the natural ecosystems, generate economic benefits, and improve the quality of life for local communities.

Another elusive objective is the creation of sustainable sources of in-country support for the long-term management of the Parks in Peril sites. The continued dependence of many of the sites on large infusions of international support is disturbing, particularly in light of anticipated declines in USAID funds available for conservation. Very few sites have designed realistic long-term financial strategies, and fewer yet have put them into practice by establishing diverse and replicable sources of local, state, and national private and public funding. While it is encouraging that other bi- and multilateral organizations, such as the Global Environmental Facility, are following the lead of USAID by investing in biodiversity conservation, there are still too few in-country commitments to sustain and expand national protected-areas systems.

The basic strategy of the Parks in Peril program is to buy time for Latin American and Caribbean countries that are now increasingly realizing the importance of their natural areas as the foundation for sustainable growth. The Nature Conservancy is committed to working with in-country conservation organizations to build organizational capacity and commitment to safeguard their nations' natural heritage. The core of the Conservancy's approach is to build partnerships to ensure conservation of large landscape areas preserving the best examples of natural diversity in the Latin American and Caribbean region. With its in-country partners, the Conservancy will focus on priority ecoregions and portfolios of sites, using them to improve the partners' skills and understanding of the ecological, economic, and societal relationships necessary to sustain their conservation into the next generation. Using these new skills, the Conservancy and its partners will extend their experiences to other organizations at the local, state, national, and international levels, forming diverse constituencies to change policies, guide financial resources to sustain conservation efforts, and replicate successes.

This compendium of case studies is an initial effort to share the experiences of the Parks in Peril program and extend conservation as widely as possible. Beyond the current list of Parks in Peril sites, there is another barefoot campesino with a machete standing by the side of a muddy road trying to save our world's incredible natural heritage. It is the courage and dedication of its local partners that will continue to inspire the Conservancy to conservation action.

2. The Parks in Peril Network: An Ecogeographic Perspective

Roger Sayre, Jane Mansour, Xiaojun Li, Timothy Boucher, Stuart Sheppard, and Kent Redford

The earth is increasingly recognized as a human-dominated system (Vitousek et al. 1997a, b). In response to this domination, approximately 5 percent of the land surface of the planet has been set aside in protected areas with a variety of management goals and structures. More than twenty-five thousand protected areas are known to exist (McNeely, Harrison, and Dingwall 1994). These protected areas often contain a wealth of the world's biodiversity (WCMC 1992), and some are managed primarily for biodiversity conservation (IUCN 1978).

Representation of distinct natural communities in conservation strategies and protected-area networks has become a fundamental conservation goal of many conservation organizations (Dinerstein et al. 1995; The Nature Conservancy 1997), multilateral lending institutions (Dinerstein et al. 1995), and international development agencies (BSP et al. 1995). Representativeness assessments are analyses that characterize the extent to which protected areas contain, or "capture," unique habitat types and other elements of biodiversity (Pressey et al. 1993). These assessments are implemented in a relatively straightforward geographic information system (GIS) analysis where protected areas are overlain on landscape units (communities, land types, cover classes, ecoregions, biomes, etc.). When implemented to evaluate the adequacy of existing reserve systems for biodiversity conservation, the approach is commonly referred to as gap analysis (Scott, Tear, and Davis 1996). Representativeness assessments at the continental or global scales have quantified protection of biomes (McNeely, Harrison, and Dingwall 1994) and ecofloristic zones (Green et al. 1997).

A great many parks and other protected areas have been established in Latin America and the Caribbean for a variety of reasons. Conservation of natural resources or biodiversity is a common theme in the park establishment process, but the extent to which actual conservation is achieved by setting land aside in reserves is difficult to measure. The area of land under protection is, however, relatively easy to quantify. While the areal extent of protected-area coverage may not be related to protected-area management, it is a surrogate measurement for conservation commitment.

One small subset of these Latin American and Caribbean protected areas is the Parks in Peril network. PiP is a site-based conservation program in which an active management presence is under development at sixty paper parks to improve their conservation impact. These parks have been mapped in a geographic information system, enabling a spatial analysis of their contribution to total protected area of countries and regions, and a gap analysis of their protection of ecoregions.

This chapter is presented in two sections. The first presents an ecogeographic analysis of the Parks in Peril network, describing the contribution of PiP sites to the overall protection of countries and geographic regions, and comparing that protection with coverage provided by other protected areas. The second describes an ecoregional classification system for Latin America and the Caribbean and characterizes the representation of these ecoregions in PiP sites. These analyses provide an ecogeographic context for the distribution of PiP sites and place in perspective the case studies that follow in this volume.

The PiP sites are a small subset of all the protected areas that exist in Latin America and the Caribbean (1,901), yet they provide meaningful protection for some of the region's highest-priority ecoregions. A comparative analysis of the protection of ecoregions by PiP sites relative to protection by all sites would be useful, but an adequate master database of Latin American and Caribbean protected areas that would permit such an analysis does not exist. The PiP gap analysis offers a landscape-level, network-wide PiP perspective; subsequent chapters will focus on single-park dynamics.

Parks in Peril and Other Protected Areas

The Nature Conservancy, in partnership with in-country conservation organizations, works at sixty protected areas, PiP sites, throughout Latin America and the Caribbean (map 2-1 and table 2-1). Approximately half of the PiP sites receive direct support from the Latin America and Caribbean Bureau of USAID. Descriptions of the original USAID-supported sites can be found in the *Parks in Peril Source Book* (The Nature Conservancy 1995). For the purposes of this chapter, we refer to this group of sites as AID/TNC PiP sites (the Conservancy was previously known as TNC). The other protected areas of the PiP program receive primary support from a wide variety of sources, including The Nature Conservancy, USAID in-country missions, other government agencies, foundations, and individual donors. We will refer to these sites as TNC PiP sites.

While thorough assessments have been conducted of the status of protected-area coverage in Latin America and the Caribbean and in the world's other major biogeographical regions (IUCN 1992; McNeely, Harrison, and Dingwall 1994), we review some of that information in this chapter in order to assess the relative contribution of coverage by PiP sites. The studies cited earlier identify more than 1,900 protected areas in Latin America and the Caribbean (see table 2-2). This number includes all sites in management categories I–VIII of the classification developed by IUCN (1978). The eight IUCN management categories are:

I. Scientific Reserve / Strict Nature Reserve

II. National Park

III. Natural Monuments / Natural Landmark

IV. Managed Nature Reserve / Wildlife Sanctuary

V. Protected Landscape

VI. Resource Reserve

VII. Natural Biotic Area / Anthropological Reserve

VIII. Multiple-Use Management / Managed Resource Area

In 1988, the IUCN's Commission on National Parks and Protected Areas undertook a thorough review of these management categories and developed a new six-category classification system (see McNeely, Harrison, and Dingwall 1994). Protected-area data are not yet available using this new classification, so our analysis was conducted using the original categories.

For purposes of the analysis, we have summarized information for two

Parks in Peril Sites

classes of management categories (I–V and VI–VIII). From a biodiversity conservation perspective, protected-area coverage in categories I–V is of greater interest, as the objectives of those categories are more closely allied with conservation, while categories VI–VIII are devoted to protecting natural resources for future use and protecting the lands of traditional peoples.

Table 2-1. The protected areas of the Parks in Peril program

Site Number	Site Name	Country	Status
1	Rio Bravo Conservation and Management Area	Belize	AID/TNC
2	Maya Mountains/Marine Transect	Belize	TNC PiP
3	Amboró National Park	Bolivia	AID/TNC
4	Noel Kempff Mercado National Park	Bolivia	AID/TNC
5	Tariquia National Fauna and Flora Reserve	Bolivia	AID/TNC
6	Guaraqueçaba Environmental Protection Area	Brazil	AID/TNC*
7	Pantanal National Park	Brazil	AID/TNC*
8	Grande Sertão Veredas National Park	Brazil	TNC PiP
9	Serra do Divisor National Park	Brazil	TNC PiP
10	Cahuinari National Park	Colombia	AID/TNC
11	Chingaza National Park	Colombia	AID/TNC
12	La Paya National Park	Colombia	AID/TNC
13	Sierra Nevada de Santa Marta National Park	Colombia	AID/TNC
14	Corcovado National Park	Costa Rica	AID/TNC
15	Talamanca-Caribbean Biological Corridor	Costa Rica	AID/TNC
16	Guanacaste National Park	Costa Rica	TNC PiP
17	Morne Trois Pitons National Park	Dominica	AID/TNC
18	Jaragua National Park	Dominican Republic	AID/TNC
19	Madre de las Aguas (Armando Bermúdez Nat'l. Park/ Valle Nuevo Scientific Reserve)**	Dominican Republic	AID/TNC
20	Parque del Este National Park	Dominican Republic	AID/TNC
21	Machalilla National Park	Ecuador	AID/TNC
22	Podocarpus National Park	Ecuador	AID/TNC
23	Antisana Ecological Reserve	Ecuador	TNC PiP

(continued)

Table 2-1. The protected areas of the Parks in Peril program (*continued*)

Site Number	Site Name	Country	Status
24	Cayambe-Coca Ecological Reserve	Ecuador	TNC PiP
25	*Galápagos Marine Park*	*Ecuador*	*TNC PiP*
26	Galápagos National Park	Ecuador	TNC PiP
27	Maquipucuna Ecological Reserve	Ecuador	TNC PiP
28	El Imposible National Park	El Salvador	TNC PiP
29	Sierra de las Minas	Guatemala	AID/TNC
30	Cerro San Gil Ecological Reserve	Guatemala	TNC PiP
31	Maya Biosphere Reserve	Guatemala	TNC PiP
32	Honduran Mosquitía (Río Plátano)	Honduras	AID/TNC*
33	Cusuco-Merendón National Park	Honduras	TNC PiP
34	Blue/John Crow Mountains	Jamaica	AID/TNC*
35	*Montego Bay Marine Park*	*Jamaica*	*TNC PiP*
36	Calakmul Biosphere Reserve	Mexico	AID/TNC
37	El Pinacate & Gran Desierto del Altar Biosphere Reserve	Mexico	AID/TNC
38	El Ocote Ecological Reserve	Mexico	AID/TNC
39	El Triunfo/La Sepultura Biosphere Reserves	Mexico	AID/TNC
40	La Encrucijada Coastal Wetland Reserve	Mexico	AID/TNC
41	Ría Celestún and Ría Lagartos Wildlife Refuge	Mexico	AID/TNC
42	Sian Ka'an Biosphere Reserve	Mexico	AID/TNC
43	*Sea of Cortez*	*Mexico*	*AID/TNC**

44	*Sierra Madre Occidental*	*Mexico*	*AID/TNC**
45	Cuatro Ciénegas	Mexico	TNC PiP
46	*Laguna Madre*	*Mexico*	*TNC PiP*
47	Bosawas Natural Resource Reserve	Nicaragua	TNC PiP
48	Darién Biosphere Reserve	Panama	AID/TNC
49	Panama Canal Watershed (Chagres and Soberanía National Parks)**	Panama	AID/TNC
50	*Bastimentos Marine National Park*	*Panama*	*TNC PiP*
51	Defensores del Chaco National Park	Paraquay	AID/TNC*
52	Mbaracayú Forest Nature Reserve	Paraquay	AID/TNC
53	Bahuaja-Sonene National Park	Peru	AID/TNC
54	Paracas National Reserve	Peru	AID/TNC*
55	Yanachaga-Chemillén National Park	Peru	AID/TNC
56	Pacaya-Samiria National Park	Peru	TNC PiP
57	*Salt River Bay*	*U.S. Virgin Islands*	*TNC PiP*
58	*St. Croix Marine*	*U.S. Virgin Islands*	*TNC PiP*
59	Aguaro/Guariquito National Park	Venezuela	TNC PiP
60	Canaima National Park	Venezuela	TNC PiP

* Proposed sites for AID/TNC Parks in Peril.

** These sites include more than one protected area but are treated administratively as one PiP site.

Italics refer to marine sites or PiP sites for which area has not been mapped/defined, and are not included in the analyses.

Table 2-2. Summary of PiP sites and total protected-area coverage by country. The last column (% all) refers to the total area of the country protected by all sites. The bottom row provides total figures for protected area. The percentage coverage columns for categories I–V, VI–VIII, and all protected areas were calculated by summing all protected area and dividing it by the total terrestrial area of the countries and islands listed.

#	Country	#AID/TNC	#TNCPiP	#All PiP	# I–V	% I–V	# VI–VIII	% VI–VIII	# All Parks	% all
1	Belize	1	1	2	10	12.7	15	17.9	25	30.6
2	Costa Rica	2	1	3	25	12.2	59	20.3	84	32.5
3	El Salvador	0	1	1	5	0.9	0	0	5	0.9
4	Guatemala	1	2	3	17	7.6	10	7.7	27	15.3
5	Honduras	1*	1	2	38	4.8	10	11	48	15.8
6	Nicaragua	0	1	1	21	6.4	10	5.8	31	12.2
7	Panama	2	1	3	15	16.9	10	18.3	25	35.2
8	Anguilla	0	0	0	0	0	0	0	0	0
9	Antigua-Barbuda	0	0	0	3	15	0	0	3	15
10	Aruba	0	0	0	0	0	0	0	0	0
11	Bahamas	0	0	0	6	8.9	0	0	6	8.9
12	Barbados	0	0	0	1	0.6	0	0	1	0.6
13	Bermuda	0	0	0	2	232	0	0	2	232
14	British Virgin Islands	0	0	0	6	10.2	0	0	6	10.2
15	Cayman Islands	0	0	0	13	31.4	0	0	13	31.4
16	Cuba	0	0	0	57	7.8	5	6.5	62	14.3
17	Dominica	1	0	1	2	9.9	3	12.7	5	22.6
18	Dominican Rep.	3	0	3	18	21.6	0	0	18	21.6
19	Grenada	0	0	0	0	0	1	1.8	1	1.8
20	Guadeloupe	0	0	0	2	11.8	0	9.1	2	20.9
21	Haiti	0	0	0	3	0.3	0	0	3	0.3
22	Jamaica	1*	1	2	2	0.1	45	8.1	47	8.2

23	Martinique	0	0	0	4	66.2	1	0.1	5	66.3
24	Montserrat	0	0	0	0	0	0	0	0	0
25	Neth. Antilles	0	0	0	2	9.7	3	10	5	19.7
26	Puerto Rico	0	0	0	18	2	1	4	19	6
27	St. Kitts-Nevis	0	0	0	1	10	0	0	1	10
28	St. Lucia	0	0	0	2	3.2	0	12.1	2	15.3
29	St. Vincent-Grnda	0	0	0	2	21.3	11	0	13	21.3
30	Trinidad/Tobago	0	0	0	9	3.4	0	0	9	3.4
31	Turks-Caicos	0	0	0	20	39.7	0	0	20	39.7
32	U.S. Virgin Islands	0	2	2	3	16.8	0	0	3	16.8
33	Mexico	9*	2	11	60	5	18	1.3	78	6.3
34	Argentina	0	0	0	100	3.4	19	1.4	119	4.8
35	Bolivia	3	0	3	26	8.4	17	14	43	22.4
36	Brazil	2*	2	4	214	3.3	286	13.5	500	16.8
37	Chile	0	0	0	65	18.2	0	0	65	18.2
38	Colombia	4	0	4	79	8.2	263	63.6	342	71.8
39	Ecuador	2	5	7	15	24.1	52	6.3	67	30.4
40	French Guiana	0	0	0	0	0	1	2.1	1	2.1
41	Guyana	0	0	0	1	0.3	0	0	1	0.3
42	Paraguay	2*	0	2	19	3.6	1	0	20	3.6
43	Peru	3*	1	4	22	3.2	19	6.6	41	9.8
44	Suriname	0	0	0	13	4.5	1	0.4	14	4.9
45	Uruguay	0	0	0	8	0.2	3	0.1	11	0.3
46	Venezuela	0	2	2	104	30.2	4	30.5	108	60.7
	Totals	37	23	60	1,033	6.33	868	12.32	1,901	18.65

* Includes proposed new AID/TNC sites.

Sources: IUCN 1992; McNeely, Harrison, and Dingwall 1994.

Table 2-2 provides an overall summary of the protected areas by country, including the total number of protected areas, totals grouped by category class (I–V and VI–VIII), and the percentage of the country's area under protection. Table 2-2 also provides a breakdown of the distribution of both AID/TNC and TNC PiP sites. The sources of information for this summary of national protected-area coverage are *Protected Areas of the World* (IUCN 1992) and *Protecting Nature: Regional Reviews of Protected Areas* (McNeely, Harrison, and Dingwall 1994).

It should be noted that marine sites are included in IUCN data, but percent protected-area coverage was calculated using the terrestrial area that constitutes a country. This has minimal consequence for larger countries but can result in inflated protection figures for smaller island nations. This is most notable in Bermuda, with two relatively large marine protected areas that result in a total protection of 232 percent. The high percent protection on some islands is, however, accurate. In Martinique, for example, 66.3 percent of the island area, is in fact in terrestrial protected areas. For clarification, see *Protected Areas of the World* (IUCN 1992), in which IUCN provides a full list of sites, both marine and terrestrial, with their areas.

The five countries with the greatest number of PiP sites (Table 2-2) are Mexico (11) and Ecuador (7) and Brazil, Colombia, and Peru, each of which contains 4 PiP sites. The five countries with the greatest number of protected areas are Brazil (500), Colombia (342), Argentina (119), Venezuela (108), and Costa Rica (84). The top five countries in total percent protection, however, differ. Excluding Bermuda and Turks/Caicos (see reasons relating to marine parks, above), these countries are Colombia (71.8 percent), Martinique (66.3 percent), Venezuela (60.7 percent), Panama (35.2 percent), and Costa Rica (32.5 percent). Note that the majority (63.6 percent) of the coverage in Colombia is in categories VI–VIII. Vast tracts of land in Colombia are in declared forest or anthropological reserve areas, though the degree to which these tracts (or category VI–VIII sites in any country) may be protected is unknown. The case is similar in Venezuela, where approximately half of the protected area total (30.5 percent) is in categories VI–VIII.

Overall protected-area coverage by type is summarized by geographic region in figure 2-1, which shows actual coverage in hectares for the Caribbean, Central America, Mexico, and South America. The large differences in the total areas of the four regions make interregional compar-

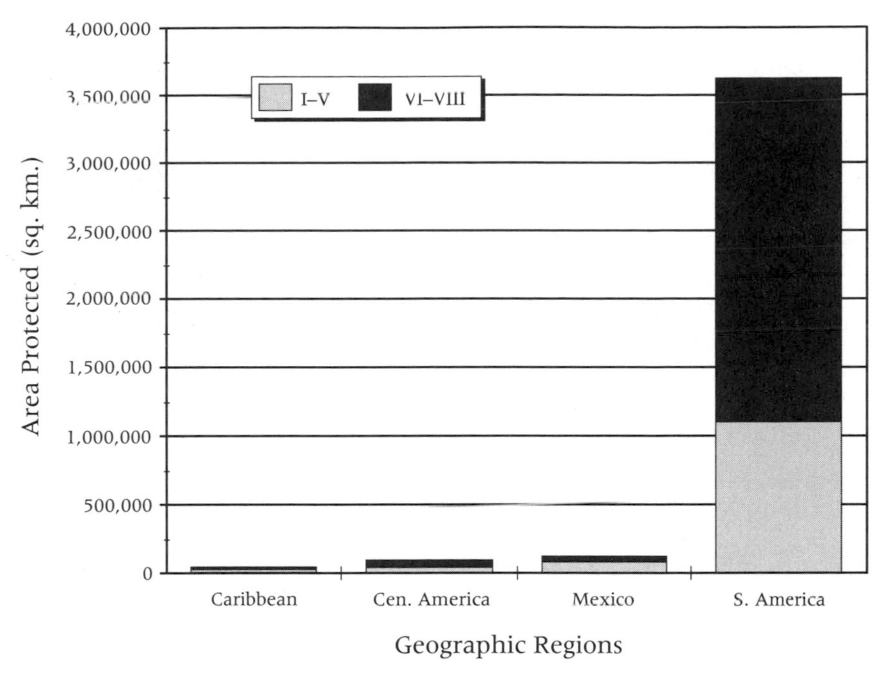

Figure 2-1. Total protected-area coverage in square kilometers for four geographic regions in Latin America and the Caribbean. Total are grouped for IUCN management categories I–V and VI–VIII. *Source:* McNeely, Harrison, and Dingwall 1994.

isons of protection amounts difficult. It is evident, however, that more than two-thirds of protected-area coverage in South America is in categories VI–VIII, while protection in the other regions is more uniformly distributed between the two management category classes.

Figure 2-2 displays these same data as a percentage of each region's total area, a more appropriate graphic for comparison of regional coverage. South America has the highest total percentage of area under protection, at 20.1 percent, approximately two-thirds of which are found in resource and anthropological reserves. Central America follows South America closely, with 19.2 percent total coverage, with approximately half in each of the management category groupings. With the exception of El Salvador, all Central American countries have in excess of 12 percent total protection, with three countries (Belize, Costa Rica, and Panama) in excess of 30 percent. In the Caribbean, total coverage is significant. Of the total designated protected area (13.4 percent), 84 percent is found in Cuba and the Dominican Republic. This is not surprising, as those two

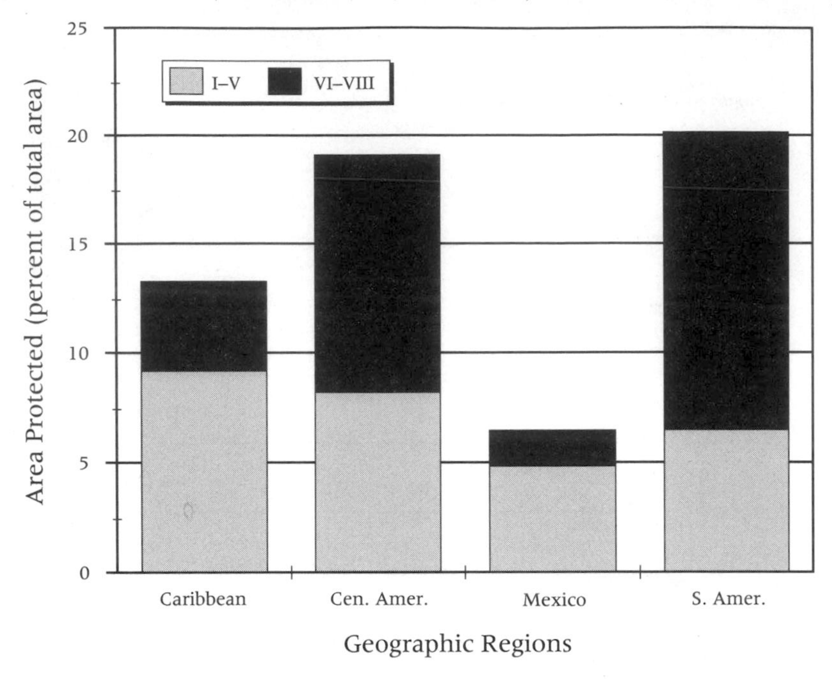

Figure 2-2. Percent of total terrestrial area in protected areas for four geographic regions in Latin America and the Caribbean. Totals are grouped for IUCN management categories I–V and VI–VIII. *Source:* McNeely, Harrison, and Dingwall 1994.

countries account for more than half of the terrestrial area of the Caribbean; however, there are many islands with few or no terrestrial parks. Mexico is the least protected of the regions, with total coverage at only 6.3 percent.

Figure 2-3 presents the contribution of the Parks in Peril program to the overall park-based protection of Latin America and the Caribbean. For this analysis, IUCN categories have been combined, and PiP sites (AID/TNC and TNC PiP sites combined) have been highlighted. The PiP contribution to total regional coverage is highest in Central America (8.5 percent), and under 2 percent in the other three regions. PiP sites constitute roughly half of all sites in Central America. These two figures are not surprising given that the sixty PiP sites represent only 3.2 percent of the 1,901 total protected areas.

Figure 2-4 shows the distribution of PiP sites across IUCN management categories. Parks in Peril contribution to overall protection is concen-

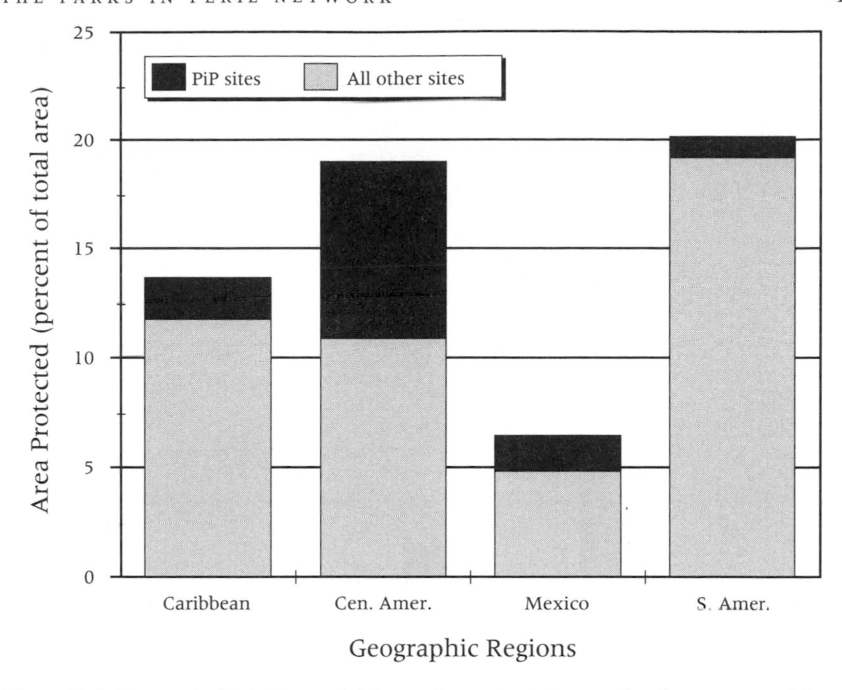

Figure 2-3. Percent of total terrestrial area in protected areas for four geographic regions in Latin America and the Caribbean. Data are grouped to highlight PiP component of protected-area coverage relative to all other protected areas. *Source:* McNeely, Harrison, and Dingwall 1994; The Nature Conservancy, unpublished data.

trated in category II, national parks (thirty-one sites), which constitute 57.8 percent of the area in the PiP program. The next most represented category in PiP as percentage of area is category VIII, multiple-use and managed resource areas (17.9 percent), representing five sites. Categories IV (ten sites), V (three sites) , I (four sites), and III (one site) make up 11.2 percent, 7.8 percent, 4.8 percent, and 0.5 percent, respectively, of the area found in all PiP sites. As stated earlier in the chapter, biodiversity conservation objectives are most closely related with IUCN management categories I–V, yet the second-largest category of protection (by area) within the PiP program is category VIII. One of these sites, Pacaya-Samiria National Reserve in Peru, accounts for 68 percent of the PiP area in that category, and at 2.1 million hectares, it is the second-largest PiP site. Moreover, Pacaya-Samiria has (as do many other category VIII sites) a core protected area in addition to multiple-use zones.

Figures 2-5 and 2-6 provide differing regional perspectives on PiP

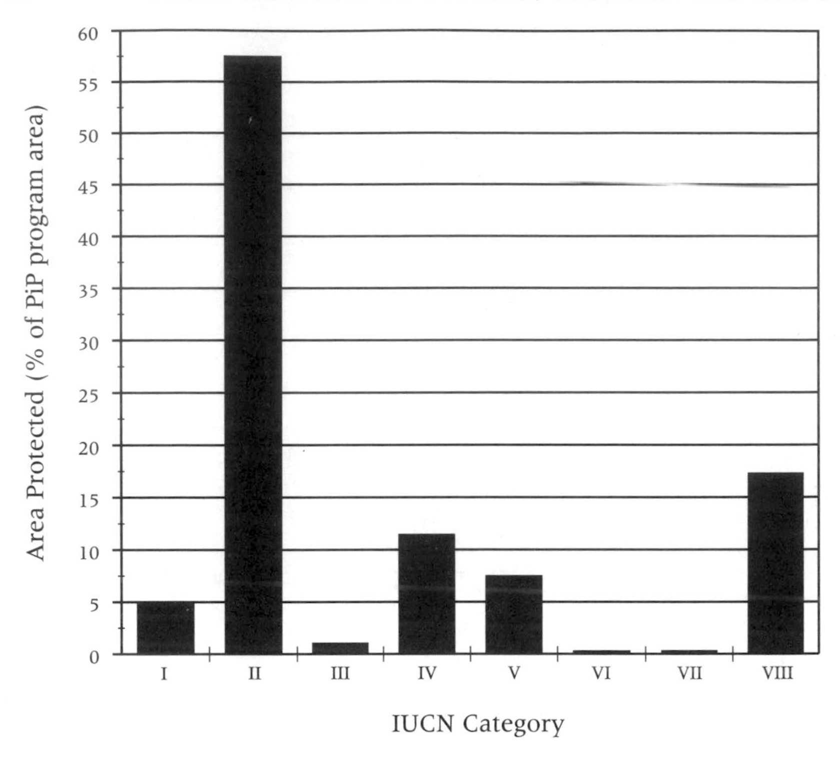

Figure 2-4. Distribution of total area in PiP sites across IUCN categories I–VIII. *Source:* The Nature Conservancy, unpublished data.

coverage. In figure 2-5, AID/TNC PiP and TNC PiP sites are shown as a percentage of all protected areas in the region. In Central America, where the Conservancy works in all seven countries, PiP sites account for 44.3 percent of total area under protection. In Mexico, 21.8 percent of protected areas are included in nine PiP sites. With the exception of Cuba, there are PiP sites in all the larger Caribbean islands and PiP sites account for 12.8 percent of designated protection. The relatively small overall protection in South America (3.7 percent) is due to two factors. First, there is currently no Parks in Peril presence in several South American countries (Argentina, Chile, French Guiana, Guyana, Suriname, and Uruguay); and second, the large difference in scale tends to make coverage in South America appear minimal. Figure 2-6 shows actual PiP coverage in hectares in the four regions. This gives a more accurate programmatic picture of Parks in Peril; almost half of the sites (twenty-six) are in South America, covering 13 million hectares.

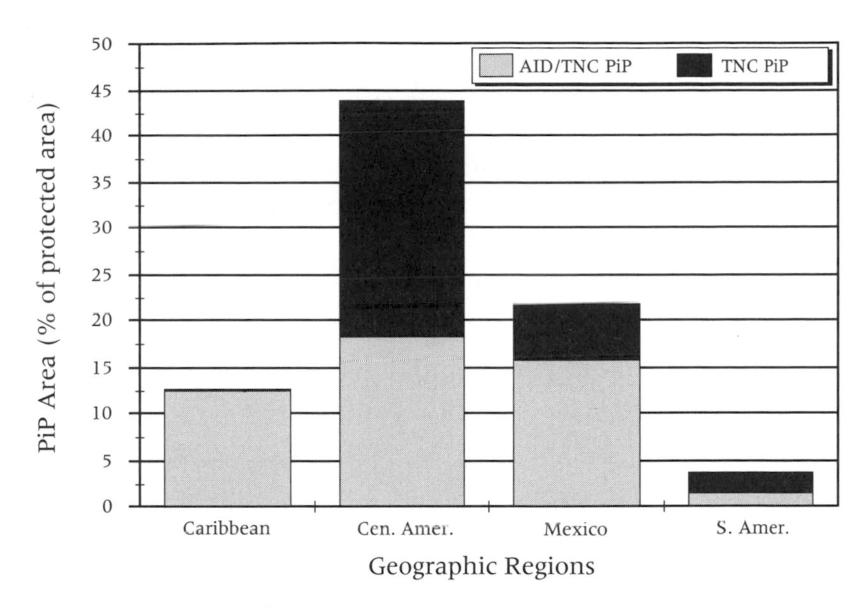

Figure 2-5. Coverage by Parks in Peril sites as a percentage of all protected areas for four geographic regions in Latin America and the Caribbean. *Source:* The Nature Conservancy, unpublished data.

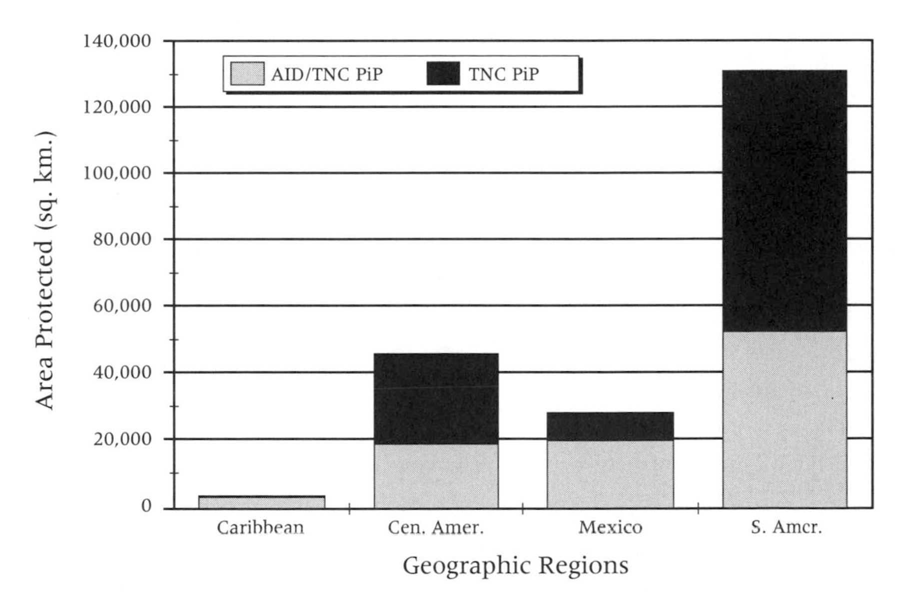

Figure 2-6. Total area, in square kilometers, of Parks in Peril sites for four geographic regions in Latin America and the Caribbean. *Source:* The Nature Conservancy, unpublished data.

These data provide an overview of the geographic distribution of Parks in Peril sites and their relative importance to the protected-area networks in individual countries and regions. They do not, however, provide insight into the representation of natural communities or ecological dynamics in this set of sites, which is a vital component in determining how they collectively contribute to the protection of biodiversity throughout the region. This discussion is presented below.

Parks in Peril Sites and Ecoregional Conservation

Ecoregions

Ecoregional planning has recently been adopted by a number of conservation organizations to focus biodiversity protection efforts and improve and systematize conservation investments. A set of ecoregions (map on next page) for LAC was developed during two recent biodiversity priority-setting initiatives commissioned by the World Bank (Dinerstein et al. 1995) and the United States Agency for International Development (BSP et al. 1995). The ecoregions were developed by staff of the World Wildlife Fund's Conservation Science Department. Staff of The Nature Conservancy have participated in the refinement and analysis of this classification. Both the World Bank and USAID have endorsed these ecoregions and are now using them for conservation planning (The Nature Conservancy 1997). The Nature Conservancy and USAID have incorporated this ecoregion classification system (and derivatives from it) into their decision-making process for conservation allocations and Parks in Peril establishment.

An ecoregion is defined as a geographically distinct assemblage of vegetation types that share a large majority of their species, ecological dynamics, and environmental conditions, and whose ecological interactions are critical for their long-term persistence (The Nature Conservancy 1997). One hundred eighty-two terrestrial ecoregions were developed for this classification. These ecoregions were aggregated into eleven major habitat types (MHTs): (1) Tropical Moist Broadleaf Forests, (2) Tropical Dry Broadleaf Forests, (3) Temperate Forests, (4) Tropical and Subtropical Coniferous Forests, (5) Grasslands, Shrublands, and Savannas, (6) Flooded Grasslands, (7) Montane Grasslands, (8) Mediterranean Coastal Shrublands, (9) Deserts and Xeric Shrublands, (10) Restingas, and (11) Mangroves. MHT distributions are largely controlled by climatic and

Ecoregions of Latin America and the Caribbean

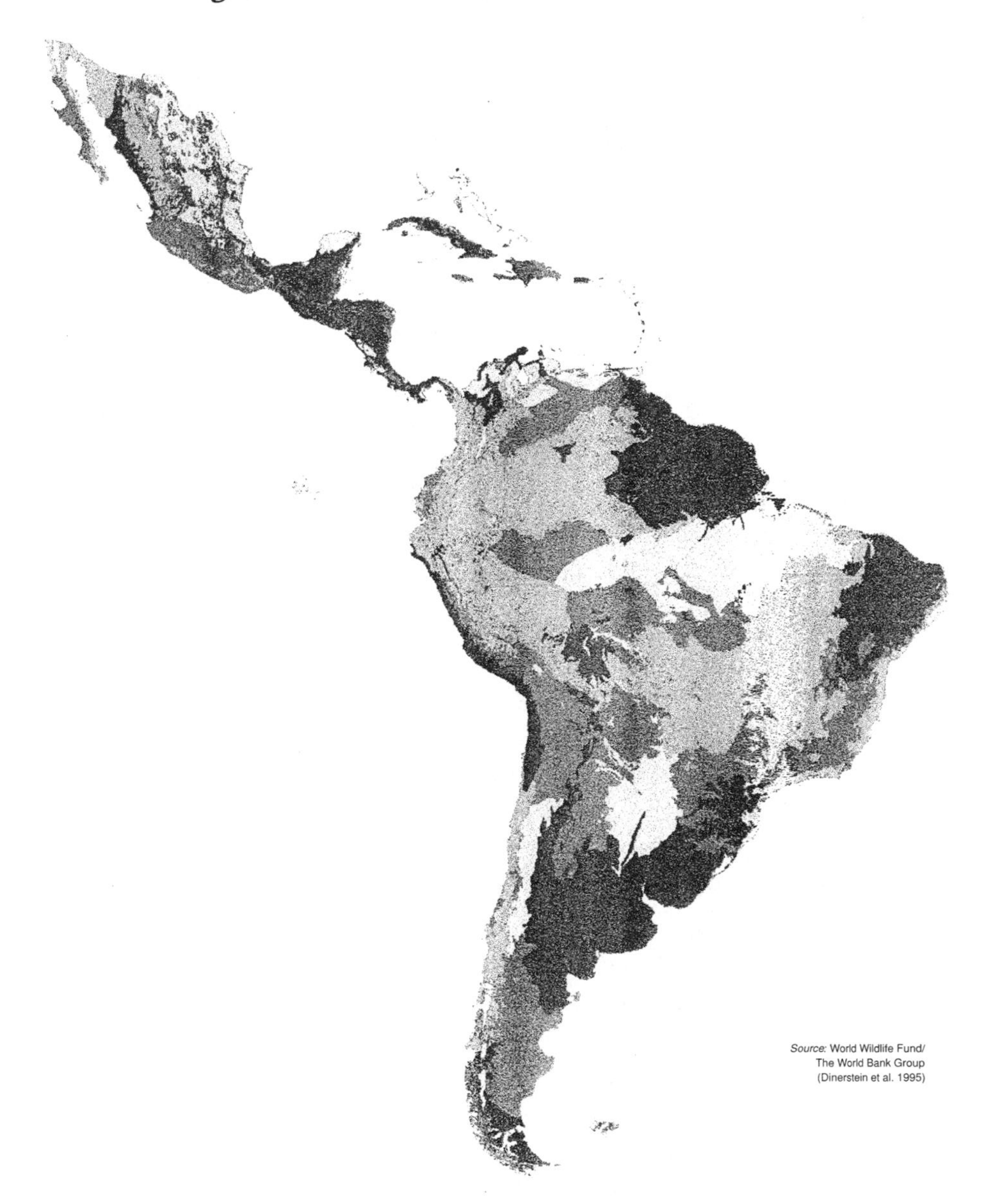

Source: World Wildlife Fund/
The World Bank Group
(Dinerstein et al. 1995)

geologic factors and are widely geographically distributed; ecoregions represent geographically discrete units of MHTs.

Ecoregions represent the original (estimated) extent of vegetation complexes, and thus are more reflective of potential vegetation, and do not describe current vegetation distributions. Ecoregions were developed from an analysis of over 250 maps and textual descriptions of the vegetation of Latin America and the Caribbean, including existing ecoregion classification systems and continental, regional, national, and subnational scale vegetation maps. Final biodiversity ranks were assigned to each ecoregion based on an assessment of an ecoregion's conservation status and biological distinctiveness. The final biodiversity priorities were: I, Highest Priority at Regional Scale; II, High Priority at Regional Scale; III, Moderate Priority at Regional Scale; and IV, Important at National Scale. For a detailed description of the ecoregions, their biodiversity priority assignments, and the source information used in their development, refer to Dinerstein et al. (1995).

Ecoregional Analysis Approaches

Digital mapping technologies allow an assessment of the areal extent of ecoregion protection afforded by PiP sites and a comparison with ecoregion protection from other protected areas. To accomplish this analysis, a geographic information system (GIS) data layer was created by digitizing the best available source maps of PiP sites. The general scale of data development ranged from 1:24,000 to 1:250,000; 1:50,000 was the most common scale. This GIS dataset of PiP sites was used for the gap analysis. The PiP protected areas were combined in a spatial overlay analysis with the ecoregion data layer, and descriptive statistics were obtained. Results are described from a number of different subsetting perspectives including major habitat types, ecoregion ranks, and type of PiP site.

Results of the Ecoregional Conservation Analysis

Including mangroves, there are 182 ecoregions throughout Latin America and the Caribbean. The number of fragments that make up these ecoregions varies from one unit to over a thousand (e.g., Subpolar Nothofagus Forests with 1,024 fragments, mainly islands). Most ecoregions have less than 20 fragments, and many have fewer than 10. The Mangroves ecoregion, widely distributed throughout LAC, is composed of 2,001 fragments. The average size of an ecoregion is 11,288,954 ha;

median ecoregion size is 3,260,002 ha. Ecoregion sizes range from a minimum of 2,267 ha (Cayman Islands Xeric Scrub) to a maximum of 207,746,220 ha (Cerrado), more than an order of magnitude difference. Variation in the size of ecoregions is an important consideration when describing their protection by parks. For this reason, this analysis emphasizes the percent protection of an ecoregion, rather than area totals.

Of the 182 ecoregions, 69 contain PiP sites. For these 69 ecoregions, the average protection afforded by the parks was 10.27 percent. From a land area perspective, the 5 ecoregions with the greatest PiP protection are: Galapagos Islands Xeric Scrub (100 percent), Santa Marta Paramo (98 percent), Quintana Roo Wetlands (85 percent), Eastern Panamanian Montane Forests (61 percent), and Santa Marta Montane Forests (45 percent). An additional 10 ecoregions have between 10 percent and 45 percent protection by PiP sites, for a total of 15 of 69 ecoregions with more than 10 percent protection.

An analysis of the frequency distribution of the percent area of ecoregions contained in PiP sites is presented in figure 2-7. It should be noted that some PiP sites protect more than one ecoregion. The majority of the ecoregions in the 0 percent to 1 percent protection class are the 113 ecoregions in which there are no PiP sites, plus an additional 22 sites in which total ecoregional area protected is less than 1 percent. The remaining

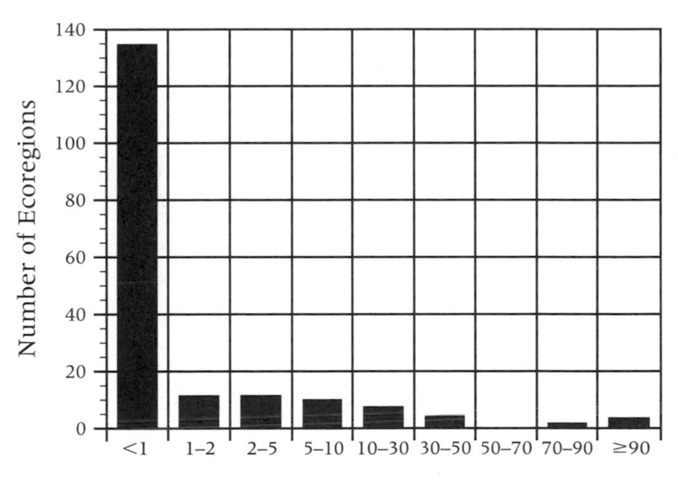

Figure 2-7. Frequency distribution of the percent area ecoregional protection by all PiP sites. *Source:* The Nature Conservancy, unpublished data.

ecoregions are distributed fairly uniformly throughout the percent protection classes, tapering off to a small number (less than 10 ecoregions) with greater than 70 percent protection afforded by PiP sites. An area target of 10 percent protection has been proposed as appropriate (McNeely and Miller 1984) for a global network of parks in terrestrial and marine regions. Using an arbitrary 10 percent protection target for ecoregion-based coverage, 15 ecoregions could be considered adequately protected by the PiP network alone.

Figure 2-8 presents the protection status of all ecoregions grouped by major habitat type. This figure illustrates the area of ecoregions, summed across ranks, protected by all PiP sites. Although PiP sites make a significant contribution to the overall protection of most MHTs; there is no PiP coverage in three of the eleven MHTs (Mediterranean Coastal Shrublands, Tropical and Subtropical Coniferous Forests, and Restingas). Total protection across ecoregion rank is highest in the Mangroves MHT (2.8 percent), followed by Flooded Grasslands (nearly 2 percent), and Tropical Moist Broadleaf Forests (1.6 percent). Relatively high PiP protection in two MHTs, Flooded Grasslands and Deserts and Xeric Shrublands, is primarily achieved by the presence of two large parks, the Pantanal in Brazil and El Pinacate in Mexico. For number of sites in ecoregions, grouped by MHT (figure 2-9), it is evident that there is a concentration of PiP sites, both AID/TNC PiP and TNC PiP, in Tropical Moist Broadleaf Forests. This result is indicative of the historical tendency toward biodiversity conservation in tropical rainforests (Redford, Taber, and Simonetti 1990) and illustrates how representation of unique habitats has not been a primary criterion for PiP site selection.

Summed across MHTs, the distribution of PiP sites into ranked ecoregions (figure 2-10) reveals that there is a greater number of priority-one ecoregions contained in PiP sites than lower-priority ecoregions. Grouped by MHT, however, and considering area protected instead of number of parks (figure 2-11), a concentration of protection in ecoregions of higher ranks is not observed, except in the case of Tropical Moist Broadleaf Forests. For many MHTs, protection is distributed throughout the priority rankings. The finding that PiP protection is not concentrated in the higher-priority ecoregions is expected, as placement of Parks in Peril sites in the past was not implemented using an ecoregional framework.

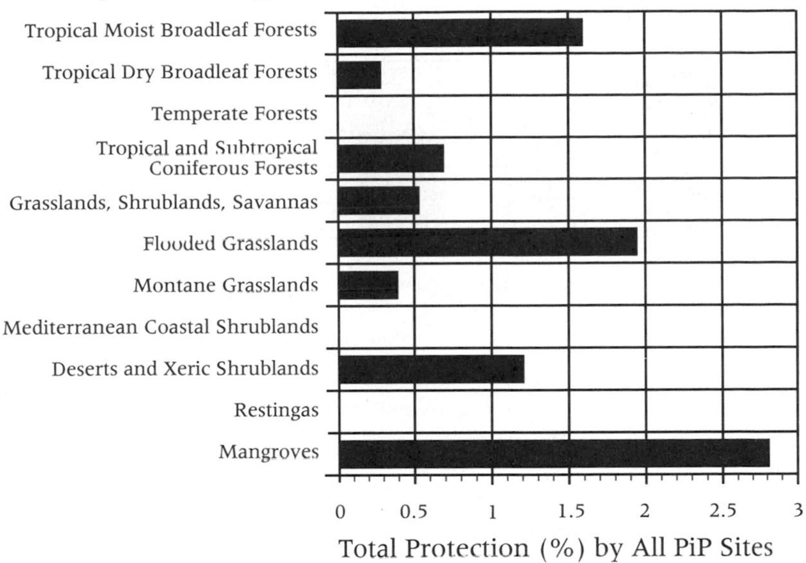

Figure 2-8. Total percent protection of ecoregions by all PiP sites ($n = 60$), grouped by major habitat type. *Source:* The Nature Conservancy, unpublished data.

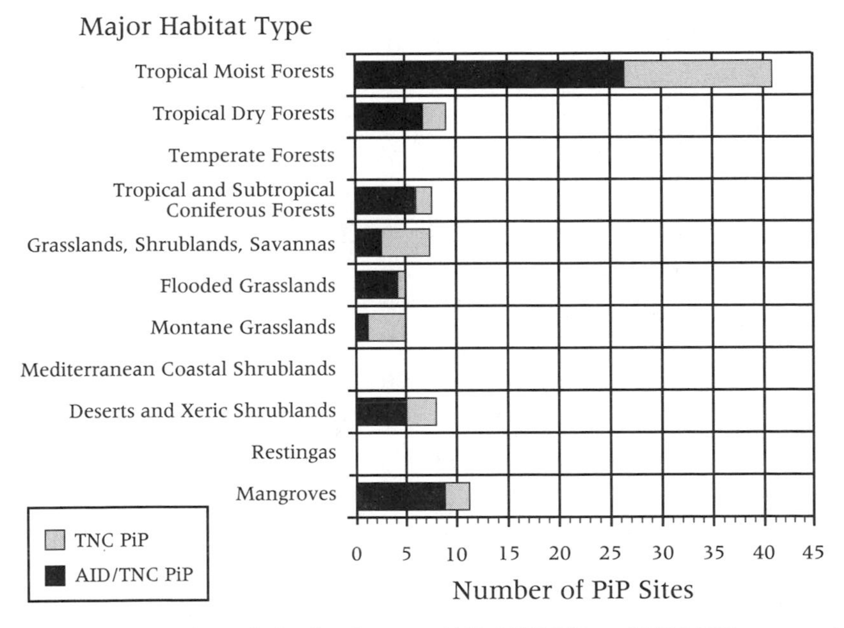

Figure 2-9. Number of PiP sites by type (AID/TNC PiP and TNC PiP), grouped by major habitat type. *Source:* The Nature Conservancy, unpublished data.

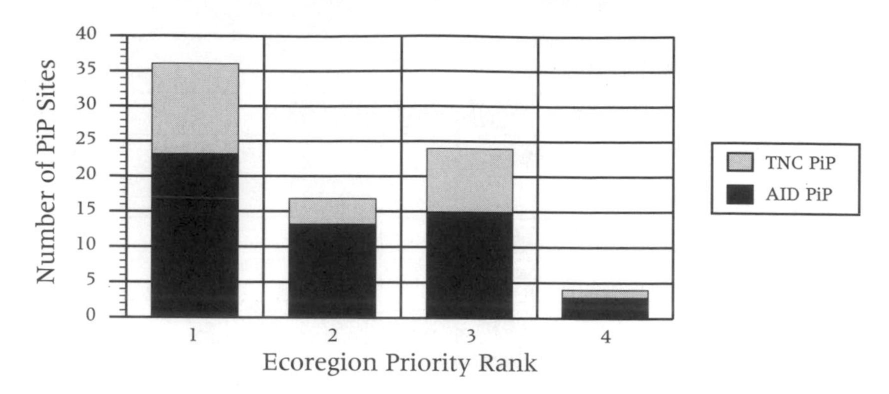

Figure 2-10: The number of PiP sites by type (AID/TNC PiP and TNC PiP), distributed across ecoregion ranks 1 to 4 (1 being highest priority). Multiple counting of PiP sites occurs when a PiP site contains multiple ecoregions. *Source:* The Nature Conservancy, unpublished data.

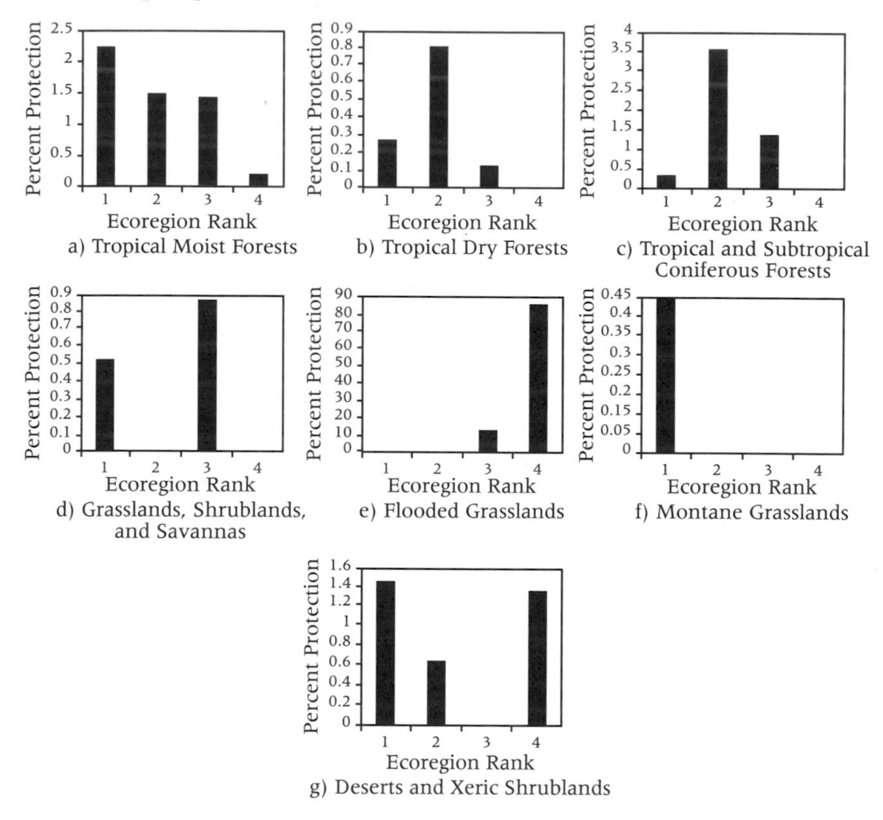

Figure 2-11. Total PiP protection (%) of ecoregions, grouped by rank, for seven of the eleven major habitat types. Restingas, Mediterranean Coastal Shrublands, and Temperate Forest MHTs are not shown, as no PiP sites occur in those MHTs. Mangroves are not shown, as they are all one rank—highest priority. *Source:* The Nature Conservancy, unpublished data.

Biodiversity within the PiP Case Study Sites

The nine PiP sites characterized in the remainder of this book are important from an ecoregional perspective. Total percent ecoregional protection by each of the case study sites is presented in table 2-3. Eight of the nine sites contain highest-priority ecoregions.

Several of the sites contain more than one ecoregion, and Amboró is unusual in containing three top-priority ecoregions. The Sierra de las Minas site does not contain a highest-priority ecoregion, but it does contain a very large portion (32 percent) of a high-priority ecoregion. Of these nine sites, the highest diversity in ecoregions is observed in Rio Bravo, with a total of four ecoregions represented.

Conclusion

The major findings of this study are as follows:

1. Total percent protected-area coverage is highest in South America (20 percent), followed by Central America (19 percent), the Caribbean (13 percent), and Mexico (6 percent).

2. PiP sites provide nearly half of the total protected-area coverage of Central America, and 22 percent, 13 percent, and 3.5 percent of the total protected-area coverage of Mexico, the Caribbean, and South America, respectively.

3. While protection of several habitat types (e.g., Moist Tropical Forest, Dry Tropical Forest, and Montane Grasslands) is significant, three out of eleven major habitat types (Temperate Coniferous Forest, Mediterranean Coastal Scrub, and Restingas) are not represented in the PiP network of protected areas.

4. Parks in Peril sites contain portions of 69 out of 182 ecoregions. Protection of ecoregions by Parks in Peril sites ranges from 0 percent (no PiP sites in the ecoregion) to 100 percent (an ecoregion completely contained in a PiP site); mean protection of the 69 ecoregions by PiP sites is 10.27 percent.

5. Parks in Peril sites are distributed throughout the range of ecoregional priority classes (I, Highest Priority at Regional Scale; II, High Priority at Regional Scale; III, Moderate Priority at Regional Scale; and IV, Important at National Scale) and are generally not concentrated in the higher-priority (I and II) ecoregions.

Table 2-3. Ecoregional protection by the nine PiP case study sites.

Site	Ecoregion	Rank	Protection (%)
Ría Celestún-Ría Lagartos	Yucatán Dry Forests	3	0.01
	Mangroves	1	2.20
Sierra de las Minas	Central Amer. Atlantic Moist Forest	3	0.16
	Central Amer. Pine-Oak Forest	3	1.14
	Motagua Valley Thornscrub	2	31.8
Corcovado	Isthmian-Pacific Moist Forests	2	1.90
	Mangroves	1	0.01
Parque del Este	Hispaniolan Moist Forests	1	0.77
	Mangroves	1	0.02
Rio Bravo	Petén Moist Forests	1	0.35
	Yucatán Moist Forests	3	0.09
	Belizean Swamp Forests	3	12.7
	Belizean Pine Forests	2	3.69
Machalilla	Ecuadorian Dry Forests	1	2.06
Podocarpus	Eastern Cordillera Real Montane	1	1.76
Amboró	Southwestern Amazonian Moist Forests	1	0.35
	Bolivian Yungas	1	5.96
	Beni Savannas	1	0.002
Yanachaga-Chemillén	Ucayali Moist Forests	1	0.13
	Peruvian Yungas	1	0.42

These area-based characterizations of ecoregion protection by PiP sites provide a quantitative conservation perspective but do not characterize the quality of conservation management in any ecoregion and must therefore be interpreted with caution. Moreover, ecoregions represent potential, not actual, vegetation, and this analysis of the degree of protection of entire ecoregions does not discriminate between altered and unaltered landscapes. Even though a park may protect a very small percentage of the total area of an ecoregion, it may be protecting the only remaining natural communities in that ecoregion.

In spite of these caveats, however, these results provide a meaningful first look at the protected-area quantity dimension of the PiP network. On a regional and national level, protected-area coverage by PiP sites (and other sites) appears substantial, often exceeding an arbitrary 10 percent land-area basis level. Ecoregion protection by protected areas in Latin America and the Caribbean is, however, currently inadequate, with many ecoregions virtually completely unprotected. A greater and more directed conservation effort is necessary to bring total percent protection of ecoregions to appropriate levels. Protection is not currently concentrated in ecoregions with relatively higher conservation priorities. Further development of protected areas in less protected major habitat types (especially Mediterranean Coastal Shrublands) is encouraged in order to achieve representation as a fundamental conservation goal. Site-based conservation planning within these protected areas should set target goals for sustaining species and natural communities over time.

If representation is identified as a fundamental goal of PiP, then efforts to place future PiP sites in Temperate Forests, Mediterranean Coastal Shrublands, and Restingas would achieve representation at the MHT level. Future incorporation of new PiP sites should also consider ecoregion priority as a criterion for establishment, and sites should be placed in higher-priority ecoregions.

Finally, the case study sites analyzed in greater detail in chapters 4 through 12 make an important contribution to ecoregion protection. Almost all of these sites contain highest-priority ecoregions, and most contain multiple ecoregions as well. One site (Amboró) contains three highest-priority ecoregions, and another (Sierra de las Minas) contains nearly a third of an entire high-priority ecoregion.

3. Analyzing the Social Context at PiP Sites

Barbara Dugelby and Michelle Libby

Just as the protected areas within the Parks in Peril portfolio are found within a wide range of ecoregions, they also are endowed with a rich socioeconomic and cultural diversity. With very few exceptions, all of the Parks in Peril sites are surrounded by human populations that in one way or another interact with and affect the natural systems of the protected area. Often the communities that surround a protected area depend on the park's resources for their livelihood. As currently carried out, many of the subsistence activities of local communities are not compatible with the ecological integrity of parks and in fact pose serious threats to the conservation of biological diversity in these areas. Working with these communities to ensure the biological integrity of the protected areas is a major emphasis of the Conservancy's work with partner organizations.

This chapter describes elements of the social context of the original twenty-eight PiP sites that receive support from the U.S. Agency for International Development (see chapter 1). First, a summary of the social context of the PiP sites is presented, including information on the populations at sites, ethnicity, and key threats. Second, the range of community-based conservation activities being implemented by the Conservancy's partner organizations is described. The chapter concludes with a general discussion of the community-based conservation work of the Conservancy's partner organizations and possible future directions of the Conservancy's community-based conservation work in Latin America and the Caribbean.

Description of Communities around Sites

The twenty-eight PiP sites are as diverse culturally and politically as they are biologically. Some protected areas have few inhabitants, while others have a large and diverse local population. For example, El Pinacate

Biosphere Reserve in the Sonoran Desert of Mexico covers 794,556 hectares but has only approximately 200 inhabitants. Similarly, Noel Kempff Mercado National Park has close to 3,500 people living around an area of 1,600,000 hectares. Conversely, sites like the Panama Canal Watershed have about 130,000 people living in and around 151,104 hectares of protected land.

In many sites, human communities are located within the boundaries of the protected area. For example, Sierra Nevada de Santa Marta National Park in Colombia encompasses over 382,996 hectares and has 17,750 people living within the park boundaries. In terms of protection status and the presence of human inhabitants, protected-area categories are not standard across countries. For example, in one country, national park status typically guarantees that no one is allowed to live within the park's boundaries. However, in another country, government agencies may tolerate or permit residents in national parks. Commonly, legislation governing the protection status awarded different types of parks does not vary much from country to country. In most countries, laws prohibit human habitation of national parks and biological reserves, for example. However, the legal framework established for protected areas often conflicts with the reality of conditions on the ground. For example, Ecuadorian law does not allow residence in national parks, yet at Machalilla National Park (chapter 9), the park's boundaries were superimposed over existing communities that have not been relocated.

The ethnic diversity of people living in and around PiP sites is as far ranging as the size of local populations. A majority of the sites have mestizo (of mixed Spanish and Native American descent) populations, yet a variety of indigenous groups and other unique ethnicities (e.g., various black-Latino groups such as the Caribe, Garifuna, Black Colombians, etc.) are found throughout the sites.

Thirteen of the twenty-eight PiP sites have substantial mestizo populations, and out of these only two, Chingaza National Park in Colombia and Machalilla National Park in Ecuador, do not have indigenous populations. Many sites are populated primarily by indigenous groups. For example, Mbaracayú National Park in Paraguay is home to the hunting-and-gathering Aché. Guaraní Indians also live near the nature reserve. Other sites, such as the Sierra Nevada de Santa Marta and Cahuinari National Parks in Colombia, have enormous diversities of ethnic groups—nine different indigenous groups are found in these two parks alone.

In addition to indigenous and mestizo groups, several sites have black populations. Afro-Caribbean live in the Talamanca Biological Corridor in Costa Rica and Black Colombians in the Darién Biosphere Reserve of Panama. Creoles live near Rio Bravo Conservation Management Area in Belize and within Del Este National Park in Dominican Republic. Tariquía National Reserve in Bolivia has a substantial mulatto population. Among the more unexpected ethnicities is a large community of Mennonites that is engaged in intensive agriculture adjacent to Rio Bravo, and a European and Asian community near Yanachaga-Chemillén National Park in Peru. Table 3-1 shows for each PiP site the size of the protected area, the population of surrounding communities or those within the site's area of influence (including buffer zones and sustainable use zones), the total number of communities, and the different ethnic groups living at that site.

Description of Threats at PiP Sites

All of the PiP sites are threatened by a range of human activities, whether rooted in national policies or local livelihood needs. Human populations in and around all of the protected areas engage in a variety of activities that create threats to the integrity of the sites. We base our description and analysis of locally induced threats on the PiP site annual evaluations, which include a section for outlining key threats to the protected area. Of the twenty-eight PiP sites, three have identified and ranked their threats and are addressing them through management actions: Yanachaga-Chemillén National Park in Peru (chapter 12), Mbaracayú Forest Nature Reserve in Paraguay, and Rio Bravo Conservation and Management Area in Belize (chapter 8). Nine sites have not yet implemented a strategy to address threats, and of these most have completed only a basic identification of threats. For this reason, we can present only the frequency with which threats are reported across the sites.

As revealed in figure 3-1, the most commonly reported (over 70 percent) local activity threatening protected areas is hunting or poaching. However, since this information is based on the frequency with which partners report on threats and not on a standardized threats analysis methodology, this figure does not reveal whether hunting is the *most* menacing threat to the ecological integrity of the sites for which it is reported. We can say only that hunting is the most frequently reported threat.

Table 3-1. Population sizes and ethnicities of PiP sites.

Name of Park	Park Size (ha)	Population in or around Park	Number of Communities	Ethnic Groups
Morne Trois Pitons	6,860			
Talamanca Biological Corridor	36,516	10,000		Afro-Caribbean, Cabécar, Kéköldi
Corcovado NP	**41,788**	**11,922**		**Guaymi**
Del Este NP	**42,000**	**2,156**	**3**	**Mulatto**
El Ocote Ecological Reserve	48,140	4,204	21	Tzotzil, Tzetzal
Chingaza NP	50,395			
Machalilla NP	55,020	11,586		
Mbaracayú Forest Nature Reserve	64,405	7,00	31	Aché, Guaraní
Rio Bravo CMA	**92,614**	**8,760**	**>9**	**Maya, Mennonite, Creole**
Pampas del Heath NS	102,109	560	3	Ese eja
Ría Lagartos/Celestún BRs	**106,883**			**Maya**
Yanachaga-Chemillén NP	**122,049**	**33,781**	**28**	**Yanesha, Amuesha**
Jaragua NP	137,445	22,519	14	
La Encrucijada BR	144,868	25,398	68	
Podocarpus NP	**146,339**	**7,356**	**105**	**Shuar**
Panama Canal Watershed	151,104	127,962		
Soberanía		59,000		
Chagres		68,000		

Sierra de las Minas BR	**236,032**	**35,000**	**110**	**Poqomchi' Q'eqchi'**
Tariquía National Reserve	246,964	4,500		Mulatto
El Triunfo BR	296,356	3,954	16	
Sierra Nevada de Santa Marta NP	382,996	17,750		Aruaco, Arsario, Kankuamo, Kogi, Tairona
La Paya NP	422,171			Coreguaje, Huitoto, Inga, Ingano, Siona
Amboró NP	**442,500**	**18,419**	**92**	**Quechua**
Cahuinari NP	574,899			Bora, Matapi, Miranya, Yacuana
Darién BR	597,166	14,000		Black Colombians, Embera, Kuna, Wounan
Sian Ka'an BR	617,515			Maya
Calakmul BR	723,482	13,000	75	
El Pinacate BR	794,556	165–200		Tohono O'Odham Nation
Noel Kempff Mercado NP	1,600,000	3,459	9	Chiquitoan

Note: Based on information as provided by partners.

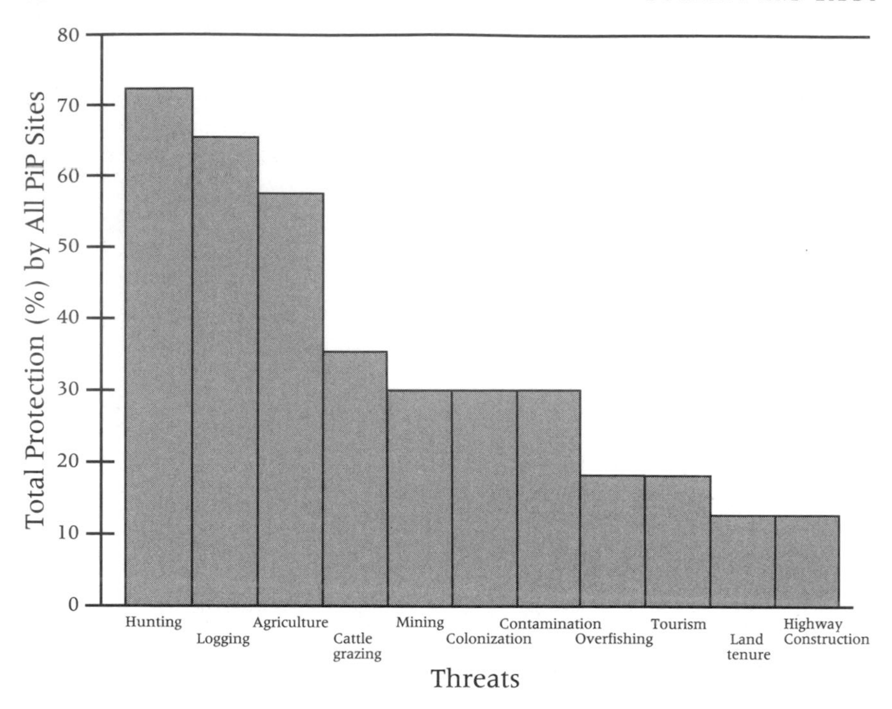

Figure 3-1. Types of threats most commonly reported by partners at PiP sites. *Source:* INE 1995.

Logging activities follow hunting as the second most prevalently reported threat, with over 60 percent of the sites reporting detrimental timber harvesting activities. Agricultural encroachment is reported to affect approximately 55 percent of the sites and is followed by cattle grazing, which is reported to affect 34 percent of the sites.

Not all of the threats reported for the central PiP sites are included in figure 3-1. Many threats reported are common only to a small number of sites. However, they do merit mention since for certain sites they represent the primary threat to an area. Inadequate or conflicting national policies affecting natural resource management, conservation, and agricultural practices were listed as threats at several sites. Another common problem is infrastructure development, particularly in areas that are either heavily populated or rapidly becoming ecotourism destinations. This problem is acute in many protected areas in the Caribbean and in Central America, as the remaining natural areas in these regions are being encroached upon by hotels, roads, and dams. Fires, whether caused by slash-and-burn practices or intentionally set, and the extraction of rare plants are threatening park integrity at several sites. Finally, drug production, processing, and trafficking, particularly of marijuana and coca products, are problematic for several South American sites.

Threats can also be very site specific, as is the case in La Encrucijada Biosphere Reserve in Mexico, where the current main threat to the area is the Chiapas Coast Hydrologic Program implemented by the National Water Commission and financed by the World Bank. Similarly, in the Darién Biosphere Reserve in Panama, a pending threat is the planned extension of the Pan-American highway through the Darién Gap to Colombia, which would expose the reserve to increased logging pressure and colonization.

Community-Based Activities Implemented by Partners

Partners have had differential success when attempting to integrate park management and conservation activities into the lives of the surrounding communities. Before true integration of local communities with site management can begin, partner organizations and park administration staff must win the trust and cooperation of community members. Establishing good relations with communities as a basis for designing and implementing conservation activities that engage local residents is not an easy task. Partners use a variety of strategies for establishing good relationships and winning the trust of local communities, and the success rate of these techniques varies greatly from site to site.

Below we review the range of strategies and activities conducted by partners for involving local residents in protected-area planning and management. We divide these strategies and activities into four general categories: local awareness building and environmental education, natural resource management, compatible economic development, and protected-area management.

Local Awareness and Environmental Education

Local awareness-building activities form the majority of the community initiatives conducted by partners. These initiatives often consist of environmental education activities, the most common of which are community meetings, workshops, and presentations; slide and video shows; and environmental fairs. Environmental education programs are underway at twenty-seven of the twenty-eight PiP sites. Some of the environmental education programs are better established than others, but none has developed a method to measure the impact they are having in the communities or on threat mitigation.

One of the most advanced environmental education programs is implemented by Programme for Belize (PfB) at the Rio Bravo Conservation and Management Area in Belize (chapter 8). One of the main goals of PfB is to promote concepts of sustainable development and the interrelationship of conservation and economic development through education and outreach. A majority of the outreach and environmental education activities are directed toward target schools and neighboring communities and have been carried out successfully with the help of other local organizations that are interested in helping PfB attain its educational goals.

ANCON, the partner NGO for the Darién Biosphere Reserve in Panama, has built an environmental education center where groups can stay and learn about land-use patterns, reforestation, wildlife management, agroforestry, and park ranger training. Similarly, in the Panama Canal Watershed, ANCON created the Río Cabuya Agroforestry Demonstration Farm, where peasants and farmers from all over the watershed can attend free training sessions regarding economic development issues and sustainable development, or be trained in reforestation, agroforestry, and soil conservation techniques. Fundación Neotrópica in Costa Rica also has developed a Tropical Center for Youth for environmental education and community outreach activities near Corcovado National Park (chapter 6).

Several partner organizations have established formal outreach-extension programs for the purpose of establishing good relations with surrounding communities. For example, the Outreach and Environmental Culture Program of El Ocote Biosphere Reserve in Mexico has established a presence in sixteen reserve communities. Here, the reserve personnel maintain qualitative data on their relations in six communities within El Ocote. The information includes the degree of confidence achieved, degree of community participation, and impact and disposition of communities to work with the partner organization staff.

The community outreach program at El Pinacate Biosphere Reserve in Mexico has increased the surrounding communities' knowledge of the reserve and their interest in participating in the conservation of the reserve. The partner organization Centro Ecológico de Sonora (CES) has formed a consultative working group composed of representatives from the region's population centers to help with the development of the reserve's management plan. Participants include communal landholders, private landowners, representatives from government and nongovernment institutions, and representatives from the Tohono O'Odham Nation.

Several sites have been unable to design community outreach programs that successfully garner the support of surrounding communities. Common characteristics among such sites are high levels of in-migration to the area, often leading to high numbers of disenfranchised people distrustful of outside organizations. This migration can be spawned by a variety of issues, such as government policies encouraging colonization, or military instability, forcing people to leave their homes and look for safe areas. The latter has been the case in Sian Ka'an, where instability in the Chiapas region forced many to migrate; high numbers of migrants have settled in the region surrounding the Sian Ka'an Biosphere Reserve. One of the main obstacles impeding better community relations at La Paya National Park in Colombia has also been people's lack of trust in outside institutions.

Natural Resource Management

Improved community relations often are sought by partners through the promotion of development or natural resource management projects. Activities that blend elements of traditional natural resource management activities with compatible and sustainable development activities are commonly referred to by partners as sustainable resource management projects. All sites except three have developed some type of sustainable resource management projects. The majority of these are small projects involving a particular community or a select group of individuals. Very few projects actually involve community organizations, or are undertaken in cooperation with major community-based institutions. Livestock breeding initiatives, such as poultry raising and iguana or paca breeding, are common sustainable resource-use activities. Other popular activities are beekeeping, aquaculture, agroforestry, and agroecology.

In Machalilla National Park in Ecuador (chapter 9), the partner organization, Fundación Natura, is carrying out projects that illustrate the mix between natural resource management and compatible development activities. For example, the community of Casas Viejas traditionally has been dedicated to subsistence agriculture, raising of goats, and sale of timber. With support from PiP, a beekeeping project has been initiated.

The partner hypothesizes that through this project, villagers will obtain increased income from the sale of honey and perceive incentives to protect the forest, as it provides the habitat necessary for the survival of the bees.

A similar example is that of vegetable ivory artisanry. Again in Machalilla, communities have begun to collect the seeds of a palm species native to the park that contains an extremely hard nut, used in the early part of this century for the manufacture of buttons. The palm nut, also referred to as vegetable ivory for its hard consistency and color, traditionally has been sold in bulk (by weight) by the communities to outside buyers. Fundación Natura is working to provide training to local residents in carving figurines from the nut, to be sold as handicrafts, adding significant local value to the product through local processing and design.

Compatible Economic Development

Several sites use compatible development activities as the vehicle for building trust and good relations with communities. These activities are geared toward community economic development and improving the quality of life of local residents, although they may or may not have a conservation benefit. For instance, Fundación Amigos de la Naturaleza (FAN), the partner NGO for Noel Kempff Mercado National Park in Bolivia, has always provided local communities with health services and emergency evacuation assistance. FAN supports monthly community visits by doctors and/or dentists and provides communities with medical supplies.

Instituto de Historia Natural (IHN), working at El Triunfo Biosphere Reserve in Mexico, makes provisions to include environmental health activities in its community development program. With the help of rural clinics, community health committees, and communal land authorities, several communities have received talks on health and prevention of intestinal diseases and family planning. In addition, this program supported the installation of latrines in several communities. Fundación Neotro´pica has trained local people as guides for ecotourists visiting Corcovado National Park in Costa Rica. Through support from PiP, many sites currently are conducting ecotourism feasibility and marketing studies or are in the process of designing ecotourism management plans, many of which will include training the locals to manage and benefit from ecotourism activities.

At Sierra Nevada de Santa Marta in Colombia the partner organization Fundación Pro-Sierra Nevada de Santa Marta (FPSN) has been working with the communities since 1986, six years before the inception of the PiP program. In those six years, through the implementation of projects in education, health, and agriculture, they have succeeded in establishing good relations with the mestizo and indigenous communities. FPSN has emphasized vigorous but low-key extension activities using nonconfrontational, nonpartisan policies and has maintained a dialogue with all groups in the area.

Local Involvement in Protected-Area Management

Prior to receiving support from the PiP program, some partners and protected-area staff had never engaged locals in protected-area management. Others, however, like IHN at El Triunfo Biosphere Reserve in Mexico and the Talamanca-Caribbean Biological Corridor Commission (CCBTC) in Costa Rica, had facilitated an active role of local residents in park management activities long before the PiP project began.

In El Triunfo, communities and individual landholders have been involved with the reserve since the area was officially declared a protected area in 1972. When the area was declared a biosphere reserve in 1990, IHN, which has managed the park since its creation in 1972, formed a local advisory board made up of county officials, landowners, and presidents of communal farms. This local advisory board is currently a key element of reserve management.

The CCBTC is a coalition of thirteen member organizations constituted by organized civil society and local landowners that have been working in the area for over five years. This commission believes that setting up ecologically sustainable and profitable activities in the communities is the best way to get community support and to ensure long-term conservation.

Among the twenty-eight PiP sites it appears that the most popular way of engaging local people in park management activities is by hiring them as local park protection or extension staff. Almost all sites hire locals as park guards and rangers, and at some sites these guards or rangers also have the role of community extensionists working primarily in educating the communities about the park and its purpose. In addition to hiring local community members as outreach extensionists, partners often hire local residents as accountants and field biology assistants. Women are commonly employed as secretaries and cooks but are not confined to those roles. Another common way of employing community members is through contract work for construction of stations, trail maintenance, and demarcation of boundaries.

Conclusion

Creating a strong constituency of local supporters who are involved in planning and implementing community-based projects that successfully mitigate threats to a protected area is a long and slow process (Wells and Brandon 1992). This process most commonly begins with an attempt on the part of outside organizations and park staff to develop good relations with local residents, winning local support for the park and educating locals about the benefits and management needs of the site. As a means of winning local support, NGOs often initiate projects that produce a direct and immediate benefit to locals—responding to the needs most strongly expressed by communities, not necessarily related to conservation or the protected area. Projects typically address issues such as health, sanitation, and general community development.

NGOs assume that after gaining the trust of locals through investing time and resources in their needs, they can enter a phase in which they move closer to issues more directly related to the management of the protected area, such as natural resource management. Focusing on issues of natural resource use allows partners and community residents to address and begin reducing locally induced threats to the protected area, such as hunting, forest extraction, agricultural practices, and fishing. As conservation programs evolve, locals typically play an increasing role in designing and implementing projects and in park management as a whole. Ultimately, locals are fully engaged not only in park management, but also in integrated monitoring programs that provide important feedback on the impacts of their activities.

The description presented in this chapter of the social context of PiP sites and the range of community-based activities currently conducted by the Conservancy's partners reveals that at many sites, partner organizations are still in the early stages of community work. Partners invest substantial amounts of financial resources in local awareness-building activities. Only a handful of partners have been able to advance their work with local communities to the point of linking community-based projects to locally induced threats, and even fewer have initiated monitoring programs to gauge the impacts of those activities. The Conservancy's next challenge is to help partners and protected-area staff move beyond the initial stages of local constituency building to actually reducing and eliminating locally induced threats to parks.

PART II

Nine Neotropical Parks

RÍA LAGARTOS AND RÍA CELESTÚN SPECIAL BIOSPHERE RESERVES

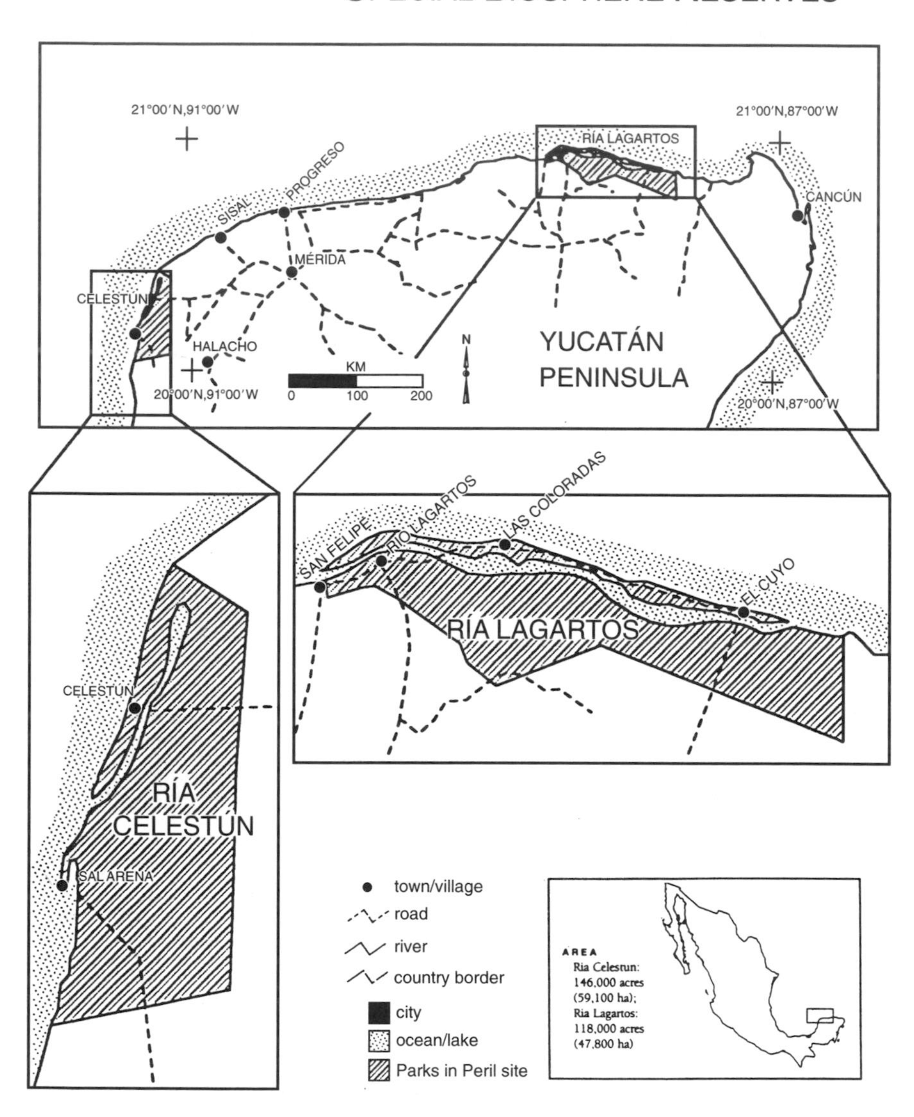

4. Mexico: Ría Celestún and Ría Lagartos Special Biosphere Reserves

Joann M. Andrews, Rodrigo Migoya Von Bertrab, Susana Rojas, Armando Sastré Méndez, and Debra A. Rose

Ría Celestún and Ría Lagartos are biosphere reserves located on Mexico's Yucatán Peninsula. Ría Lagartos is not too distant from Cancún, one of the more popular destinations for North American tourists. Rapid social, economic, and infrastructure changes transformed Cancún from a fishing village to an international tourism destination. While the magnitude of change is less elsewhere on the peninsula, the social context around Ría Celestún and Ría Lagartos is undergoing change as well. Migration into or near both reserves is relatively high, there is increased development of infrastructure such as roads, and there are relatively rapid changes underway in how people earn their livelihood. Ecological changes, both natural and humanmade, are underway as well, and there is often a clear relationship between these ecological changes and the social changes that result. For example, Hurricane Gilbert opened new outlets to the sea, making coastal fishing and fishing within the Celestún lagoon more accessible to fishermen. However, overfishing of once abundant species has converted other fishermen into part-time tour boat operators. This theme of relatively rapid change is perhaps the most dramatic element demonstrated by this case. Changes in tourism policy, land use, population levels, resource use, employment, and government policy are also demonstrated here.

Park Establishment and Management

Ría Celestún and Ría Lagartos were declared wildlife refuges in 1979 to protect estuarine wetlands on the northern coast of the states of Yucatán and Campeche. Coastal dunes, estuaries, mangroves, hummocks, and scrub forests protect a diverse fauna. These include 304 avian species catalogued to date (Correa and Barrón, cited in Arellano 1994), among them resident and migratory waterfowl and the largest mainland flock of American flamingos (*Phoenicopterus ruber ruber*) in the Americas. Also present are a number of protected species including Morelet's crocodiles (*Crocodilus moreletti*), marine turtles (*Eretmochelys imbricata, Chelonia mydas*), jaguar (*Panthera onca*), ocelot (*Felis pardalis*), and three endangered palms locally known as *nakax* (*Coccothrinax readii*), *chit* (*Thrina radiata*), and *kuká* (*Pseudophoenix sargentii*). Both areas were upgraded in 1988 to the status of special biosphere reserves, a national category created for parks not internationally recognized by the Man and the Biosphere program but modeled after the MAB biosphere reserve concept.

Primary responsibility for park administration was transferred in 1992 to the National Institute of Ecology (INE), now within the Ministry of Environment, Natural Resources, and Fisheries (SEMARNAP). In early 1991, the Conservancy initiated a PiP project for Ría Lagartos/Ría Celestún with Pronatura Península de Yucatán (PRONATURA) as its local NGO (nongovernment organization) partner. Together with the Center for Investigations and Advanced Study (CINVESTAV) of Mérida, PRONATURA cooperates with both state and federal offices of SEMARNAP in reserve management and conservation of these reserves.

The Ría Celestún Reserve is located on the northwestern tip of the Yucatán peninsula, with a declared area of 59,130 hectares in the states of Yucatán and Campeche. Isla Arena, a small island that forms part of the Ría Celestún Reserve, is in the state of Campeche. The coordinates included in the decree are not sufficiently complete, however, to allow accurate delimitation of the reserve's boundaries. A similar error exists in the decree for the establishment of the Ría Lagartos Reserve on the eastern border of the state of Yucatán. The official area of the reserve is 47,820 hectares, whereas the true boundaries encompass an area of 55,350 hectares. The 1993 management plan for Ría Lagartos therefore creates a core zone of 20,996 hectares and a buffer zone of 34,354 hectares. Decrees amending the boundaries of both reserves are expected to be issued in 1996 (see "Postscript" at end of chapter).

Although both reserves are critical to the life cycle of the American flamingo, they are located at opposite ends of a long expanse of coastal wetlands. The state reserves of El Palmar, to be managed jointly with the Celestún reserve under an agreement signed in February 1995, and

Dzilam de Bravo have been created in the coastal zone and serve as buffer zones for the two federal reserves. Much of the central coastal zone, however, is threatened by rapid coastal development. A proposal has therefore been put forward by the Wetlands Board of Trustees of the Yucatán, a committee formed by state and federal agencies, NGOs, and the State University of Yucatán, to create a coastal bioreserve for the protection of coastal and wetland areas along the entire coastal zone of the state.

Well-enforced protection has been spasmodic in both reserves, mainly because of the budgetary limitations of the Mexican government. PRONATURA during this period has provided stop-gap protection in the form of equipment and supplies for guards. The Ría Celestún Reserve still lacks a management plan and a reserve director. In 1995, a Technical Advisory Committee (TAC) was formed to produce a plan reflecting adequately the major concerns of all sectors interested in the reserve.

Ría Lagartos is the only Mexican site listed by the Ramsar Convention for the Protection of Wetlands of International Importance. It has fared only slightly better than Celestún. The reserve is now included in the $25 million Mexico Protected Areas Program of the Global Environment Facility that was agreed upon in 1992. As preconditions for GEF funding, a Technical Advisory Committee for Ría Lagartos was created in 1993, a management plan for the reserve was published also in 1993, and an interim operational plan was adopted in late 1994. The management plan did not take into account several priority issues of the local community and is considered inadequate. As a result, a new management plan was undertaken in 1995.

Because of delays in development of the management and operational plans, administrative reorganization, and personnel changes, the release of GEF funds was delayed until mid-1995. A bridge loan by the World Bank and care packages from PRONATURA permitted the Mexican government to continue to employ a reserve director and thirteen staff hired in the anticipation of funding availability. In the interim, the components covered by the program have expanded to include incremental personnel costs and productive activities in buffer zone areas as well as infrastructure, equipment acquisition and maintenance, and conservation inventories (A. Demayo, pers. comm.). For the first year of GEF funding, SEMARNAP initiated projects in livestock ranching, ecotourism, and environmental education (INE 1995).

Land and Resource Tenure

Serious problems over the definition of boundaries exist for both the Celestún Special Biosphere Reserve and the Ría Lagartos Faunal Refuge. The Celestún Special Biosphere Reserve is located in the municipalities of Celestún, Yucatán, and Calkiní, Campeche. The boundaries of the reserve are not clearly defined, and reliable information on land tenure within the reserve has not been researched. Private property is located on the beach, and some areas of the reserve are being considered for subdivision for the construction of summer homes (Arellano 1994). One study (Arellano 1994) gives approximate information on land holdings (figure 4-1).

The 1979 decree creating the Ría Lagartos Faunal Refuge did not define legal land tenure, nor were lands included within the reserve's boundaries expropriated by the federal government, despite the existence of four human settlements within the reserve. The reserve's boundaries again have not been clearly delimited, and land ownership and land-use patterns within the reserve are not well known, although reserve personnel are currently interviewing landowners in order to map landhold-

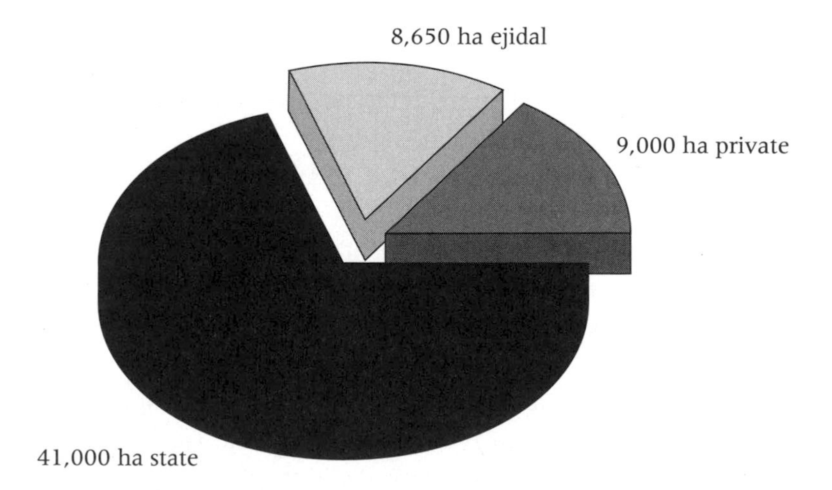

Figure 4-1. Types of land holdings in the Ría Celestún Reserve for the area of the reserve located within the state of Yucatán.

ings within the reserve. The 1995 management plan (INE 1995) gives some information on land holdings (figure 4-2). Population estimates of four communities for 1993 and 1995 are shown below in table 4-1.

Because the physical limits of the reserve have not been defined by the Ministry of Agrarian Reform, the sale and cession of lands considered to be within the reserve have been problematic (Abundes and Gamiño 1991). Extensive cattle ranching on the southern edge of the reserve has taken advantage of poorly delimited boundaries to extend into what the park authority considers to be areas included within the reserve; in some cases, grazing lands have extended to the edge of the lagoon itself. The present reserve staff has removed livestock from several critical areas, and efforts have been made to initiate the documentation of land tenure and land use in order to clearly delimit and demarcate the reserve's boundaries and prevent further incursions. However, the area affected and the

Table 4-1. Population Growth in Four Communities

	Ría Lagartos	San Felipe	Las Coloradas	El Cuyo	Total
Population in 1993	1,690	1,254	829	802	4,575
Population in 1995	3,300	2,700	500	1,500	8,000

Sources: INE 1993; R. Rubio, director, Ría Lagartos, pers. comm.

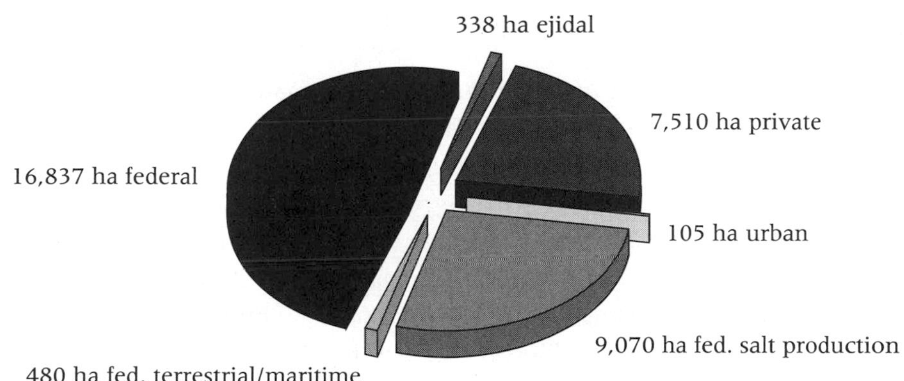

Figure 4-2. Types of land holdings in the Ría Lagartos Reserve.

number and identities of the landholders involved remain unknown. Land uses on private and *ejidal* (Mexican communal landholding) property have not to date been reconciled with prohibitions on agricultural use contained in the General Law of Ecological Equilibrium and Environmental Protection.

Resource Use

The GEF Small Grants Program administered by UNDP has agreed to fund four pilot projects for the development of sustainable resource uses within both the Ría Celestún and the Ría Lagartos reserves. In Ría Celestún, GEF is financing a project to develop blue crab aquaculture by the Association for Social Solidarity (SSS, Sociedad de Solidaridad Social). In Ría Lagartos, projects funded were blue crab aquaculture and shrimp aquaculture.

Celestún

No information is available on resource use in Isla Arena; the following discussion therefore refers only to the population of Celestún.

The town of Celestún, within the boundaries of the Celestún Reserve, is located some 40 meters from the beach and 1.3 kilometers from the estuary. The community now covers an area of approximately 2.5 square kilometers and is rapidly expanding to the south and into the wetlands east of the town (Abundes and Gamiño 1991). The town is experiencing rapid population growth, although population estimates vary widely (see table 4-1). One reason for the wide disparity in population estimates is the influx of migrant workers during the octopus season (August–December) and salt harvest.

Despite the wide variation in population estimates, the community's population appears to have stabilized recently. Several factors are responsible for past growth, including the completion of a highway from the city of Mérida to Celestún in the 1970s, the collapse of the sisal industry in the state, and state and federal government efforts to promote fisheries as a means of absorbing unemployed agricultural labor from the interior. Ninety-six percent of the immigrants live on the periphery of the port. The immigrants are generally contracted at low salaries, experience limited admission into fishing cooperatives, and are forced to build their homes in marginal areas by filling wetlands with waste and sand. A number of temporary structures have been constructed of wood and corrugated tin and black paper by fishers who reside in Celestún during the octopus and lobster seasons.

Before the development of the region's fisheries in the 1970s, agricultural production, fishing, and timber extraction were the region's predominant economic activities. Agricultural production is now minimal in Celestún; some ten individuals cultivate rice and beans in combination with fishing or another economic activity. Livestock rearing is similarly limited by poor coastal soils, with some five ranches employing an estimated fifteen individuals. Copra production is also currently minimal due to the destruction of most of the coconut palms in the region by lethal yellowing. Seasonal employment is available to residents as guides for hunters visiting the area of El Palmar during the duck season, primarily from November through March; no information is available on the importance of direct hunting and game consumption to the community. Subsistence and commercial use of sea turtles and their eggs continue despite their protected status. Collection of firewood and of wood for construction and repairs to homes, boats, and fishing gear also takes place within the reserve (*Programa de desarrollo social y ordenamiento ecológico* 1995). Although this activity is permitted, no information is available on its importance.

Fisheries

Fisheries currently employ the majority of the economically active population, and the port is the second-largest fisheries producer in the state of Yucatán. The fishery has grown rapidly, from a total of 1,584 fishers with a total of 391 small vessels and 3 large vessels in 1986, to a total of 2,569 fishers with a total of 584 small vessels and 11 large vessels in 1991. Direct and indirect employment in the fishery is estimated to involve more than 90 percent of the population (Arellano 1994). Many fishers are recent arrivals from the declining agricultural sector of the interior of the state and have no training or experience in fisheries. As a result, boat safety and resource management have suffered.

The principal fisheries species include white grunt, sea trout, mullet, sardine, anchovy, red snapper, and gray snapper. Immigrants often enter the fishery by netting in the lagoon, and an important part of the nearshore catch consists of small fish known locally as "sardines" that may actually be juveniles of other species. A fishmeal plant operated in Celestún until the late 1980s but was forced to close due to quality-control problems. Most fisheries production is now destined for the fresh seafood market, with the catch immediately transported on ice to Progreso. Most fishers are independent, hired by permit holders, or part-time, with only some 10 percent associated with cooperatives. There are currently five cooperatives with a total of 171 members (74 of whom are inactive) and nine rural associations with a total of 86 members (53 of whom are inactive). Filleting fish, peeling shrimp, and cleaning blue crab are important secondary activities associated with the fishery and are usually the responsibility of women and children (*Programa de desarrollo social y ordenamiento ecológico* 1995).

Total fisheries production has remained relatively constant in recent years, but per capita production has declined; market changes make the economic impact of declining per capita production difficult to assess. Several important species such as octopus, lobster, blue crab, and shrimp are considered overfished as a result of the entry of new fishers and failure to comply with closed seasons, size limits, and gear restrictions, but no recent fisheries assessments are available. A survey was conducted in July 1995 to identify the number of fishermen and boats active in the lagoon and in the open sea and to identify those with and without required permits, but the results of the survey have not as yet been made available. Overexploitation of crocodiles and of adult sea turtles and their eggs have been the primary factors in the decline of these species in the region; human exploitation is now prohibited but continues on a small scale. PRONATURA has initiated several efforts to protect sea turtles, including beach patrols, relocation of nests, and research and education.

Tourism

Tourism has become an increasingly important source of income in the community. Four hotels with a total of thirty-six rooms have been constructed in the town of Celestún, in addition to the rental of summer houses and room rental by three private homes, while five restaurants offer locally caught seafood. Visitation to the lagoon alone is estimated to total nearly thirteen thousand individuals annually in the estuary, with peak visitation in July, August, December, and January (Arellano 1994). Detailed information on employment and revenues generated by tourism to the rest of the reserve, including the town of Celestún, is currently being collected. However, much of the reserve's visitation is oriented toward observation of flamingos and other aquatic birds in the interior of the lagoon and therefore provides part-time employment to an increasing number of fishers in Celestún.

Tourism services are offered by two SSSs, Paraíso Escondido, with seventeen members and ten boats, and Santa Cruz Cambalam, with seventeen members and thirteen boats. Both acquired their boats through credit provided by the Mexican government's Solidaridad Program in 1991, and the cooperatives share a small docking area that allows direct access to the lagoon from the highway entering Celestún from Mérida. Following the organization of the cooperatives, PRONATURA and the Secretary of Urban Development and Ecology (SEDUE) collaborated in providing training on natural history of the estuary and of the flamingos and methods of tour operation to minimize disturbance to the flamingos, and posted interpretive signs within the estuary. Boat operators belonging to the two cooperatives charge a fixed rate of $130.00 per boat to be transported into the lagoon, with launches holding a maximum of eight persons. Key attractions to be seen by boat are flamingos, mangrove "tunnels," fresh water springs, beaches, the abandoned settlement of Real de Salinas, and a "petrified" mangrove forest.

Despite the rapid growth of tourism in Celestún, a number of continuing problems are evident. Although rotation of tours has been arranged both within and between the two cooperatives, competition for visitors has occasioned conflict. In addition, a number of fishers who are not members of the SSSs offer services directly from the beach, often at a rate below that charged by the two cooperatives, creating resentment on the part of cooperative members. Furthermore, only a small proportion of the visitors stay overnight in local hotels or consume meals in local restaurants, and the benefits of tourism currently appear to be confined to a small number of boat operators and hotel and restaurant owners.

Only recently have initial studies on the impacts of tourism on park conservation been systematically assessed, and although upwards of one hundred boats have been observed within the estuary in a single day, the capacity of the area to support current levels of visitation or future increases is not known. However, the passage of tour boats close to flamingo colonies in order to view their flight, and in proximity to nesting areas of other avian species, is likely to cause some disruption to resident wildlife (Arengo and Baldassarre 1995). Additional impacts include those of improper trash disposal (Arellano 1994). Interpretative materials and opportunities for environmental education remain inadequate.

Salt Production

Small-scale artisanal salt production continues to employ a significant proportion of the population, particularly recent immigrants and seasonal laborers, although more exact data are not available. The production of salt was the most important economic activity in Celestún until the 1940s, when the market was flooded by production by the Salt Industry of the Yucatán (ISYSA) in nearby Las Coloradas. In the early 1980s, production was renewed through a program of economic diversification designed to alleviate underemployment in the henequen (*Agave fourcroydes,* plant fiber for rope and coarse fabric) sector. In the mid-1980s, four SSSs, one cooperative, and three private concessionaires were involved in salt production within the reserve, with a total of seven hundred evaporation pools having been formed (Batllori 1990a). In 1988, the National Institute of Statistics, Geography, and Information (INEGI) reported total employment of 106 individuals working 103 days of the year. In 1994, a total of ten Societies of Salt Employees (SES) existed, with a total of 190 members, although employment varies widely depending on environmental conditions (Arellano 1994; *Programa de desarrollo social y ordenamiento ecológico* 1995). Salt production has led to localized habitat modification and destruction, particularly disruption of water circulation through the construction of containment walls, damage to coastal dunes, salinization, and damage to mangroves and their use in barrier construction (*Programa de desarrollo social y ordenamiento ecológico* 1995).

Ría Lagartos

The community of Ría Lagartos is the largest located within the reserve. The majority of immigrants originated in livestock zones of the peninsula and have settled along the principal highway to Tizimín, constructed in the 1960s, or have obtained land through filling in estuaries and wetlands. The primary occupation of the population of Ría Lagartos is fishing, as described next.

Fisheries

Fishers in Ría Lagartos are organized into the Cooperative Manuel Zepeda Peraza, consisting in 1991 of 128 members with a total of 79 vessels, and the Cooperative of Fishermen of Ría Lagartos, with 147 members and a total of 71 vessels. Infrastructure in the community in 1993 consisted of three reception centers, an ice plant, a dock of twenty-five meters in length, and a total of some 280 boats. The most important fisheries species are grouper, octopus, lobster, shrimp, sharks, mojarra, mullet, and drum (Abundes and Gamiño 1991; INE 1993). Most fishing is conducted in the open sea with the exception of the fall–winter season of *nortes,* when bad weather forces the fishermen to stay inside the lagoon.

San Felipe was a small fishing community until the 1970s, when immigration was encouraged by the construction of a highway from Tizimi´n to San Felipe, electrification, and development of fisheries production, which attracted immigrants from the henequen zone. More than half of the households in San Felipe raise small domestic livestock on their patios, but few grow vegetables because of poor water drainage (Abundes and Gamiño 1991). The majority of the economically active population is employed in the fishery, and most fishers are members of the United Fishermen's Cooperative of San Felipe, which had 168 members in 1991. Infrastructure in 1993 consisted of a total of 110 boats, a freezer, and a warehouse. The most commonly caught species are octopus, lobster, grouper, snapper, and shrimp (Abundes and Gamiño 1991; INE 1993).

The economically active population of El Cuyo, the size of which varies both by season and depending on the estimate, has an estimated total population of between 802 individuals (1993) and 1,500 (1995), the majority of whom are employed in fisheries for grouper, octopus, lobster, shark, skipjack tuna, and Spanish mackerel. Some one hundred individuals are organized in a single fishing cooperative; in 1993, a total of 311 boats were registered. This community is the most marginalized of the four populations located within the reserve and is experiencing extremely high rates of unemployment (Abundes and Gamiño 1991; INE 1993).

In addition, fishing is a growing industry in Las Coloradas; in 1991 it employed twenty members of the Rural Cooperative Society Las Coloradas with 8 vessels; twenty-five members of the Cooperative Association of Fisheries Production with 10 vessels; and an additional twenty-five independent fishers. The principal species caught are scale fish, shark and ray, octopus, and lobster (Abundes and Gamiño 1991).

Tourism

Two boat owners' cooperatives also exist in Ría Lagartos for tourism within the lagoon, although visitation is much less developed than that in Celestún due to the greater difficulty of access to the lagoon (Arengo and

Baldassarre 1995). There is a single hotel and two restaurants providing services to visitors; currently, a new hotel is being constructed in San Felipe. The reserve administration plans to construct two facilities for the provision of information to tourists. As yet, however, no mechanism has been established for the collection of visitor entrance fees to provide resources for reserve management. Under an ecotourism project financed by GEF, reserve staff plan to promote tourism to the reserve by encouraging the formation of a single tourism organization, constructing three interpretative signs, and providing education to tour operators regarding the need to conserve the reserve's resources (INE 1995).

Salt Production

The population of Las Coloradas, estimated at 828 people in 1993 and 500 in 1995, is currently expanding eastward toward El Cuyo. The majority of the economically active population (60 percent of the male population) is employed by the company Industria Salinera de Yucatán, S.A. (ISYSA), Mexico's second-largest producer of salt. Many of the employees of ISYSA have received land and housing from the company. Salt extraction in the Ría Lagartos area dates from the early Mayan civilization. In the 1950s, much of the area of the reserve was purchased by a single family, and large-scale commercial extraction through solar evaporation was initiated. Damming of large areas of the lagoon to create evaporation pools, pumping of fresh and salt water, discharge of untreated wastewater directly into the ocean, road construction, removal of dune vegetation, and destruction of mangroves have resulted in the opening of breaches into the lagoon, altered water quality, and decreased water circulation, leading to the inundation of flamingo nesting areas (Moan 1992; INE 1993).

Other

Cattle ranching has also created severe impacts within the reserve. As of 1995, an estimated seven thousand hectares had been cleared for grazing in the southeastern portion of the reserve. In some cases, cattle grazing has extended nearly to the edge of the estuary. At least eight ejidos are involved in cattle raising, while private ranchers are organized in the Regional Ranching Union of the Western Yucatán. Expansion of cattle grazing is often achieved through the renting of ejidal lands by private ranchers (INE 1993).

Collection of wood for fuel and construction within the reserve is permitted by residents of Ría Lagartos and is known to occur, but no information is available on its scale and impacts. Information on hunting of terrestrial wildlife is similarly lacking, although hunting of crocodiles is believed to have led to their near eradication from the reserve named for them. Illegal taking of sea turtles and their eggs continues, as in Celestún, although research programs and beach patrols supported by PRONATURA, the U.S. Fish and Wildlife Service, and fisheries agencies have helped to reduce this activity.

Organizational Roles

Pronatura Península de Yucatán (PRONATURA) is a private, nonprofit organization based in Mérida that works on conservation issues throughout the Yucatán. It was originally formed as a regional chapter of Pronatura, a Mexico City–based NGO that is Mexico's oldest environmental group. In 1990, PRONATURA became independent of Pronatura. Since 1991, PRONATURA has been the local counterpart organization of the Conservancy–sponsored PiP program for Ría Celestún and Ría Lagartos. Together with the CINVESTAV of Yucatán, the NGO has signed an agreement of cooperation with federal agencies charged with reserve management. The NGO's management and conservation activities in the reserves have focused on the provision and maintenance of basic protection infrastructure, research, and education and outreach. More recently, the NGO has become involved in fostering alternative production activities. The NGO has contracted a fisheries extensionist to conduct fisheries assessments and support local projects designed to reduce the impacts of overfishing and has recently obtained funding from the Ford Foundation to design and implement community development projects in Ría Lagartos Reserve.

PRONATURA has installed and maintained signs in both reserves, equipped guard stations with radios, provided and maintained transportation (vehicles, boats, and bicycles) for park staff, and provided in-kind support to reserve staff to complement low salaries. In the past few years, however, the government has begun supporting these activities. Research supported by PRONATURA has played an important role in providing technical and financial support needed for park management and conservation, including inventories of fauna and flora found within the two reserves. Overflights organized with Project Lighthawk to monitor the reserves in early 1995 determined that some seven thousand hectares located within Ría Lagartos had been converted to pasture for cattle grazing. The NGO has also supported sea turtle research in both reserves. PRONATURA, together with CINVESTAV of Yucatán is implementing the Biodiversity Data and Environmental Monitoring System. This is funded by the North American Waterfowl and Wetlands Council through the U.S. Fish and Wildlife Service. The system has provided basic scientific and socioeconomic information on the western coastal region of Yucatán.

PRONATURA's Department of Environmental Education has developed educational materials, including a video on Ría Celestún, organized student exchange programs in El Cuyo, Celestún, and Ría Lagartos; and fostered the creation and development of a conservation youth group, the Ecological Group of Celestún. PRONATURA has long been active in providing education and training courses to tourism operators in both reserves and to ecotourism operators in Mérida. In 1994, PRONATURA contracted the services of an ecotourism specialist to develop a plan for ecotourism in the Celestún Reserve (Arellano 1994), and in 1995–96, PRONATURA will assist in the development of tourism infrastructure in both reserves. In Celestún, the NGO will assist in the construction of a dock; the government will construct an interpretive trail for a spring (*ojo de agua*) at Baldiosera and an interpretive center for visitors to the lagoon. A formal training course was held in 1995, in cooperation with Ducks Unlimited of Mexico (DUMAC), Grupo Ecológico Celestún (GECE), and SEMARNAP, for members of the two tourism cooperatives in Celestún.

PRONATURA has also worked to establish good relations with other organizations involved in reserve conservation and management. The NGO is a member of the TAC for Ría Lagartos and was chosen to head the Secretariat of a planned TAC for Celestún and charged with the design of a management plan. In spite of the initial resistance from Campeche state representatives, the committee has been formed and PRONATURA is proceeding with the development of the management plan. PRONATURA also currently serves as a representative to the governing body for the El Palmar State Reserve. Although few resources are currently available to protect this state reserve, PRONATURA's presence in both contiguous reserves helps to link conservation plans of El Palmar with those of Celestún.

Linkages between Parks and Buffer Areas

A variety of different activities are underway at both of the reserves to begin to promote sustainable use of resources within the reserves and alternatives to current patterns of overexploitation of certain resources. At both sites, at least some of these projects have been hampered by technical problems. Work to resolve land tenure issues at both sites also offers one way of establishing stronger linkages between management of the protected areas and community development.

Celestún

Given the threats posed to the Celestún Reserve by rapid development, and the difficulty of stimulating local initiatives, attempts to engage both the community and government agencies in urban planning appear to be priorities. In 1995–96, PRONATURA began compiling information on land tenure in the Celestún Reserve. This research is critical to efforts to develop the management plan and to define core and buffer zones for the reserve, needed to control future development of the area. Again, these activities offer numerous opportunities for informal outreach. Also critical is commitment by SEMARNAP to reserve protection, starting with the hiring of a director and support staff in the reserve. In this regard, PRONATURA is working with SEMARNAP to develop a financial plan to pay personnel from locally generated funds.

In Celestún, PRONATURA has worked to strengthen the Grupo Ecológico Celestún (GECE), a conservation group for youth, ages eight to thirty, that has been active in beach cleanups and educational activities. In July 1995, with a change in the municipal government, one of the founders and most active members of GECE was selected to head the Municipal Ecological Office until 1998. PRONATURA has also provided environmental education and training for tourist guides in the community. In 1995 it focused on the needed infrastructure for the development of ecotourism. In general, social and economic conditions in Celestún have not been conducive to the formation of strong local organizations or local leadership. Immigrants are considered by long-term residents to be responsible for many of the natural resource management problems evident in the reserve, particularly overfishing in the estuary. Yet because immigrants are generally excluded from existing cooperatives or other producers' organizations, they are also excluded from discussion of problems and potential solutions.

In 1995, PRONATURA undertook a five-year Ecotourism Program for Celestún Reserve in order to increase the educational and economic benefits of tourism and to minimize its environmental impacts. The NGO is consulting with SEMARNAP in the development of appropriate regulations on tourism use, and as mentioned previously, is assisting the cooperatives in the construction of a wooden dock. A training course for guides was provided by the NGO in 1995. However, due to lack of organization and of information on their participation, sectors of the population other than the tourism cooperatives have not been consulted in planning or decision making, and there are currently no programs underway to address this issue.

In 1994, the GEF Small Grants Program administered by UNDP provided a grant of US$16,129 to the Association for Social Solidarity of Ría Celestún for a project for rustic aquaculture of blue crab. The stated purpose of the project was to produce blue crabs for repopulation of mangroves and for human consumption (GEF 1995). Initial efforts resulted in the capture of 500 kg of juvenile crabs, which when mature produced a total of 1.5 tons for sale to restaurants and residents. Based on these results, the facility is being expanded. However, there have been major problems:

- the project was not accompanied by ongoing technical assistance and extension;
- technical problems identified early in the project by fisheries agents were not addressed by the cooperative;
- the development and conservation benefits of the project remained unclear, as the primary use of blue crab in the region is bait for the octopus fishery, which already appears to be overexploited; and
- the much greater importance of the octopus fishery was reflected in the unauthorized diversion of funds from the project for the construction of a storage facility for octopus.

The cooperative's long-term commitment to the maintenance of the project is uncertain.

Responding to these problems, PRONATURA has begun working with the SSS Ría Celestún to redirect and strengthen the focus of the project. The NGO contracted a fisheries extensionist to provide continuous onsite assistance and education to familiarize fishermen with the life cycle of the species and emphasize the importance of a closed season and minimum size limits. The project is also working to achieve closed-cycle production for repopulation of the lagoon as well as for human consumption. The Ford Foundation is currently providing funding for the development of infrastructure needed to complete the productive cycle. Although the long-term commitment of the community to the project is not yet assured, PRONATURA considers the project to be an important strategy for education regarding alternative methods of sustainable production and has provided for continuous on-site training and extension.

Ría Lagartos

The Ría Lagartos Reserve, with an estimated resident population of eight thousand, has, like Celestún, been subject to rapid population growth due to immigration encouraged by the possibilities for employment in the fishery or in ISYSA. Immigration flows have been much less dramatic than in Celestún because of the more limited growth capacity of its fisheries. As a result, local and regional organization has been much more extensive and effective. For example, in San Felipe, the majority of fishers belong to a single cooperative that provides a number of social services such as unemployment and health insurance. The municipal government has also promoted measures to slow the rate of immigration and has requested improvements in urban planning from the TAC.

In addition, fishing cooperatives in San Felipe, Ría Lagartos, Las Coloradas, and El Cuyo have formed a single federation, Fishermen of the Western Yucatán State, that has become the single most important organized force in fisheries in the state. In the spring of 1995, the cooperatives initiated the protection of a marine zone extending from Isla Cerritos some six kilometers seaward in order to protect what is considered to be a refuge or nursery for key fisheries species. A protected fisheries area was recently created within the lagoon at the initiative of the cooperative, with hook-and-line the only fishing method permitted within the area. Assistance in marking and enforcing the area was solicited and obtained from the local fisheries inspector and from the navy.

Fishermen in the four communities are eminently aware of the impacts of overfishing. The area's capture fisheries, however, have not been subject to a thorough evaluation or to further conservation measures, internal or external. The opportunity does remain to engage fishers in productive dialogue regarding possibilities for resource conservation. Given the demonstrated capacity of local organizations to respond effectively to new information regarding conservation needs, fisheries assessments demonstrating opportunities to conserve economically important species are likely to be an effective means of stimulating local initiatives.

In 1994–95, the GEF Small Grants Program provided assistance for two pilot aquaculture projects in Ría Lagartos. A blue crab aquaculture project was initiated by the Cooperative Fishing Society Manuel Cepeda Peraza, with UNDP providing assistance for training and CINVESTAV providing technical assistance under contract to SEMARNAP with GEF funds. However, preliminary technical studies were not conducted prior to initiation of the project. Following the death of all the crabs introduced into the constructed enclosure, the following findings were made:

- water chemistry and salinity were not conducive to blue crab production, and
- blue crabs were not found naturally in significant numbers.

A second project was initiated for shrimp aquaculture by the Cooperative of Fishermen of Ría Lagartos. A team of ten has already completed its training course, and a site has been selected for the construction of twelve hectares of tanks, of a total of one hundred hectares planned for development. SEMARNAP, with funding from the GEF Mexico Protected Areas Program, has also initiated pilot projects for shrimp aquaculture in raised tanks in the communities of Las Coloradas and El Cuyo. The project in Las Coloradas is designed primarily for training and research, while the project in El Cuyo involves raised-tank aquaculture administered by an ejido of eighty families. According to the terms of agreement negotiated between the reserve manager and the ejido, 60 percent of project earnings will be retained by the ejido, while 40 percent will benefit the reserve.

The cooperatives have expressed considerable interest in shrimp aquaculture projects supported by external funding, but the technical and financial demands of these projects are likely to exceed local capacity while distracting attention from more feasible options. A further local sustainable development initiative supported by the GEF Small Grants Program has proposed ranching of the Morelet's crocodile for reintroduction into the reserve and for human consumption. Again, however, the proposal appears to have been made without relevant background studies considering economic feasibility issues. There was no consideration of:

- local technical capacity
- availability of resources for infrastructure maintenance

- local capacity to provide feed and disease control
- environmental impact issues.

The Morelet's crocodile is a protected species in Mexico and is listed as endangered in appendix I of Convention on International Trade in Endangered Species (CITES), which prohibits its commercial trade.

In 1995, with GEF financing, reserve staff initiated a pilot project to foster intensive managed cattle ranching in the zone in order to reduce the potential area affected by grazing, allow recuperation of degraded areas, and improve production by ejidos and private ranchers. The project proposes to create four production units of twenty-four hectares each on lands provided by local landowners, providing training, soil studies, and infrastructure. A system of electrical fences is to be used to allow rotational grazing through a succession of fenced enclosures in which native forage plants would be cultivated. Following consumption of available forage in one enclosure, cattle would be transferred to the next to allow the restoration of vegetation. The strategy also attempts to foster alternative land uses, such as semi-intensive production of white-tailed deer and reforestation of cleared areas with species compatible with fruit and timber production (INE 1995). According to reserve staff, an additional objective of the project is to facilitate the redefinition of reserve boundaries. By reducing the land area currently required for livestock rearing, reserve personnel hope to regain areas of the reserve currently used for grazing.

Opportunities for strengthening local participation in reserve protection are increased by the 1995 hiring of a corps of young, bright, and enthusiastic, albeit inexperienced, park guards in the Ría Lagartos Reserve, made possible in part by GEF funding. The park director and staff are committed to community education and outreach efforts, and park guards have been assigned individual responsibilities for developing conservation and development projects targeted at plant conservation, fisheries, and ranching. Support for research needed to direct the efforts of reserve staff as well as NGO extensionists is likely to be an efficient means of maximizing their effectiveness. Particularly important to this process is improved information on the socioeconomic profiles of the communities within the reserve and on fisheries status and conservation.

Promising programs have also been initiated to ensure long-term park protection through land-use mapping and boundary redefinition in Ría Lagartos. SEMARNAP, PRONATURA, the National Institute of Statistics,

Geography, and Information, and Biocenosis, an NGO, are collaborating in mapping the reserve in order to correct its boundaries. This action will facilitate the prevention of incursions by neighboring ranchers on reserve lands. Part of this effort involves interviewing residents by reserve staff in order to define land tenure within the reserve. This program presents numerous opportunities for community outreach, including solicitation of informal input into the development of a new management plan for the reserve. Moreover, the program could easily be extended, in conjunction with boundary redefinition and outreach efforts directed at neighboring cattle ranchers, to landholders in adjacent areas, in order to monitor land-use changes in buffer areas and to begin negotiating agreements with neighboring residents to respect the redefined reserve boundaries.

Conflict Management and Resolution

The Technical Advisory council, established in Ría Lagartos in 1993, and a council likely to be formed in the near future in Celestún, are currently the only institutions available or potentially available for local participation in management and conservation decisions or for conflict management and resolution. The councils may in time serve important functions in facilitating the airing of grievances and communication among members in the case of more localized conflicts, such as those between organized and independent fishermen and tour operators. The history of the councils, however, suggests that they may be established primarily for the transmission of and mobilization of support for federal policies and programs. A more important function may therefore be in encouraging communication and coordination among federal agencies with authority within the reserves.

Although the majority of Celestún's land area (56 percent) is in the state of Campeche, nearly all of the human population of the reserve is located within Yucatán state, with the exception of Isla Arena. The island, which has a small resident fishing community of some one thousand inhabitants and roughly one hundred vessels, is the subject of a prolonged dispute between the citizens of Celestún and Isla Arena. Conflicts between fishermen of the two communities arise each year during the early summer, when income from fishing activity is low and fishers from both communities compete in the waters adjacent to the other. During the conflict, fishers in Celestún reiterate the long-standing but entirely unfounded claim that Isla Arena belongs to the state of Yucatán. With the beginning of the octopus fishery in early August, the conflict is forgotten. In 1995, however, particularly severe conflict led to violence and the burning of boats, and Isla Arena stated its "independence" by buying and selling merchandise almost entirely in the state of Campeche. The conflict hindered efforts by the newly formed TAC for Celestún to develop a management plan for the reserve; it awaits the Campeche state government's agreement to the plan. Representatives of fishing cooperatives from both Campeche and Yucatán have attended the meetings of the council, but it is not yet clear whether and to what extent the council will be able to contribute to the resolution of the conflict.

Large-Scale Threats

Celestún

Rapid population growth in Celestún has been accompanied by severe impacts, particularly in the estuary and wetlands. The construction of highways and of a permanent outlet to the ocean has disrupted water flows. A shortage of available land for housing construction has forced the expansion of community settlements largely through filling of wetlands with solid waste. Furthermore, although 89 percent of Celestún's population receives running water piped in from a cenote located some ten kilometers from the community, more than half of the population lacks sewage service or septic tanks. Those facilities that exist permit filtration into the aquifer, leading to contamination of wetlands and water sources (*Programa de desarrollo social y ordenamiento ecológico* 1995). Untreated water is discharged directly into the estuary, and solid waste is dumped in empty lots and wetlands. Eutrophication of the principal estuary and of smaller bodies of water is evident. Port development in nearby Progreso has led to intrusion of highly saline water into the estuary, which is gradually extending in the direction of Celestún.

In addition, new immigrants to the community continue to enter the fishery, increasing the potential for overexploitation of marine species. Noncompliance with regulations on catch and gear is the norm, in large part due to lack of experience or capacity on the part of new entrants. Both permanent and seasonal immigration have undermined the capacity of local organizations to respond to these threats.

To date, the increasing importance of tourism has compounded these problems rather than generating the resources needed to address them. Increasing visitation to the reserve is not generally linked to environmental education, nor have revenues been generated to support reserve conservation and management. A number of actual or potential impacts of tourism have been noted, including additional development of the coastal zone, increased stress to limited infrastructure, and disturbance to flamingos and other nesting birds by boat operators and tourists.

Ría Lagartos

In Ría Lagartos, cattle grazing, salt mining, pollution, inappropriate tourism development, and potential overfishing appear to be the most important large-scale threats to reserve protection. Cattle ranching has been associated with deforestation and habitat degradation not only on the limits of the reserve but also within it, in some cases extending nearly to the edge of the estuary. Industrial salt production has heavily impacted a large area of the reserve, including part of the area currently designated as the core zone. As is the case in Celestún, in Ría Lagartos damming of large areas of the lagoon to create evaporation pools, pumping of fresh and salt water, discharge of untreated wastewater directly into the ocean, road construction, removal of dune vegetation, and destruction of mangroves have led to the opening of breaches into the lagoon, altered water quality, and decreased water circulation. In Ría Lagartos it has also resulted in inundation of flamingo nesting areas.

As in Celestún, coastal ecosystems have been subject to significant modification throughout the reserve. The construction of highways from Tizimín to Ría Lagartos and El Cuyo has obstructed water flow and resulted in the destruction of large areas of mangrove. Urban development has resulted in the destruction of dunes and vegetation on the barrier island. Inadequate sewage disposal and the reclamation of wetlands by filling them with solid waste have created problems similar to those in Celestún. In the late 1970s, a permanent inlet was constructed in front of Ría Lagartos. Sedimentation in the inlet and in the lagoon itself has resulted, making boat passage difficult in the western end of the estuary. Annual dredging is now necessary to keep the inlet and estuary navigable. Additional impacts from saltwater intrusion through the inlet into the estuary have not been investigated.

National Policy Framework

Many of the impacts evident in the Celestún and Ría Lagartos reserves are the results of national and local development strategies that have proceeded without regard to the protected status of these areas. In the 1970s and 1980s, a nationwide effort to develop fisheries production as a means of employing landless or land-poor campesinos (peasants) encouraged development of related infrastructure such as roads, docks, and storage facilities in these coastal reserves as elsewhere along the Yucatecan coast. Until the exhaustion of key fisheries and the bankruptcy of both cooperatives and state agencies became evident in the early 1990s, artisanal fisheries in particular were encouraged through credits to purchase boats and gear and to work in cooperatives to catch high-value species such as shrimp, lobster, and conch.

Mexico's environmental protection agencies have been unable to entirely curtail the serious aforementioned impacts. Other agencies have shown spasmodic commitment to the protection of these areas since the declaration of Ría Celestún and Ría Lagartos as wildlife refuges in 1979. Examples in Ría Lagartos include:

- the authorized expansion of ISYSA operations despite their potential impacts
- the Federal Electric Commission clearing a right of way for power lines in Ría Lagartos without consulting reserve administration, with the result that a significant number of flamingos and other birds have been electrocuted
- the Banco de Crédito Rural's offer of credit to campesinos for the establishment of farms and ranches within the reserve's limits in the 1980s.

In recent years, this situation has been aggravated by frequent reorganizations in the Mexican ministry responsible for environment and park protection. The first Mexican agency responsible for environmental matters was the Secretary of Urban Development and Ecology (SEDUE), founded in 1985. In 1992, environmental affairs were transferred to the Secretary of Social Development (SEDESOL). In 1995, the Secretary of the Environment, Natural Resources, and Fisheries (SEMARNAP) was created to consolidate environmental concerns. These many shifts involve policy as well as personnel and contributed to a three-year delay in the release of funds approved by the GEF for protected-areas management in Mexico. Bureaucratic obstacles seem to have limited the capacity of the Mexican government to ensure long-term financing of efforts currently underway in Ría Celestún and Ría Lagartos. No trust fund or other financial mechanism has been established to ensure their continuation following termination of GEF assistance.

The rapid exhaustion of available funds is particularly worrisome due to a shift in the attitudes of government agencies toward natural resource conservation and protected-areas management. While the creation of SEDUE in the early 1980s ushered in a period of strict controls on or prohibition of human activities impacting wildlife and habitat, the creation of SEMARNAP in 1995 has led to a greater emphasis on community development programs to the detriment of conservation goals. Many of the projects intended to protect natural areas and wildlife are difficult to link directly to conservation. This appears to be true of some of the projects approved for GEF funding in Ría Lagartos, including shrimp aquaculture and crocodile ranching.

Another shift in government conservation strategies has involved the decentralization of both regulation and administration. Authority for resource management and environmental protection has in many cases been transferred from the offices of SEDUE, SEDESOL, and SEMARNAP to state delegations of federal agencies, to state governments, and to nongovernment organizations. Decentralization of state authority for protected-areas management has opened several new channels through which both nongovernment organizations and local communities can influence protected-areas management. It has also created new opportunities for linkages between protected areas and between parks and adjacent areas, as in the agreement for joint management of the Celestún and El Palmar reserves.

Transboundary Issues

Celestún Reserve is located within the municipalities of Celestún in the state of Yucatán and Calkiní in the state of Campeche. Although recent efforts have been made to develop coordinated management and conservation efforts by incorporating the TAC, border and fisheries conflicts between the states of Yucatán and Campeche continue to make a bistate management plan a real challenge.

Conclusion

The complexity of issues facing Ría Lagartos and Ría Celestún is indeed staggering. Both of the reserves face both large- and small-scale threats from multiple sources. Yet many of these threats are also interrelated, adding enormous complexity to the task of addressing each of them. Policy changes at the national government level that appear to be innocuous, such as encouraging a switch from employment in henequen to employment in fishing, have had unforeseen negative consequences—overfishing of once abundant areas within the two parks. Changes in the Mexican land-use system of ejidos, communally managed land areas, will no doubt have substantial implications for how some of these areas use and manage their lands. Cattle ranching, which has been prevalent in some areas, especially near Ría Lagartos, shows no signs of diminishing. Tourism offers both potential benefits and sources of revenue for reserve financing and management, as well as benefits to local communities; yet organizing the tourism in such a way that it can provide these benefits is a difficult task, especially given the rapid potential for the reserves, particularly Ría Lagartos, to become so popular that the tourism is no longer environmentally appropriate.

There is also political complexity in dealing with the diverse government agencies involved in setting policies for this area. PRONATURA has worked with a variety of municipal policies, multiple states and their agencies, and national-level agencies and policies. All of these have shifting demands and agendas, yet their actions, as described above, can often have unintentional consequences that affect the reserves and the communities that surround them. Addressing the threats, and the rapid set of changes that are underway on the Yucatán, represents a serious management challenge and requires a tremendous amount of coordination and cooperation. Yet failing to accomplish these objectives means the loss of a wide array of migratory waterfowl and wildlife, as well as the only feeding and breeding habitats for greater flamingos in the Americas.

Postscript

A number of positive changes have taken place at Ría Lagartos and Celestún. When the reserves were decreed, the coordinates included in the decree were not sufficiently complete to allow for accurate delimitation of boundaries. A new proposal has gone to the government for both Ría Lagartos and Celestún that would result in a new decree with precise information for boundary demarcation and reserve status. While both of the areas were in a category called special reserves, the status they would have in the future is unknown, but they might possibly be biosphere reserves. One complicating factor is that Celestún straddles two states. The state of Campeche wants all of its coastal zone to be a biosphere reserve known as Los Petenes. PRONATURA would like to see Celestún stay intact as one uniform area, rather than have separate categories and management regimes for the two sides.

When the case study was written, there were a total of twenty-three boats and thirty-four members organized into two cooperatives providing tours of the lagoon at Celestún. There are now seven cooperatives registered, with five cooperatives for a total of sixty-five boats providing tours of the Rías and two cooperatives providing tours of the beach. The price for rides has increased from 130 pesos per boat to 200 pesos per boat. A study that looked at the distribution of tourism in Celestún and the carrying capacity concluded that the existing number of boats appears to be the maximum. PRONATURA is working to determine how many can be in what areas during peak times and the best routing of boats to minimize disturbance to wildlife, primarily flamingos, and submerged vegetation (Leslie 1997).

A document containing an analysis of threats found that for Ría Lagartos, the primary threats are: (1) cutting of native vegetation; (2) fragmentation of habitat; (3) reduction of aquatic species; and (4) reduction of mammals and birds. Factors that were found to contribute to these are population increase, summer home construction, increased road construction, overfishing, the advance of cattle ranching, and the expansion of the salt industry. In Celestún, the trends were similar, with principal threats of: (1) cutting of native vegetation; (2) reduction of aquatic species; (3) alteration of water flows; and (4) habitat fragmentation. These are due to the same problems as in Ría Lagartos.

Glossary

CINVESTAV—Centro de Investigaciones y Estudios Avanzados (Center for Investigations and Advanced Study)

Cooperrative Association of Fisheries Production—Sociedad Cooperative de Producción Pesquera

DUMAC—Ducks Unlimited of Mexico*ejido*—Mexican communal land-holding system

GECE—Grupo Ecológico de Celestún, A.C. (Ecological Group of Celestún)

GEF—Global Environment Facility

General Law of Ecological Equilibrium and Environmental Protection—Ley General de Equilibrio y Protección al Ambiente
INE—Instituto Nacional de Ecología (National Institute of Ecology)
INEGI—Instituto Nacional de Estadística, Geografía, e Informática (National Institute of Statistics, Geography, and Information)
ISYSA—Industria Salinera de Yucatán, S.A. (Salt Industry of the Yucatán)
MAB—Man and the Biosphere Program
Municipal Ecological Office—Regiduría Ecológica del Municipio
National Institute of Statistics, Geography, and Information—Instituto Nacional de Estadística, Geografía, e Informática
nortes—storms from the northeast
PRONATURA—Pronatura Península de Yucatán
Rural Cooperative Society—Sociedad Cooperativa Rural
SEDESOL—Secretary of Social Development
SEDUE—Secretaría de Desarrollo Urbano y Ecología (Secretary of Urban Development and Ecology)
SEMARNAP—Secretaría de Medio Ambiente, Recursos Naturales, y Pesca (Ministry of Environment, Natural Resources, and Fisheries)
SES—Societies of Salt Employees (Sociedades de Salineros)
SSS—Sociedad de Solidaridad Social (Association for Social Solidarity)
TAC—Technical Advisory Committee (Comité Técnico Asesor)
UNDP—United Nations Development Program
Wetlands Board of Trustees of the Yucatán—Patronato de Humedales de Yucatán

Sierra de las Minas Biosphere Reserve

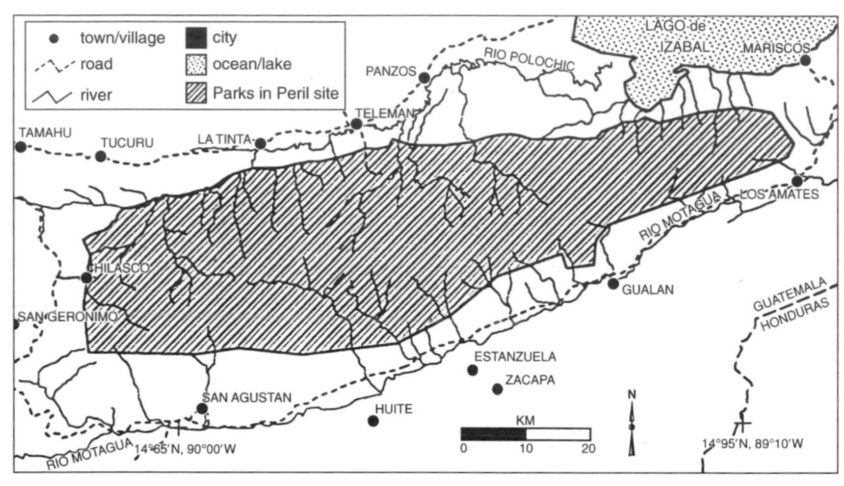

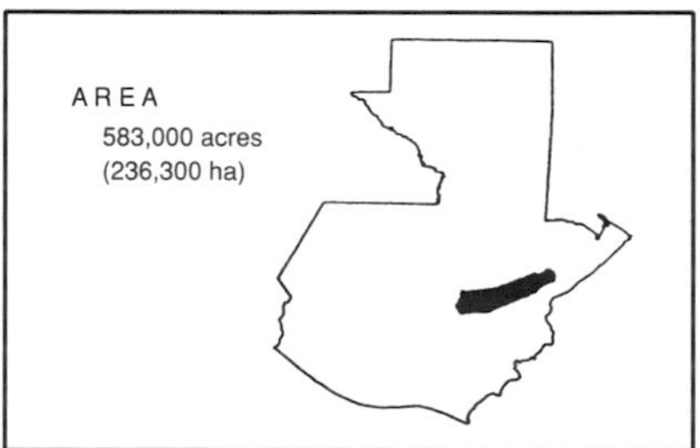

5. Guatemala: Sierra de las Minas Biosphere Reserve

Andreas Lehnhoff and Oscar Núñez

Sierra de las Minas is a spectacular mountain chain in eastern Guatemala that rises from a few meters above sea level to more than three thousand meters in altitude. Bordered on the north and south by the Polochic and Motagua river valleys, Sierra de las Minas extends about 130 km in length and ranges from 10 to 30 km in width.

In many senses, Sierra de las Minas is an area of contrasts. Its steep slopes rise like an upthrust that towers between the Motagua and the Polochic geologic faults, where the North American and Caribbean continental plates collide. Its soils are highly susceptible to erosion due to the steepness of the slopes (40–80 percent), which are principally formed by the oldest Paleozoic rocks in Central America (Campbell 1982).

Sierra de las Minas has a great diversity of climates and rainfall patterns. Due to the rain shadow caused by the height of the range, the central part of the arid Motagua valley receives less than five hundred millimeters of rain a year and is considered the driest area in Central America. In contrast, the Polochic valley, located only a few kilometers away, receives more than forty-five hundred millimeters of rain annually.

The diversity of climates, the biogeographic position and isolation, and the dramatic differences in altitude—which vary between 15 and 3,015 meters above sea level—have contributed to Sierra de las Minas' great biological diversity and high degree of endemism. Its vegetation, comprising various forest types, is a mixture of Nearctic and Neotropical associations. It contains at least fifteen species and six genera of conifers, which represent an invaluable source of seeds for potential reforestation projects throughout the world. The western peak of the range is covered by approximately 600 square km of montane cloud forest, the largest remaining extension of this ecosystem in Guatemala and in Central America. This forest has been described as a spectacular example of the coexistence of conifers, oaks, and diverse populations of *lauraceae* with abundant associations of mosses, ferns, and epiphytes (CDC 1993).

Sierra de las Minas is home to 885 species of terrestrial vertebrates, 70 percent of all those reported for Guatemala. There are approximately 400 species of birds, including some threatened species such as the quetzal (*Pharomacrus moccino*), the harpy eagle (*Harpia harpyja*), the peregrine falcon (*Falco peregrinis*), and the horned guan (*Oreophasis derbianus*). The reserve is also home to 5 species of felines: the puma (*Felis concolor*), the jaguar (*Panthera onca*), the jaguarundi (*Felis yagouaroundii*), the ocelot (*Felis pardalis*), and the margay (*Felis wiedii*). Other mammals include the spider monkey (*Ateles geoffroyi*), the black howler monkey (*Alouatta pigra*), the mantled howler monkey (*Alouatta palliata*), the red brocket deer (*Mazama americana*), the collared peccary (*Tayassu tajacu*), and the white-lipped peccary (*Tayassu pecari*).

Sixty-three river basins originate in the forests at the peaks of Sierra de las Minas, of which twenty-nine flow into the Motagua River and thirty-four are tributaries of the Polochic River. These rivers provide cool water and influence the microclimatic conditions of the valleys. They are used widely for irrigation, agroindustry, hydroelectricity, and domestic water supply. Especially in the arid and densely populated Motagua valley, these sources of water are vital to the health and economy of the region.

Park Establishment and Management

On October 4, 1990, the Guatemalan national congress enacted law 49-90, declaring Sierra de las Minas a protected area under the management category Biosphere Reserve. This represented the culmination of an intense process that included the preparation of studies, meetings, and negotiations led and promoted mainly by the Fundación Defensores de la Naturaleza (Defensores), a private Guatemalan nonprofit organization.

The process of creating the Sierra de las Minas Biosphere Reserve (SMBR) was intense and relatively rapid. In the mid 1980s, several biologists from Guatemala's Universidad del Valle who had been studying the ecosystems and species at Sierra de las Minas suggested to Defensores the idea of promoting the creation of a natural reserve in the central part of the range (M. Dix 1991, pers. comm.).

After several orientation visits and after gathering more information on the site, Defensores decided to promote the creation of a legally declared protected area at Sierra de las Minas. With technical and financial support from the National Environment Commission (CONAMA) and the World Wildlife Fund, in 1988 Defensores formed a multidisciplinary team of scientists and experts to conduct a detailed technical study of the ecological, social, and economic conditions of the area. As a result of this study, a formal proposal was developed for the establishment of SMBR.

To familiarize the public with this study and proposal, Defensores produced a series of audiovisual and printed materials about Sierra de las Minas, with which it launched a broad information campaign throughout the country. Meanwhile, it consulted different groups concerned with the area. Among those giving their most enthusiastic support to the idea of creating the reserve were the local authorities of the arid Motagua valley, motivated primarily by the need to protect the water sources that originate in the mountain forests, upon which the valley is economically dependent. In a consultative workshop with authorities and local leaders, the seventeen local mayors who were gathered signed a request to Congress asking for the legal establishment of the reserve. This strong local support would later be a determining factor in the acceptance of the proposal by members of Congress.

Naturally, the proposal also encountered some opposition. The Forestry Guild of the Chamber of Industry, an association of the organized logging sector in the country, was strongly opposed to the declaration of Sierra de las Minas as a protected area, fearing that it would completely prohibit use of the forest. Defensores decided to meet directly with the group to explain the proposal to them in detail and to resolve possible differences. After intense negotiations, a proposal was drafted that was mutually acceptable, especially regarding the conditions and regulations of use of the forest in the sustainable-use, buffer, and recovery zones of the reserve.

After overcoming that obstacle, the proposal was formally presented to and approved by the National Protected Areas Council of the Presidency of the Republic (CONAP). CONAP was then charged with following the proposal through the government, and a few months later, in a public event celebrated at the National Palace on June 5, 1990 (World Environment Day), the president of Guatemala signed the bill and sent it to Congress. During the next few months, the bill to create SMBR was analyzed and approved by the appropriate legislative committees and finally was enacted into law by Congress.

Nevertheless, a few weeks after the declaration, a group of landowners with property within the reserve's limits who disagreed with the agreement between Defensores and the Forestry Guild took legal action before the constitutional court to request it to repeal this law. The group argued that the legal establishment of the reserve was imposing limitations that violated their constitutional right to use their private property. This action unleashed a six-month legal battle that was finally won by CONAP and the groups defending the creation of the reserve. Thus the constitutional court ratified the legal declaration of the reserve, setting an important precedent in support of conservation. Later, in January 1993, the SMBR obtained international recognition when it was included in UNESCO's international network of biosphere reserves.

Zoning

The SMBR is 236,300 hectares and occupies the greater part of the Sierra de las Minas range. It was divided according to the management zones defined by the Guatemalan legislation for biosphere reserves (Protected Areas Law and Bylaws). The four management zones are: the core zone, sustainable-use zone, buffer zone, and recovery zone.

The Natural or Core Zone

Located in the heart of the reserve, this zone contains the highest peaks of the range. It covers 105,000 hectares, approximately 45 percent of the total protected area. According to its law of creation (Law 49-90) and the protected areas bylaws (Ac. Gub. 739-90), the core must be used for preservation of the natural environment, conservation of biological diversity, preservation of water sources, scientific research, and ecotourism in suitable areas, if these activities do not negatively affect the ecosystems. Extractive activities and human settlements are not allowed, with the exception of already approved logging permits on private land, which can continue to operate after passing an evaluation and adjusting to the legal regulations. In the SMBR there was only one such permit issued, on the properties called Montaña Larga/El Cóndor, and it was canceled by the government in 1995 due to noncompliance with the approved management plan. Currently, about 95 percent of the core zone is covered by pristine, natural forest, both montane tropical cloud forest and humid subtropical forest. The other 5 percent is being used for migratory subsistence agriculture practiced by seven small Q'eqchi' and Poqomchi' communities. Four of these settlements are located within the core zone, while the others live outside and have only their agricultural fields in the core.

Sustainable-Use Zone

Extending over 34,600 hectares, this zone represents approximately 14 percent of the reserve. It forms a relatively narrow ring around the core zone and contains primarily forest cover or soils suited only for forests. In this zone the sustainable use of the forest and other natural resources is allowed, but the establishment of permanent human settlements is not (Ac. Gub. 739-90, Defensores 1992).

Buffer Zone

The outer ring of the reserve is called the buffer zone. It consists of 91,800 hectares, or 39 percent of the reserve. It includes about 110 human settlements, which include villages, ranches, and co-ops. The objectives of this zone are to achieve the sustainable use of natural resources, especially those of traditional activities, including agriculture, agroforestry, and limited livestock raising, to benefit both the local communities and the natural resources.

Recovery Zone

The conifer forests of the Teculután river valley in the central zone of the reserve were severely depleted by a logging company several years ago. For this reason, the area was designated a recovery zone. It spans 4,200 hectares, representing approximately 2 percent of the reserve (Congreso de la República 1990; Defensores 1990). The current vegetative cover in the sustainable-use, buffer, and recovery zones is as follows: approximately 65 percent of the surface of the sustainable-use zone has forest cover, and the remaining 35 percent is used for annual and permanent cultivation of cardamom and coffee. About 30 percent of the buffer zone and 50 percent of the recovery zone are also covered by forest. The natural cover in these three zones is dominated by mixed conifer forests (30 percent), mixed broadleaf and conifer forests (30 percent), pure conifer forests (20 percent), and broadleaf forests (20 percent).

The boundaries and zoning of the reserve were determined by the team of scientists that completed the original technical study. The coordinates—including both external limits and internal zoning—were established by the team with the help of the Guatemalan Geographic Institute, based on aerial photography from 1987, topographic maps at a scale of

1:50,000, and field verifications. These coordinates were included in the law that created the SMBR. Among the parameters used to determine the reserve's limits were forest cover, particularly old-growth cloud forests and humid subtropical forests; habitats for endangered species such as the quetzal and horned guan; current land use and location of human settlements; and land tenure. Due to the extent and inaccessibility of the area and the lack of available information—particularly on land tenure—the law allows the possibility of modifying the limits of the core zone in the master plan, as long as this modification would not reduce the area of the core zone by more than 15 percent. This provision was included to allow for adjustment of the design of the SMBR when further information becomes available.

Since the SMBR's creation in 1990, local people's awareness of the existence of different use zones has steadily increased, and Defensores has learned to effectively foster this awareness. This is illustrated in the case of the Lato River basin, where during the reserve's first year in existence, Defensores began demarcation without providing enough information to the local population. This caused discontent, as the communities did not understand the significance of the use zones in the reserve. As a result of this experience, Defensores began to actively share information about the reserve and its use zones through environmental education programs, community meetings, training, and technical assistance to farmers and local peoples. This strategy has been highly effective and has regained the trust of local people, the majority of whom came to accept the use zones and their rationale, as well as the physical demarcation of the most pressured and accessible sites.

Administrative Structure of the Reserve

Law 49-90 designated Defensores as the executive secretary of SMBR—that is, the administrative body—under the supervision of CONAP and a board of directors (more details to follow in section entitled "Organizational Roles," p. 121).

In order to assure its proper management, Defensores divided the SMBR into three administrative districts—Chilascó, Motagua, and Polochic—which in turn were subdivided into sectors. Initially, ten sectors were established, increasing to eleven in 1994 when one was split in half. The division into sectors was necessary in order to have land units of manageable size to allow for effectively dealing with the particular needs and conditions of each hydrographic basin in the Sierra. This administrative structure was approved by CONAP in the first SMBR Master Plan for 1992–97. In order to design the districts and sectors, Defensores took into account several factors besides the division of the hydrographic basins, such as existing access roads, ethnic distribution and social relations, and the limits of the five departments and thirteen municipal jurisdictions that share the SMBR. Each of the districts and sectors includes part of the core, sustainable-use, and buffer zones (Defensores 1992).

The SMBR has a director and two deputy directors (one responsible for land and protection and the other for sustainable use). They supervise both the personnel in charge of thematic programs such as protection, environmental education, research, and forest management and the coordinators of the reserve's three districts. The district coordinators in turn are responsible for supervising the implementation of management programs, as well as ensuring coordination with other local stakeholders, such as development boards, municipalities, landowners, and regional delegations of governmental entities. Each district coordinator has a strategically placed office in a community close to the respective district, and many sectors also have field stations.

Land and Resource Tenure

Most of the Sierra de las Minas is inappropriate for annual crops or livestock due to its soils, climate, and topography. It is appropriate for the protection of biodiversity, forest management, and some permanent crops such as coffee and cardamom. Both land use and tenure differ greatly on the north and south of the SMBR because of historical differences and cultural and socioeconomic characteristics. The total population of the SMBR is roughly thirty-five thousand, of which ten thousand are ladinos (people of mixed or Spanish origin), twenty thousand Q'eqchi', and five thousand Poqomchi'. While in the north there remain vast forested areas—mostly contained in properties that cover more than a thousand hectares—accessibility to the south side has allowed for greater human intervention. Thus the crest of the Sierra conserves only a narrow fringe of forest.

Colonization

On the northern slope, inhabited mainly by Maya descendants, there is a growing demand on the part of the local population for land for both annual cultivation (such as corn and beans) and permanent crops (coffee and cardamom). The need for agricultural land is driven by the accelerated population growth and the lack of economic alternatives for the rural population. Consequently, the communities from the lower part of the range have a direct impact on the forested land higher up, as they try to convert it to agricultural fields. This generally results in the slashing and burning of the forest.

Communities obtain the land in a variety of ways. In some cases they rent or buy it from private landowners, or they receive the land from landowners as payment for labor. Often communities organize invasions of public or private land with the intent to force negotiation. The occupants almost always clear the forest immediately to demonstrate their possession of the occupied area.

In contrast, the southern slopes of Sierra de las Minas—inhabited principally by ladinos—is not attractive to colonists or immigrants who want to practice subsistence agriculture. This is because the land tenure structure is relatively clear, the majority of lands are private property, and there are practically no suitable lands left for colonization. The migration trend has actually been the opposite. As the population has grown naturally, the low agricultural productivity of the southern slopes has caused out-migration, both temporary and permanent, of young, productive people to other parts of the country or even to the United States. Nevertheless, the agricultural frontier continues to be advanced toward the forested areas by subsistence and livestock farmers, particularly in municipal lands like those of San Agustín Acasaguastlán.

Tenure and Possession Security

In most parts of the SMBR, particularly on the north side, there is insecurity about tenure and possession of land and other natural resources, due to the following reasons:

- The country lacks a current, reliable, and coherent land survey to clearly determine property limits. The mosaic of properties currently registered in the SMBR has dramatic overlaps and imprecisions.

- As in other regions of the country, the properties are poorly demarcated due to registry imprecisions, incongruence of real and registered possession, difficult topographic conditions, and, in many cases, the impenetrable forest.

- Traditional users of natural resources lack legal security guaranteeing their use rights. There have been few legal mechanisms developed thus far in Guatemala to protect the rights of traditional natural resource users.

Indigenous or tribal lands do not exist in Guatemala as they do in other countries in the Americas. During the colonial period, the majority of the indigenous population was concentrated and settled in what were called Indian villages, where the Spanish Crown assigned lands to them (generally six leagues in circumference). This situation remained until 1871—fifty years after Guatemala's independence—when the new liberal government instituted the so-called *censo efitéutico* (a special census). This procedure deprived traditional indigenous people of their lands since lands that were not registered in a peremptory fashion were considered unoccupied and available to the government, which would section and sell the land to individuals. Thus traditional indigenous property was practically abolished and exists today only in the form of municipal and some titled community lands (A. Duarte 1996, pers. comm.).

Within the existing legal framework, the most effective strategy for ensuring traditional user rights is to allow people to acquire ownership of the land. Nevertheless, given the lack of adequate, available national lands in Sierra de las Minas, there have been several cases in which the Instituto Nacional de Transformación Agraria (INTA) has entitled rural communities with lands unsuitable for agriculture. This results in the destruction of the forest without solving the problem of rural poverty. In recent years, better coordination and communication between INTA and Defensores has stopped this inadequate titling in the SMBR.

Public and Private Property in the SMBR

An estimated 45 percent of SMBR land is public, 50 percent is private, and 5 percent is municipal. However, due to the outdated and imprecise land survey, these data are approximate. In 1991, the INTA passed to CONAP fifteen national properties within the SMBR, most in the core zone, to be managed according to the reserve's objectives. As the reserve administrator, Defensores made all possible efforts to protect and manage these properties. Nevertheless, until 1995, CONAP had not physically demarcated this land, resulting in unclear boundaries, which made it difficult to manage effectively.

Most of the private properties located in the core zone are large and covered with pristine forest. The majority of the landowners are well-to-do families from the capital or private companies. Because of their inaccessibility and rough landscapes, these properties have been conserved for many years. On the northern slope, in the sustainable-use and buffer zone, most of the co-ops and private farms are very extensive. On the south side, properties are generally smaller, particularly in the most arid zones, where the productive lands with access to water have been subdivided into smaller and smaller pieces.

Defensores has acquired about twenty-three thousand hectares of private land in the SMBR to ensure conservation of vitally important areas located primarily in the core zone. Defensores usually acquires land by direct purchase, with the exception of one case in which a private company donated the use rights for thirty years of twenty-five hundred hectares of forest in the SMBR. Defensores' land acquisition strategy is consistent with protected-areas legislation, which establishes that the land in the core zone should preferably be owned by either the government or Guatemalan nonprofit conservation organizations.

Resource Use

Most of the local people living in and around Sierra de las Minas depend directly on the available natural resources. Defensores conducted a detailed study in 1993 called "Diagnosis for Human Integration into the Sierra de las Minas Biosphere Reserve" (Margoluis and Gálvez 1993). The study was carried out in more than fifty ladino (nonindigenous) and Maya communities in the western part of the range, and compiled extensive information about the local population's perceptions and knowledge about natural resources. The following resources stand out as the most used:

Timber and nontimber forest products. A large variety of timber and nontimber forest products are used by the local population for the construction of houses and furniture. This includes wood in the form of logs, beams, and boards, which are sawed in simple pit sawmills or, at times, with a power saw. A variety of palms, reeds, and vines are collected from the forest and used for roofs and walls (Margoluis and Gálvez 1993).

In some populations, women make large baskets using as the primary material vines, palms, and reeds, which are collected by family members in the reserve's broadleaf forest. The species of reed called locally *vara de canasto* (*Cusquea* sp.), which grows in the cloud forest in the western part of the SMBR, has been affected by this extraction. Nevertheless, according to studies in 1995, the growth still exceeds the community's extraction (Flores 1997).

More than 90 percent of the local inhabitants of SMBR use firewood for cooking, and this wood usually comes from the natural forest. Among the species preferred as firewood and for other multiple uses are the pine (*Pinus* sp.), the oak (*Quercus* sp.), and the *cuje* (*Inga* sp.), although the order of preference differs between ladinos and Mayas. In the last few years, there is growing interest in reforesting with multiple-use trees as the distance to areas where firewood is collected increases (Margoluis and Gálvez 1993).

Edible and medicinal plants. The local population uses a great variety of plants for food and medicine. Most individuals have a general knowledge of the medicinal qualities of some plants, though healers have a more profound knowledge. In some Maya communities, local knowledge of medicinal plants has been lost with the introduction of Western civilization and the communities' disassociation from the forest.

Hunting. For many years the local population of the SMBR—both ladino and Maya—have hunted wildlife. Nevertheless, since the SMBR is a protected area and hunting is illegal in the core zone, this has become a very delicate topic and it has not been possible to gather reliable information to quantify the scope and impact of this activity (Margoluis and Gálvez 1993). Local people have many motivations for hunting wildlife. They do it to complement their diet or to obtain medicinal substances from certain animals. They often hunt animals that damage their crops, and other times hunt for recreation. There has also been news of sport hunting groups from the cities that contract locals to serve as their guides. Among the preferred game in the SMBR are various types of turkeys such as the so-called *pajuil* (*Penelopina nigra*) and also the threatened horned guan. Hunted mammals include the white-tailed deer (*Odocoileus virginianus*), red brocket deer, paca (*agouti paca*), white-lipped peccary, and collared peccary. However, during the last five years there has been a significant recovery of many of the game species—particularly the horned guan—in sites where reserve personnel have been present (C. Méndez 1994, pers. comm.).

Subsistence crops. Most of the indigenous and ladino communities in the SMBR continue to plant corn and beans as subsistence crops even when they have other sources of income. The family diet is based on these staples and complemented by other minor crops such as pepper, bananas, sugar cane, citrus fruits, squashes, and herbs, as well as domestic animals.

For corn and beans, local farmers generally use the *milpa* system of crop rotation. This means that after farming the land for one season, the land must "rest" for a year or two. During this time the land becomes covered by secondary vegetation called *guamil* or *guatal*. To prepare the land, almost all farmers use *rozas*, controlled burns, which can easily turn into forest fires if the necessary precautions are not taken. The farmers of the SMBR agree that their harvest yields very little corn. This is due to the steep topography, poor soils, and insect pests, as well as the lack of knowledge and application of appropriate farming techniques (Margoluis and Gálvez 1993). Particularly on the north side, this low farming productivity has helped advance the agricultural frontier toward forested areas.

Small-scale cattle ranching. The ladino population of the SMBR's southern watersheds practice small- and medium-scale cattle ranching. The livestock, usually owned by large cattle ranchers of the Motagua valley, roam free for three or four months a year in the high part of the mountain. There they graze in semiforested areas, such as open pine and oak forests. This affects the regeneration of the understory, since the livestock not only eat from the young trees but also compact the ground on the slopes, contributing to soil erosion. Burning the pasture at the end of the dry season to promote quicker regeneration creates another negative environmental impact. The burns usually occur without adequate precautions to protect the forest. On repeated occasions fire has advanced to the high part of the mountain and caused forest fires.

Water. The communities and populations in the valleys surrounding Sierra de las Minas depend in many ways on the reserve's water. In the arid Motagua and San Jerónimo valleys south and west of the range, this water is essential to daily life and the regional economy. Communities rely on it not only for domestic uses but also for supplying ranches and plantations. The agriculture and agroindustry of both slopes and valleys also depend on the mountain streams to supply a number of gravity, sprinkling, and dripping irrigation systems. A series of ranches and local communities have small hydroelectric plants to generate energy locally; the Río Hondo hydroelectric power plant supplies electricity to the national electricity system. In addition, the carbonated and alcoholic beverage industry in the Motagua valley depends on underground water, which is extracted from deep wells.

However, during the last fifteen years the deforestation of the watersheds has caused a decrease in river flow, thus reducing the surface of land that can be irrigated and negatively affecting local ecosystems. Some of the streams are polluted by fecal waste and agrochemicals. The underground water levels have also dropped in the valley, and several industries have had to invest significant amounts of money in deepening their wells.

As in other arid zones, there are frequent conflicts over water rights, particularly between users from the upper and lower parts of the watershed. These conflicts have been over both water quantity and quality. A fee has never been charged for the use of this resource, and hence local municipalities neither have the necessary funds nor feel responsible for investing in the conservation of the watersheds or springs. Defensores has done an ecological and socioeconomic study of the water coming from Sierra de las Minas, which will be used as the basis for designing and executing a campaign to raise local awareness of the need to take concrete steps toward conserving water sources (Brown et al. 1996).

Commercial use. The natural resources of the Sierra de las Minas have been exploited since the colonial period. Primary activities include agriculture, logging, and mining.

Logging. Sierra de las Minas contains some species of timber that are very attractive to industrial operations. The preferred species all pertain to the genus *Pinus*. One of the most highly valued species is the white pine (*Pinus ayacahuite*), which is used for furniture and construction. However, the *Pinus caribea, P. maximinoi,* and *P. tecunumanii* are also highly valued for their commercial timber.

The difficult topographic conditions, fractured geology, and inclement climate in the Sierra have restricted large-scale forest exploitation due to the high road construction and maintenance costs, which reduce the profit of the operations. In spite of this, for many years the forests of Sierra de las Minas supplied wood to sawmills located in the arid Motagua valley on the southern side. Twelve of these sawmills were still operational in 1995.

With the exception of a few well-managed private natural forests and plantations, commercial logging in Sierra de las Minas has not been on a sustainable base or with a long-term vision. This has contributed significantly to the loss of forest cover, additionally causing the genetic degradation of the forest due to the practice of extracting the most valuable and best-formed species first.

The inappropriate forestry practices can be illustrated by some recent examples. In the Teculután river valley, a tributary of the Motagua river, which forms in Sierra de las Minas, the Maderas El Alto sawmill felled extensive areas of forest during the 1970s and 1980s. This destroyed natural ecosystems and seriously affected the water sources in the zone. Upon establishment of the SMBR, this deforested area was designated a recovery zone.

During the same period, at the beginning of the 1980s, the Guatemalan government began constructing a paper pulp plant in the Motagua valley, called CELGUSA, with the intent of supplying it with timber from the Sierra de las Minas and other nearby areas. However, the plant never became operational, in part because sustainably produced sources of timber to feed the plant were never developed.

Another case, which occurred during the first half of the 1990s, was the destructive exploitation of the oldest part of the cloud forest in the core zone of the reserve on a property called Montaña Larga / El Cóndor. Because it had a logging permit issued before the establishment of the reserve, by law it was allowed to continue operating until this expired. However, the permit holder committed some serious violations of the approved management plan, thus provoking a series of public denouncements in the national press. After investigating and verifying these violations, the government finally canceled the permit.

The local communities of the Sierra have received few benefits from the commercial logging. It has created only very limited employment and income for field workers, compensating little for the loss and degradation of natural resources. The municipalities of the zone, always in need of funds to provide public services, have likewise been unable to charge significant taxes or fees for the extraction of wood in their jurisdiction.

Commercial crops. Various regions of Sierra de las Minas have ecological conditions that are ideal for growing commercial crops. On the towering, rainy hillsides of the north side—in the Polochic river basin—coffee has been grown for more than a century, and cardamom was introduced several decades ago. The plantations occupy a belt of land that runs along the north side at an intermediate level of elevation, essentially within the limits of the SMBR in the sustainable-use and buffer zones. At first, coffee and cardamom were grown on large plantations, many of which were established by European immigrants. However, as the years passed, there has been a significant increase in small local producers, of both Maya and ladino origin. Recent scientific investigations indicate that small-scale cardamom and shade-coffee plantations are compatible with the conservation of biological diversity, if and when they maintain a high canopy diverse in species and structure (Greenberg et al. 1995).

At the western end of the SMBR, around the village of Chilascó, broccoli and cauliflower were introduced as commercial export crops in the second half of the 1980s. This has significantly raised the income of local people, but it has also had negative impacts on health and the environment due to the massive and inappropriate use of agrochemicals.

In the SMBR, commercial crops represent the most important source of cash income for local inhabitants, in the form of employment generated by the major plantations or the income small farmers receive for their harvests. For this reason, it is vital to ensure that these sources of income are maintained, but within a sustainable ecological framework.

Marble. In the central part of Sierra de las Minas, since the 1940s, the Guatemármol company has had a concession to quarry marble on public lands. This medium-scale activity is limited to an area of only approximately fifty hectares, which contains exploitable marble blocks, and its environmental impact has been relatively small. In addition, the presence of the marble company may have even helped to protect the natural forest in this zone by controlling access to the Sierra.

Organizational Roles

The Congress of the Republic of Guatemala took an unprecedented measure in the history of biosphere reserves in Latin America when it designated Fundación Defensores de la Naturaleza, a private Guatemalan nonprofit organization, to be, by law, the administrator of the SMBR. Defensores performs this function under the supervision of the National Protected Areas Council of the Presidency of the Republic (CONAP). CONAP chairs the reserve's executive board, which also includes representatives from various local governments and indigenous communities, as well as landowners. Until 1995, however, this board had not been formally constituted, due to lack of clarity regarding the mechanisms for designating landowners and indigenous peoples as representatives.

In Guatemala the management of protected areas may be delegated to specialized entities, including nongovernment organizations (NGOs). All protected areas in the country are governed by the Protected Areas Law, and all managing organizations are under the overall supervision of CONAP. CONAP is also responsible for the authorization and supervision of natural resource extraction in protected areas where the law and area management plan allows such action and the administrative entity approves it.

As a Guatemalan nonprofit conservation organization, Defensores was eligible to be charged with the management of a protected area. Defensores' founders and board, a group of Guatemalan conservationists and influential business leaders and philanthropists, had made the conservation of Guatemala's biodiversity the organization's mission. For the implementation of programs, over the years, the organization recruited a staff of eighty-five qualified and highly committed technical, administrative, and field personnel, which represent both genders and a wide array of professional disciplines and ethnic and geographic origins. This human diversity has allowed Defensores to collaborate effectively with a wide range of stakeholder groups, and it has proven to be a critical factor when operating in the complex context that is Guatemalan society.

In its administrative capacity for the SMBR, Defensores carries out all the programs and subprograms described in the master plan and the annual operational plans approved by CONAP. These include the protection of wilderness areas, rural sustainable development, environmental education, research, and administration.

Another important function of Defensores has been interinstitutional coordination within the SMBR, since in this area a multitude of jurisdictions and institutional interests overlap, from both central and local government agencies, such as INTA, DIGEBOS (the Forest Directorate), the Ministry of Energy and Mines, CONAP, and development councils and municipalities. With all of them, Defensores exchanges information, facilitates institutional coordination, and plans joint efforts. It also organizes regular workshops, meetings, and field trips with decision makers from different levels—local, regional, and national—to keep them informed and help them design and implement the protection and sustainable management of the reserve and its natural resources.

In addition, Defensores maintains strong working relations with local people, both within the SMBR and in the area of influence. It provides technical assistance, training, and environmental education to men, women, and children alike. These efforts usually focus on themes related to natural resource management and protection, such as sustainable agriculture, community forestry, women's participation, and environmental education. In 1995 Defensores launched a pilot project to strengthen local and community groups in the SMBR that were interested in working on different aspects of conservation and sustainable resource management. In this way it expects to increase the local capacity and sense of responsibility for the conservation of natural resources in this valuable protected area. (These activities are described in greater detail in the next section.)

When Defensores detects violations of the laws governing the reserve, such as illegal timber exploitation, poaching, or invasion of wilderness areas, it gives notice to the government police forces and the district attorney to enforce the law. In such cases, Defensores, as the protected area's managing organization, acts as guardian and not as police, thus its field employees and resource guards do not carry arms. During the last five years there have been several law enforcement cases within the SMBR. In each case it was difficult to get the authorities to act quickly and efficiently. Defensores has repeatedly received proposals to assume the role of police in the SMBR but has not accepted this responsibility as it is incompatible with its role as a promoter and technical advisor to local communities.

Independent of its administrative function at the SMBR, Defensores has purchased nearly twenty-three thousand hectares of private lands covered with forest and other pristine ecosystems, mainly in the core zone of the SMBR, for protection and conservation. Most of the funds for these acquisitions were raised through tropical forest "adoption" campaigns carried out in Sweden and the United States, in which thousands of children and adults contributed money. In addition, in 1993, a private landowner donated the use rights for thirty years of a forested area of about twenty-five hundred hectares in the SMBR. Guatemalan law supports this land acquisition strategy, stating that the core zone of the biosphere reserves should preferably be public lands or the property of Guatemalan nonprofit conservation NGOs. Land acquisitions in the SMBR have secured the conservation of examples of unique and threatened ecosystems in the core zone that otherwise could have been pillaged, including a unique tract of ancient cloud forests on the Pinalón peak. On the other hand, the land purchases have been a form of indemnifying many owners of forested properties that had lost interest in keeping them due to the logging restrictions in the core zone. Additionally, land ownership has provided Defensores with more legitimacy in the eyes of the other landowners, who now see it as just another neighbor.

The government contributes funds to the SMBR for the salaries of ten field resource guards, but all other financial, human, and material resources for the management of the SMBR have been provided by Defensores. Funds are raised privately in Guatemala and abroad through campaigns and technical and financial cooperation projects. In order to ensure a continuous budget for the SMBR, Defensores is working actively to develop innovative mechanisms for financial self-sufficiency.

Defensores has made a great effort to carry out a series of participatory planning processes to establish the program, management, and development priorities of the SMBR. The outcomes have been the Defensores Strategic Plan, the SMBR Master Plan, and the annual Operations Plans (Defensores 1992, 1995). So far these processes and documents have been successful in ensuring a coherent, orderly, and coordinated implementation in the reserve.

In terms of implementation, Defensores has established strategic alliances with a wide array of national and foreign organizations, both public and private. It also has received technical and financial assistance from international entities with whom it shares common objectives. This support is mainly for the management and protection of the SMBR, but it has also strengthened Defensores' institutional capacity as the reserve administrator. Among the most important cooperating organizations are:

World Wildlife Fund (WWF). WWF has provided technical and financial assistance to Defensores since 1988, when it supported technical study in which Sierra de las Minas was originally proposed as a protected area. Since then it has provided constant support for reserve management, particularly in the fields of environmental education, sustainable agriculture, and community forestry. Part of this technical cooperation has been provided through COSECHA, an organization that specializes in soil conservation and agricultural outreach to small-scale farmers. WWF considers SMBR to be an example of an integrated conservation and development project in Central America.

The Nature Conservancy. The Conservancy has cooperated technically and financially with Defensores and the SMBR since 1990, when the reserve was included as a site of the Parks in Peril program. The assistance offered by The Conservancy was fundamental in the first phase of the management and consolidation of the SMBR; it focused mainly on wilderness-areas protection, acquisition and stewardship, ecological research, and the institutional strengthening of Defensores as the administrative body of the reserve. Financial resources have come from many different sources, including the Parks in Peril program, which is largely funded by USAID, the PACA Project, and the Adopt-an-Acre campaign.

MacArthur Foundation. MacArthur has played an essential role in the initial institutional development of Defensores as the manager of the SMBR by providing critical funding for core operations and supporting the development of mechanisms to enable it to become financially self-sufficient.

Liz Claiborne /Art Ortenberg Foundation. For many years this foundation has provided important funds to Defensores, through both WWF and the Conservancy, primarily financing activities to promote community-based conservation, sustainable rural development, and land tenure security for indigenous people in the SMBR's buffer zone.

CARE. Through the PACA Project, CARE supported Defensores from 1992 to 1995, mainly in the areas of environmental education, sustainable rural development, and community forestry in the SMBR.

RARE Center for Tropical Conservation. For several years this organization has provided technical assistance to Defensores, particularly applied research in the SMBR, which includes a study on the resplendent quetzal and a hydrological study.

Peace Corps. Since 1990 Defensores has hosted several groups of Peace Corps volunteers in the SMBR. The number of volunteers has grown each year from three in 1990 to fifteen in different parts of the reserve in 1995.

U.S. Forest Service. Through the Sister Forest Program the U.S. Forest

Service has provided technical assistance to Defensores since 1993. Among the areas of cooperation are the prevention and fighting of forest fires and ecological studies in the SMBR.

Linkages between Park and Buffer Areas

The extensive wilderness areas of the SMBR not only hold unique natural values, but also provide vital goods and services to the communities of the surrounding hillsides and valleys. Defensores has thus recognized since the beginning that the wilderness areas would survive in the long run only if the local population that lives and works in the sustainable-use and buffer zones of the reserve actively participated in its protection and management. This implied that the communities should come to sustainably manage their natural resources and assume responsibilities for the protection of wilderness areas. It was clear that this would be possible only if the communities' quality of life were to improve during this process.

The first step in helping the communities of the SMBR identify with and become involved with the objectives of conservation and sustainable development was to hire a group of local townspeople as resource guards and outreach workers. They had a good knowledge of the area, language, and local culture, considerable interest in and knowledge of natural resources and sustainable agriculture, and the necessary leadership qualities. Technical personnel for the districts were also contracted in the region. This strategy proved successful, since it facilitated the adoption of the conservation objectives in the field, and was particularly effective in the reserve areas inhabited by Q'eqchi' and Poqomchi' groups. These personnel were trained in workshops, courses, and exchange trips and received regular technical assistance from Defensores and other entities. These local field personnel have been key to Defensores' ability to carry out stewardship and extension activities.

After two years of work at SMBR, Defensores recognized the need for more detailed baseline information in order to help determine its program priorities and to measure its progress in the field. This was vital because of the social complexity—with populations of ladino and Maya origin (Q'eqchi' and Poqomchi')—and enormous biological diversity of the reserve. Thus, in 1992–93, Defensores commissioned the following studies:

- a detailed socioeconomic diagnosis carried out through participatory techniques to establish the needs of the communities, their perceptions about natural resources, and their development expectations
- a first rapid ecological assessment of the SMBR.

Additionally, in 1995, Defensores commissioned a comparative study on the advance of the agricultural frontier using aerial photography from 1964, 1990–91, and 1995.

To select its focus areas and communities, as well as its types of action, Defensores applied criteria such as:

- the distance of the community from and threat of the community to important wilderness areas
- the need of the population for sustainable management of its natural resources
- the type of natural resources and practices of the community
- the potential for success and receptiveness to the support of Defensores.

The annual operating plans for SMBR are prepared based on an annual evaluation and participatory planning process organized by Defensores. For example in 1995, 1,320 people from forty-five communities of the reserve participated in workshops to evaluate the activities already carried out and the programming of future activities. Usually, all of Defensores' technical and field personnel participate in this process, which lasts nearly a month. Hence, the communities have direct influence and responsibility in evaluating and programming the protection and management activities of the reserve.

Each year Defensores' work has expanded its coverage in the reserve, and in 1995 management programs were underway in all eleven sectors. Among the main programs are wildland protection and stewardship, which include regular patrols to control poaching, illegal logging, and land invasions. It also includes the demarcation of the more accessible and threatened areas and the prevention and combat of forest fires. Resource guards received binoculars and training for the identification and monitoring of wildlife. This training of local personnel has been provided mainly as part of Defensores' research projects, such as the ecological study of the resplendent quetzal and a study on birds in coffee plantations. The information gathered by the regular monitoring is being processed monthly in a database created to produce updated lists of wildlife.

Another program focuses on providing sustainable agricultural assistance to farmers in forty-nine of the eighty rural communities surrounding the core zone. This component includes both agroforestry and agriculture activities. This is to assist rural communities, particularly the indigenous communities on the north side, to make a change from migratory to sedentary agriculture and to apply sustainable practices. The process generally followed by Defensores consists of three phases. The first phase focuses on improving crop productivity through soil conservation, using green manure, minimal tillage, contour planting, soil conservation barriers (for example, stone retention walls or plant barriers), use of compost, and application of organic pesticides. It also includes the establishment of small communal nurseries to plant trees for watershed protection and as a future source of firewood. The second phase of the process is centered mostly on assisting local communities in improving or securing their land tenure situation in order to avoid their continual migration to other wilderness areas. Among the strategies are support in land purchase negotiations with private owners, orientation of INTA (the land titling agency), and land exchanges. And finally, the third phase emphasizes the development of economic activities that provide local communities with an alternative income to subsistence agriculture.

Simultaneously, Defensores implements a women's participation component. For example, in 1995 it worked in seventeen communities on projects such as kitchen gardens, family health and nutrition, development of income-generating activities, and introduction of appropriate technologies. Another important component of Defensores' action in the SMBR is environmental education, which in 1995 was conducted in seventy-seven communities within the reserve and its influence area. It is aimed at schoolchildren, adults, and local leaders alike, and includes education and training events, field visits, and communication through the media.

Defensores has also recognized the importance of minimizing the effects of population growth and considers the demographic pressure on the reserve to be a long-term threat. Nevertheless, due to political and religious sensitivities about family planning, Defensores has decided to only inform and promote discussion about this topic among the communities.

During the last three years Defensores has stimulated the formation of various community groups in the SMBR. Among these are the Integral

Development Association of El Progreso in San Agustín Acasaguastlán (ADIPSA); the Chilascó Waterfall Committee, which focuses on ecotourism; and the Association of Environmental Teachers of the Sierra de las Minas Biosphere, created to promote environmental education in the reserve. In 1995, Defensores initiated a pilot project to further develop local organizations in two watersheds of the SMBR.

In general, the rural inhabitants of the SMBR understand conservation of natural resources, land stewardship, and rural development in a holistic way, integrating protection of wilderness areas, natural resource use, and their own quality of life. They have always depended on natural resources for survival, particularly water, soil, forest products, and wildlife. In addition, the indigenous communities that live on the northern slope of the Sierra still consider some of the peaks to be sacred sites. In spite of this, the local people usually adopt ecologically sustainable practices only when they really understand and perceive the direct benefits of these practices, both individually and collectively. A good example is provided by the reforestation activities, which they are particularly enthusiastic about because of both their interest in maintaining the water sources and their need for firewood. Defensores' activities have thus become more and more integral, and its field personnel have had to adopt a broader vision of the management of SMBR and operate as interdisciplinary teams within each district and sector.

Until now Defensores' management approach of the SMBR has been relatively successful, as suggested by the significant growth of community involvement in forest protection and sustainable natural resource management activities, particularly in the adoption of sustainable agriculture practices. Other positive indicators have been fewer and less severe forest fires and a reduced number of invasions of the wilderness areas on the northern slope. However, the agricultural frontier has not been stabilized. Accomplishing this objective will depend first and foremost on finding economic alternatives to subsistence agriculture, stabilizing the population growth, and ensuring that communities increase and maintain respect for the reserve's wilderness areas.

Conflict Management and Resolution

As in all protected areas, conflicts have occurred in the SMBR. Most are related to use, property rights, and management of natural resources and lands. Defensores has used creative methods to confront these problems and has in general been successful. However, some of the disputes have such deep and complex causes that resolution will be a lengthy process.

Logging in the Core Zone

As mentioned previously, a verdict by the Constitutional Court established that all landowners in the SMBR had to adjust to the use regulations established for each of the management zones in the reserve. This means that no extraction of timber and nontimber products is allowed in the core zone. A minority of landowners on the southern slope of the reserve have generated political pressure to reverse the legislation in this zone, with the intention of logging. They have also openly opposed the work of Defensores. In the case of biologically and ecologically important areas, or in the reserve's strategic points, Defensores has offered to buy the land with the support of financial campaigns organized by the Conservancy in the United States and the Children's Rainforest Group in Sweden.

However, controversies are often not resolved that easily. One of the worst conflicts to take place in the SMBR was related to the large-scale timber exploitation of Montaña Larga, a property in the core zone of the reserve owned by a wealthy logger. The logger had received a logging permit before the reserve existed and by law was allowed to continue operating until the permit expired. Defensores opposed this timber operation for many years because of its destructiveness but could not convince the government to cancel it. The situation was especially bothersome because local communities criticized Defensores—as administrator of the protected area—for allowing the logger to continue operating while they were restricted from logging.

During the five years of the conflict, two resource guards were attacked and suffered permanent injury, and personnel and board members of Defensores, including the director of SMBR and the chairman of the board, received death threats. Even so, Defensores managed this conflict with both diplomacy and persistence, avoiding direct confrontation, and even conducting conciliatory meetings with the logger, while still pressuring the government agencies, especially DIGEBOS and CONAP. In 1995 the case was turned into a public scandal by the national press, which denounced the serious legal violations and excessive exploitation of the logging operation. Under pressure, intensified by an international letter-writing campaign, the government conducted an exhaustive investigation of the case and, after proving the irregularities, canceled the permit. Months later, an accrediting bank seized the logger's property as the result of an unpaid mortgage. The Conservancy and Defensores organized a fund-raising campaign, and finally Defensores was able to acquire the property from the bank.

The cancellation of this logging permit and the later acquisition of the land were great achievements for Defensores. They demonstrated to rural communities that wealthy and influential people could not illegally exploit the reserve's forests and go unpunished. They also saved one of the oldest and most spectacular tracts of cloud forests in Sierra de las Minas.

Land Use and Tenure

An equally important source of conflict is land use and tenure, especially in the Q'eqchi' and Poqomchi' communities on the northern slope. In spite of the fact that the summit of Sierra de las Minas has never been inhabited by humans, the indigenous communities have in recent years been provoking the advance of the agricultural frontier on the northern slope of the SMBR. This is due to a combination of factors including population growth, lack of available land in the valley, land tenure insecurity, migratory farming practices, lack of economic and social development opportunities, and political violence. Defensores has used a variety of methods to address these cases, including the following:

Negotiation. In several cases where migratory farmers invaded the pristine forests of the core zone of SMBR, Defensores was able to persuade the invaders to abandon the land voluntarily. In the case of both Finca Concepción and Las Pacayas, this was achieved after patient dialogue negotiation and persuasion, in which Defensores relied on the support and mediation of local municipal authorities.

Voluntary relocation. Like other Q'eqchi' communities, during the violent years of the late 1970s and early 1980s, the thirty-family Santiagüilá community moved to the highlands of the Sierra, where they felled the forest to practice subsistence agriculture and plant cardamom. At the beginning of the 1980s, the national army moved local communities, including Santiagüilá, concentrating them together in the village of San

Lucas at the foot of the range. When the repression was reduced after 1985, the lack of available land and opportunities in the valley made them migrate to the higher slopes of the range in search of land. Most of these farmers established themselves on land located outside the core zone, occasionally using areas inside as agricultural fields. Santiagüilá, however, established itself on private land inside the core zone of the reserve. The high slopes, the humidity, and the poor soil that could not support agriculture and the remoteness and lack of roads isolated them from services like education, health, and stores. Nor could they gain titles to these lands, because they were in the core zone of the reserve.

To resolve the situation, in 1991 Defensores established contact with this community and after several meetings, both parties agreed to the voluntary relocation of the community to an area outside the reserve. Defensores would acquire property in the buffer zone or outside the reserve limits, establish a program to title land to the community, help community members move their crops and gardens, and assist them for several years until they were completely settled.

Things did not go as expected, however. The lengthy land acquisition process resulted in Defensores negotiating for three years with the landowners. During this time, the community decided not to wait and moved to a lower slope, settling on another private property. Defensores had to change its strategy, and it is trying to find a way to title the community with the land on which the community has now settled. As of 1995, the case had not been resolved, but Defensores hopes the titling process will be complete no later than early 1997. It will have then fulfilled its two objectives of ensuring the protection of wilderness areas in the core zone of the reserve, and helping the Santiagüilá community acquire rights to adequate lands, which will allow its people a better quality of life.

Another important example of voluntary relocation is provided by the Poqomchi' indigenous community of Vega Larga. This case differs from Santiagüilá because the community of thirty-two families has a collective title to 1,350 hectares of land in the core zone given to them by INTA in 1987. During the years they occupied the land, they destroyed nearly 250 hectares of ancient cloud forest to plant corn and beans. However, this did not relieve their poverty. In spite of technical and financial assistance from an organization called the Family Integration

Center, their harvests barely kept them alive due to inclement weather, infertile soil, and isolation.

After four years of conversation and negotiation, Defensores and Vega Larga agreed to a land exchange. This meant that Defensores had to find suitable agricultural land where the people could resettle, land with good access to roads and in the same departmental jurisdiction. In addition, Defensores would assist them with the move and the reconstruction of their homes, and also with reestablishing their fields. In exchange, Vega Larga would sell to Defensores the property in the core zone for the recovery of the cloud forest.

For four years, Defensores used a series of strategies to advance the dialogue and negotiation with the community. In the first stage it involved community members in conservation activities as resource guards and assigned an environmental instructor who spoke Poqomchi' to work with the women. Together with community representatives Defensores organized visits to inspect potential relocation properties. Finally, it organized a series of community meetings, which involved participation by INTA (which finally realized its error in titling the community with land unfit for agriculture), representatives of the Catholic church, the Family Integration Center, and the departmental and municipal governments. Currently, Defensores is raising funds to buy the lands and finance the move of the community to the new property. If this procedure is completed as expected, it will be an example of a voluntary relocation in which all desirable objectives are achieved: the conservation of nature and a substantial improvement in the quality of life for this indigenous group.

Legal measures. In the SMBR there have been some invasions of the core zone in which Defensores exhausted all processes of dialogue without arriving at a satisfactory solution. In such cases, it has resorted to the legal process, as it did when a group of invaders called Tierra Colorada settled on Defensores' private property located in the core zone of the reserve. When dialogue proved ineffective, Defensores filed a suit against them, and at the end of 1995, was waiting to get a court order for their eviction.

One can conclude that resolving issues involving communities in the core zone is one of the most pressing issues and one of the biggest challenges Defensores will face during the next few years. The issue of land tenure in Guatemala, especially for indigenous communities, is delicate and of particular importance during the peace process, which is in progress. For this reason, whatever strategy it uses—persuasion, voluntary relocation, titling of land currently occupied, legal methods, or others—Defensores policy is to act with patience and to exhaust all possibilities of dialogue, always looking for agreements and solutions that benefit conservation as well as improving the quality of life of the people.

Water Rights

Water use and rights is another area of dispute in the SMBR, particularly on the southern side of the reserve in the arid Motagua valley, where water is particularly scarce. Water users in the lower watershed resent the loss of quality and quantity of water caused by deforestation in the upper watershed, particularly when it is caused by logging companies. The use of excessive quantities of water for irrigation or agroindustry and water pollution are also sources of conflict.

In order to have the baseline information for further action, Defensores completed a hydrological study of two principal watersheds in the reserve, the Lato and Colorado rivers, in 1995 and 1996 with the support of various international organizations. With the results, Defensores is developing an educational campaign about the value of water and will support the strengthening of organizations and local groups to protect and manage water resources. The organization is also considering whether to assist local municipal authorities in developing a user-fee system to raise funds to protect the watersheds. One concrete result has been the formation of twenty community brigades to prevent and fight forest fires. Its members are increasingly aware of the role the forest plays in the protection of water sources, which they discussed in workshops organized by Defensores. In spite of having only minimal equipment and tools, their efforts were key in preventing forest fires during the dry season of 1995.

Large-Scale Threats

Successful management of the SMBR is limited and made difficult by two key factors: the weakness of government institutions and unsustainable farming practices.

Weakness of Government Institutions

As of 1995, government agencies with jurisdiction over natural resources were not operating in a satisfactory manner. One major problem is that during the last few years, CONAP has lost political influence and leadership ability and has faltered in its administrative capacity. Additionally, it has faced severe budget constraints.

There is also a lack of clarity regarding jurisdiction and authority among central government agencies and local authorities. This has caused problems, for example, between CONAP and DIGEBOS about the issue of forestry operations in the SMBR. In addition, some agencies have opposing institutional objectives, a constant source of problems and difficulties. For example, INTA's legal mission is to promote the titling of public land to landless peasants, while CONAP's mandate is in favor of dedicating these national lands to conservation. Defensores has spent a lot of time and energy facilitating information exchange and coordination among government agencies with jurisdiction over the SMBR.

On the other hand, the enforcement of natural resource laws has not worked adequately, due in part to the ineffectiveness of both security forces and the judicial system and the fact that the country's laws include only very weak punishments for environmental offenders.

Unsustainable Farming Practices

Slashing and burning the forest to advance the agricultural frontier places the most direct pressure on wilderness areas. This destructive subsistence farming practice particularly affects the northern slope of the reserve and is still practiced by those who are marginalized, poor, and lacking in social and economic development opportunities, which is the case for most of the Sierra's communities. The problem is exacerbated by the lack of land tenure clarity and legal security. Another important threat to the forests of SMBR is fire caused by cattle ranchers who burn the pastures, as well as farmers of annual crops.

National Policy Framework

In Guatemala there have been significant accomplishments in environmental policy during the last few decades. The country signed and ratified the main relevant conservation agreements, including the international conventions on World Heritage, CITES, Ramsar, Biodiversity, and Climate Change. It is also part of the Western Hemisphere Convention and the Central American Conventions on Forests, Biodiversity, Climate Change, and Toxic Waste.

The government administration during the period 1986–91 had an authentic political commitment to conservation, as several indicators show. For example, it established a new legal and institutional framework for the environment for both Guatemala and Central America. This includes the creation of the law for Environment Protection and Improvement and the institution charged with its implementation, CONAMA; the Protected Areas Law and its implementor CONAP; and CCAD (the Commission for Central American Environment and Development).

In 1990, during the same administration, the Maya and Sierra de las Minas Biosphere Reserves were created, which together cover about 17 percent of the surface of Guatemala. (This contrasts with the period 1991–95, during which no new protected areas were created, in spite of being proposed.) Also, political decisions from that administration strengthened the position of new protected areas, as in the 1990 case of the Basic Resources oil refinery. The Ministry of Energy and Mines had authorized the refinery's construction inside the Laguna del Tigre National Park, one of the core zones of the Maya Biosphere Reserve. After a highly publicized public debate and opposition from conservationists, the government decided to override the ministry's decision and move the project outside the protected area.

Different Guatemalan administrations have been relatively open toward international environmental organizations, although they have not always been an efficient national counterpart. Many foreign public and private organizations continue to support a variety of conservation activities throughout Guatemala. An important step forward for conservation in Central America was the Alliance for Sustainable Development initiative proposed by the Guatemalan government and adopted by all seven Central American countries in 1994. This agreement, which serves as a general development agenda for the isthmus, is based on environmental protection and the sustainable use of natural resources. The first external partner for this initiative was the United States, with whom all seven countries signed the Central America–USA Agreement known as CONCAUSA in 1995.

Nevertheless, there have been shortcomings in the development of some vital aspects of environmental management during the last ten years. For example, government funding for environmental agencies and protected areas has been truly insignificant to date. Protected-area operations have been funded largely by foreign organizations and, to a lesser extent, by private Guatemalan sources. Thus there is uncertainty about how to ensure appropriate long-term financing for the country's legally declared protected areas.

The Guatemalan government has also not been particularly proactive or visionary in exploring innovative mechanisms to finance conservation, which could compensate for the lack of ordinary budget contributions. An example of this is the debt-for-nature swap proposed by a group of national NGOs, which has not been implemented despite year-long negotiations. The government has also lagged in incorporating environmental issues into other spheres of government. Although the two administrations in power between 1990 and 1995 made the environment part of their government agendas, this inclusion was mostly nominal and rhetorical, and few concrete pro-environment and conservation measures were actually taken.

Another shortcoming has been that government agencies charged with conservation have failed to implement policies favoring biodiversity conservation, from either lack of ability or lack of interest. An example is the fiscal incentives program created by the Protected Areas Law in 1989 to allow for the establishment of private nature reserves. As of 1995, it had still not been implemented by CONAP.

Conclusion

Sierra de las Minas is a crown jewel in the natural heritage of Guatemala, combining extraordinary natural values and great social complexity. Its establishment and management as a biosphere reserve have created a unique opportunity to harmonize the conservation of its natural resources with the sustainable rural development of its population.

The preliminary results of Defensores' first five years in the SMBR seem to indicate that strategies applied were appropriate for the following reasons:

- Its objectives and institutional roles are clearly defined, which allows Defensores to assume and develop its role as leader, coordinator, and facilitator of the reserve.
- There has been a consistent effort to generate objective baseline information about natural resources and socioeconomic issues, which serve as benchmarks to measure the advance of the work.
- The work has focused directly on the social players who depend on, threaten, or have interest in the reserve's natural resources.
- Defensores has demonstrated a capacity to learn from its experiences and those of others and to continuously improve itself.
- Defensores has a long-term commitment to the reserve, allowing it to conduct lengthy programs in the SMBR to attain the desired results.

In the near future, Defensores' ability to maintain its institutional strength, credibility, and management expertise will be a key factor in determining its success in achieving management objectives in the SMBR. Especially important are the cooperative relationships with all the stakeholders involved in the SMBR's management, including local communities; landowners; the private sector; local, regional and national government; and foreign partner organizations. Only then can conflicts be minimized and work agreements be reached among all the groups to coordinate the work for the reserve's conservation and protection.

There will also be important long-term challenges for Defensores in the SMBR, including:

- ensuring financial resources for reserve management
- getting other groups or agencies to assume concrete responsibility for protection and management of areas or specific themes of the reserve (such as watersheds)
- identifying and objectively measuring the goods and services that the SMBR provides for the local population, Guatemalan society, and the planet, in order to justify its ongoing need for protection and stewardship.

Land tenure, particularly concerning indigenous communities, is going to be a high-profile political theme in the country during the next few years. This issue will be directly related to policies resulting from peace agreements and the ratification of Convention No. 169 of the United Nations International Labor Organization concerning tribal rights and lands. This will have important implications for the SMBR, particularly regarding possible voluntary relocations and titling of land to indigenous communities.

Many difficult challenges face the SMBR; nevertheless, after only five years of work, results are already encouraging and indicate that with appropriate and adaptable processes, this marvelous area can be a real example of harmony between conservation and sustainable human development.

Postscript

On June 11, 1996, the Congress of the Republic of Guatemala declared Bocas del Polochic a wildlife refuge, delegating its administration to Defensores de la Naturaleza. This new protected area is located next to the northern border of the Sierra de las Minas Biosphere Reserve and can almost be considered an extension of it. It covers the wetlands of the Polochic River delta on the western shore of the Izabal lake. In February of 1997 the site received international recognition when it was included as a wetland of international importance by the Ramsar Convention.

In addition, a series of laws creating new protected areas have been passed by the Guatemalan Congress. These new protected areas include a group of small areas in the southern part of the Petén Department; the Bisis-Cabá Reserve in Alta Verapaz, which will be managed by the Asociación Chajulense, an indigenous community association; and the Cerro San Gil Nature Reserve on the Atlantic coast, whose administration will also be delegated to an NGO. These developments are encouraging: they add to Guatemala's still incomplete protected-areas system, and they include new nongovernmental organizations responsible for the management or comanagement of the protected areas. Adding new lands is an important step in increasing the representation of the country's most important species and habitats in Guatemala's protected-area system.

There have also been some important developments involving the negotiations with communities in the SMBR. The case of voluntary relocation of the Vega Larga community has advanced very well. To raise the necessary funds, the project was included as a target site for The Nature Conservancy's Adopt-an-Acre campaign. The actual process of relocation will happen gradually during 1998 by agreement of the relevant parties.

The relocation of the Santiagüilá community had not fully been resolved by March 1998. This was because the legally registered limits of the property purchased for its relocation in fact overlapped with neighboring lands. This required that a lengthy survey be conducted. The land-surveying project will also involve titling lands of three other communities (San Marcos, Tres Arroyos, and Río Blanco) that already were occupying part of the purchased land. If there are no other unforeseen delays, the land will finally be titled to the communities in 1998.

The lawsuit that Defensores helped the district attorney file resulted in the eviction of the Tierra Colorada group based on a court ruling. To this date there has been no further encroachment on the property.

Another important development is that Defensores has diversified and secured its funding sources for the management of the SMBR with an integrated conservation and development approach. Thus, in 1997 it started two new projects, one of which is financed by the European Community and the other by GEF/UNDP. These will be of vital support while the organization strengthens its endowments and other mechanisms for financial self-sufficiency.

Glossary

ADIPSA—Integral Development Association of El Progreso in San Agustín Acasaguastlán

CCAD—Commission for Central American Environment and Development

CDC—Conservation Data Center

CELGUSA—a paper pulp plant in the Motagua valley

CONAMA—National Environment Commission

CONAP—National Protected Areas Council of the Presidency of the Republic

CONCAUSA—Central America–USA Agreement

Defensores—Fundación Defensores de la Naturaleza

DIGEBOS—The Forest Directorate
guamil or *guatal*—secondary vegetation
INTA—Instituto Nacional de Transformación Agraria (the land titling agency)
ladinos—people of mixed or Spanish origin
milpa—system of crop rotation
rozas—controlled burns
SMBR—Sierra de las Minas Biosphere Reserve

Corcovado National Park

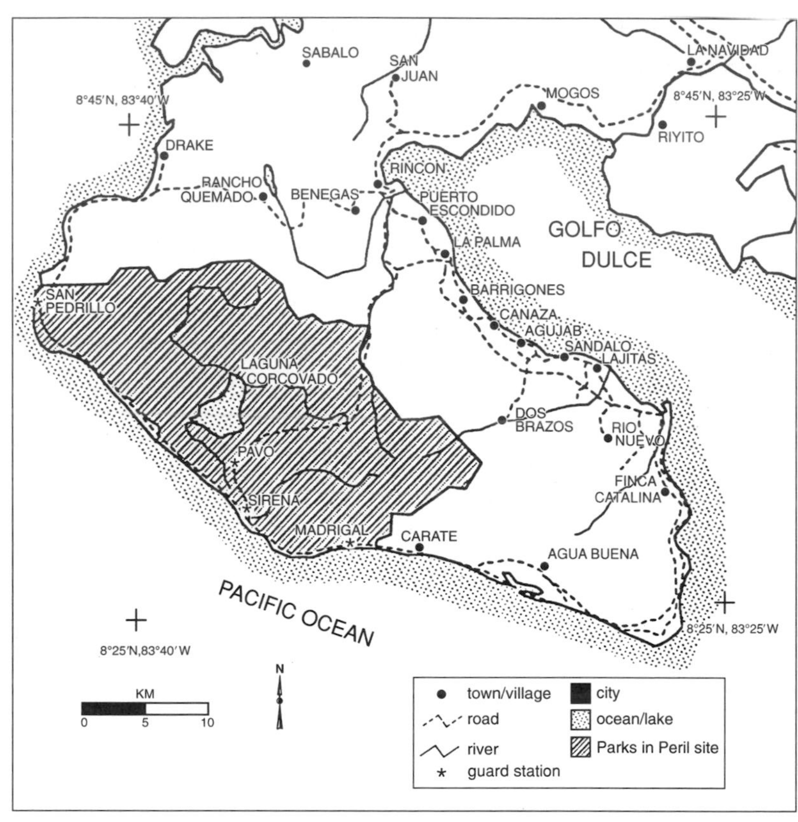

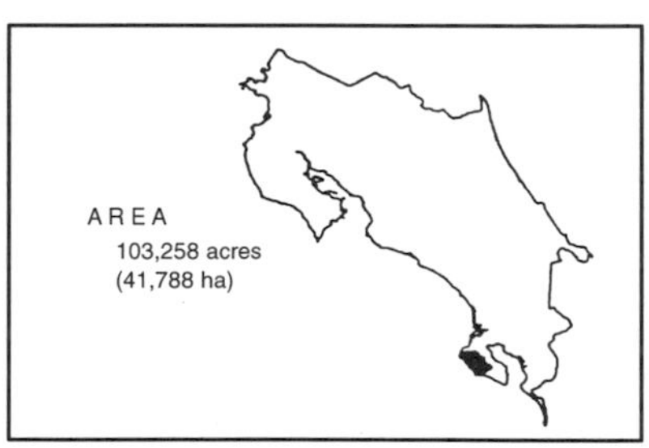

6. Costa Rica: Corcovado National Park

César Cuello, Katrina Brandon, and Richard Margoluis

The political issues . . . introduced into the OTS–Nature Conservancy discussion of Osa nature reserve acquisition would prove of vital significance during the tumultuous years of 1971 to 1975. During that time, partisans of Osa conservation would learn well that creating a scientific reserve, or any kind of nature reserve, on the Osa inevitably would require them to address . . . wider political issues.

—Catherine Christen

Corcovado National Park (CNP) and the surrounding conservation areas on the Osa Peninsula represent the largest remaining lowland rainforest on the Pacific coast of Central America. Despite the fact that nearly 80 percent of the peninsula (160,000 hectares) is legally protected in some form, rapid land-use changes and forest clearing threaten the biological integrity of the peninsula. Approximately 20 percent of the land on the peninsula, or 40,000 hectares, is not subject to any use restrictions. Despite the existing regulations, the deforestation rate for the whole peninsula was 5 percent in 1992 (Alvarez and Márquez 1992).

Corcovado Park is located in one of the wettest and most remote areas left in Costa Rica. There are over 1,513 species of trees identified thus far in Corcovado, and one estimate placed the overall number of tree species in the park and surrounding areas at 3,000 (Soto 1992). The park contains over one-quarter of all the tree species known to exist in the country, including some of the largest trees in the country's tropical forests, towering fifty meters above the ground. The fauna are also incredibly rich; to date 375 species of birds (18 of which are endemic), 124 species of mammals, 46 species of amphibians, 71 species of reptiles, 61 freshwater fish species, and approximately 8,000 insect species have been recorded in CNP (Soto 1992). CNP also protects many endangered tropical species, especially cats and reptiles.

Corcovado National Park is largely surrounded by the Golfo Dulce Forest Reserve (RFGD). This reserve, as well as other protected areas on the Osa, is of substantial ecological importance, since CNP alone is not sufficiently large to protect many of the species inside its boundaries. Therefore, this study focuses on both CNP and surrounding protected units.

The social and political context on the Osa Peninsula, particularly in terms of Corcovado Park, is complex and highly conflictive. Rapid land-use changes on the peninsula brought about by road construction and other infrastructure development, combined with greatly increased pressures due to the recent economic situation in Costa Rica, are placing serious strains on the park. Opportunities for sustainable resource extraction, such as forestry or agriculture, are limited, since most of the peninsula is hilly and susceptible to rapid erosion once forests are cleared. The Costa Rican government has been unable to implement its espoused policies of sustainable development. There has been virtually no coordination of the many institutions involved in the Osa.

What is perhaps remarkable is how well Corcovado Park and the surrounding Golfo Dulce Forest Reserve have been maintained given the history of the peninsula over the past thirty years. However, it appears that the level of pressure on the park and surrounding adjacent areas will dramatically increase in the coming years due to national-level economic problems, lack of productive land elsewhere, the country's need for wood, and increased access to and infrastructure development (such as electrification) on the peninsula. At current rates of logging and conversion, the RFGD will be destroyed within the next five to ten years. When this happens, pressures on Corcovado Park will intensify. Furthermore, species loss will be inevitable given the small size of CNP relative to the area needed to support many of the species that the park protects.

Park Establishment and Management

One of the instructive things about the founding of Corcovado National Park is that it provides great depth of understanding on how the history of a place—in this case in the last forty years—can substantially influence present-day outcomes in conservation planning and protected-area management. The conservation activities developed to protect Corcovado National Park also provide strong insights into the passing phases of conservation as a field: from direct park establishment and protection, to involvement of communities, to integrated conservation and development projects (ICDPs), to a regional perspective. One chronicler of the Osa has commented that many of the present-day animosities on the Osa are historically based; recent migrants to the Osa may be unaware of the roots of the conflicts, but they quickly learn to take sides. This section presents a historical overview of the three key competing land uses already evident by the early 1970s (subsistence farming, corporate development, and natural resource conservation) and how these three sectors affect current conservation policies (Christen 1994, 201).

By the 1930s the Costa Rican Banana Company, a subsidiary of the United Fruit Company, had begun exploring the southern zone's flatlands, including the Osa Peninsula. It established banana and African palm plantations in Sierpe, Palmar de Osa, and throughout the Golfito area. The company recruited people to work on its plantations, precipitating a wave of migration of people mostly from Costa Rica's northern Pacific regions and Nicaragua. The presence of the company marked the beginning of a long-term pattern of development focused outside the area; few profits remained in the Osa (Cuello 1997). The discovery of placer gold in the Osa in 1937 caused a rash of gold fever, furthering the rapid migration of people in search of a quick fortune. Few would find it, although most would remain there for many years as small-scale gold panners or *coligalleros* (Brenes et al. 1990). The significance for the Osa of these events was that they resulted in an increase in population, some conversion of the flatland forests to pasture lands, and the presence of the banana company in the region. However, by the 1940s, the majority of the Osa Peninsula still had its forest cover intact, except in the coastal area along the gulf (Cuello 1997). Construction of the Interamerican Highway South between 1947 and 1960 helped promote further migration toward the southern zone in general and the Osa Peninsula in particular. The new colonists came to establish themselves in the remaining unclaimed, undedicated state lands (known as *baldíos nacionales*).

The Osa Forest Products (OPF) was officially registered in Costa Rica in 1959 and granted permission for all forestry and mining activities on its holdings (Christen 1994). The manager of OPF decided to "create a tropical sustainable forestry industry centered on rotational harvests and waste-free processing of all the harvested wood into a variety of products for maximum economic returns" (Christen 1994, 74). OPF never undertook any significant logging; instead, it became mired in disputes over land title and uses. At both regional and national levels, anti-imperialist rhetoric was pervasive—and much of this sentiment at both levels was directed to OPF. The tangled history of OPF after 1971 turns from one in which there was a genuine interest in forestry and agriculture to a tale of corruption, real estate scams, and proposed tourism development. According to one historian who has chronicled the history of the Osa Peninsula, substantial conflict began in 1972, when a new OPF manager ordered guards to burn down houses and farms and shoot at peasants to scare them in order to regain control over OPF lands claimed by squatters (Christen 1994).

In 1971 hearings were conducted by the national legislature about the OPF and its actions. OPF quickly became a symbol for "land-hoarding" expatriots and their companies, who didn't use land productively and prevented Costa Ricans from using it. By 1971, as a way of demonstrating what it controlled and getting the settlers off the land, OPF began intensive road construction and forced evictions within its holdings. Armed squatters ambushed and captured OPF staff and an OPF tractor, warning that "blood would flow" (Christen 1994, 226). OPF asked the Rural Guard to come to its assistance.

The establishment of the park on the Osa is grounded in a long chain of key actions and consequenes. The proximity of OPF offices and those of the Tropical Science Center (TSC) on the same block in San José was an influential factor. The OPF manager had invited TSC staff in 1962 to use OPF land to construct a research facility in Agua Buena de Rincón. Between 1962 and 1973 over a thousand scientists and researchers visited this facility on the Osa (Christen 1994, 129–130). Many of these scientists were affiliated with the Organization for Tropical Studies (OTS), a scientific research and tropical educational center representing a consortium of U.S. and Costa Rican universities. By 1969, OTS was working with The Nature Conservancy's Western Regional Office to discern ways to acquire the land from OPF. By 1970, this Osa proposal had moved from the regional office to headquarters staff at the Conservancy, where concerns were raised over "the policy issues raised by the prospect of a United States conservation organization commencing a 'Latin American Program' with the Osa Initiative" (Christen 1994, 189).

The 1973 publication of a booklet called *The Corcovado Basin*, which described the area that later became CNP, served to unite conservationists around the idea that any protected area to be established should include the entire Corcovado basin and be run by Costa Ricans, not foreigners (Christen 1994, 192–193). Although the Rincón facility was not used after 1973, historian Catherine Christen asserts two important outcomes influencing Osa conservation patterns because of the existence of that station. First, it made it possible for researchers to travel to the Corcovado basin by foot: "Rincón station's location was crucial to the eventual conservationist focus on preserving the Corcovado basin" (1994, 194). Second, however, was that it isolated the scientific community from the Osa's residents and linked the conservationists with the greatly disliked OPF in local minds, even though the scientists deliberately tried to keep science and conservation separate from OPF.

Between 1963 and 1973, numerous conservationists, both Costa Rican and U.S.-based, began defining what they thought would be the "best" portions of the Osa to save. By 1972, "confusion and friction were not infrequent outcomes of the jockeying for position of the large number of diverse organizational entities on the Osa conservation front" (Christen 1994, 276). The tourism development plans promoted by the OPF in 1972 began to galvanize the conservationists into taking immediate actions to protect the Corcovado basin. The development of "Rincón Resorts" was to be followed by a project that would construct a road, dredge the Corcovado lagoon and connect it to the sea so it could serve as an inland marina, and construct two thousand homesites (Christen 1994, 268).

Meanwhile, by 1973 a committee of the national legislature looking into abuses of the OPF recommended expropriation of much of their land and the creation of a reserve on the Osa using some of the land. The proposal to create a reserve was largely influenced by scientists and others based outside the Osa. Toward the end of that year, an OPF guard was killed, probably by settlers. In early 1974, the second manager of OPF ran off with the down payments for homesite and resort development. OPF owners renewed their interest in their landholdings and began planning for intensive development of their holdings; logging, cattle ranching, agriculture, and tourism all figured in their plans. At the same time, representatives from WWF and the Conservancy began working closely together to intensify their efforts to raise funds to purchase the Corcovado

Table 6-1. Historical Chronology of Key Osa Peninsula Events

• 1930s	The Costa Rican Banana Company, a *United Fruit subsidiary, establishes plantations* near the Osa, causing a wave of migration from Nicaragua and North Pacific Costa Rica.
• 1937	*Discovery of gold* leads to rapid immigration.
• 1947–60	Construction of Interamerican Highway South and increased small-boat traffic lead to further migration and *settlers claiming state land.*
• 1959	*OPF is officially registered,* and government grants it permissions for forestry and mining. While no significant logging occurs, disputes arise over land titles and uses.
• 1962	OPF manager invites TSC staff to construct *research facility at Rincón*; in the next eleven years, over 1,000 scientists and researchers visit Rincón and Osa.
• 1963–73	U.S. and Costa Rican *conservationists define "best" portions of Osa to save;* friction arises among factions.
• 1969–70	OTS works with TNC's California Regional Office to acquire land from OPF. TNC headquarters staff is concerned about a U.S. conservation organization initiating a Latin American program.
• 1971	*OPF turns to corruption, real estate scams, and proposed tourism development.*
• 1971–73	*National legislature committee conducts hearings* into actions of OPF—which for many become a symbol for underutilized, foreign-owned lands. *The legislature recommends expropriation of much of OPF's land and creation of a reserve.*
• 1972	*OPF guards burn houses and shoot at peasants* under the orders of new manager to regain control over OPF lands. Intensive road construction and enforced evictions begin. Armed squatters warn "blood will flow" and capture OPF staff and tractor. OPF requests Rural Guard assistance.
• 1972	*Plans for dredging Corcovado lagoon for a marina and constructing 2,000 home sites* are part of OPF tourism development proposals.
• 1973	*"The Corcovado Basin"* is published describing the area that later becomes CNP. Conservationists agree any protected area should include the entire basin and be run by Costa Ricans.
• 1973	*OPF shuts down Rincón Research Station by force*—ironically, conservationists are linked with the disliked OPF in residents' minds.
• 1973	*OPF guard is killed,* probably by settlers.
• 1974	*Second OPF manager runs off with down payments* for home site and resort development.
• 1974	WWF and TNC begin working closely together, intensifying *efforts to raise funds to purchase Corcovado Basin*; they want assurance of Costa Rican political will to support the park's creation and long-term protection.
• 1971–75	*Population in plain swells* from five families in 1971 to 80 or 100 settlers claiming the entire plain and surrounding hillsides. Report to Parks Service stresses need for protection before 1976 dry season.

Table 6-1. (*continued*)

• 1975	OPF agrees to 1:1 land swap in face of bill to expropriate its lands.
• 1975	*CNP is established* October 31, 1975.
• 1978	*GOCR creates RFGD.* Local hostility toward protected areas increases.
• 1979	*GOCR purchases land for RFGD* from OPF and ITCO.
• 1980	*CNP boundaries are expanded,* increasing size of CNP by a third, to nearly 42,000 ha. To do this, thirty farms are expropriated; owners aren't compensated. Small-scale gold miners are evicted from park.
• 1980	*Migration to Osa* after Costa Rican Banana Company ends workers' strike by closing operations in Palmar and surrounding area.
• 1987	*GOCR creates National System of Conservation Areas* (SINAC).

Sources: Christen 1994; Cuello 1997; Vaughan cited in Christen 1994.

Basin area outright if the land-use conflicts could be resolved: "These organizations hesitated to promise funding for Corcovado without incontrovertible assurance of Costa Rican political will to support the park's long-term development and protection, not to mention its initial creation" (Christen 1994, 344).

Rapid migration and settlement intensified the need to quickly take action. While only five families lived in the plain in 1971, the population had swelled to between eighty and one hundred families by 1975. Many of those families had arrived during the dry season in 1975, and by the end of the season, "there was not one square meter in the Corcovado plain or in the nearby hills that was not marked with boundary lines and claimed by an owner" (Vaughan, quoted in Christen 1994, 362). A field report sent to the director of the Parks Service (SPN), Alvaro Ugalde, said that the park would have to be declared in 1975 for measures to be in place to protect the area during the 1976 dry season. A combination of political forces coincided to make the timing right for the establishment of CNP. OPF agreed to a land swap for its Corcovado basin lands in exchange for national land elsewhere, and Corcovado Park was officially decreed by President Daniel Oduber on October 31, 1975. Ironically, one reason that the OPF may have agreed to a 1:1 land swap was that the legislative assembly voted to expropriate its lands. However, the president, who was supportive of foreign investment and did not want to see expropriation become a precedent the legislature could exercise at its whim, had vetoed the bill.

The boundaries of CNP were expanded in 1980, increasing the size by nearly a third from 34,346 ha to nearly 42,000 ha. Another thirty farms were expropriated, and the owners were supposed to be compensated. The creation of the RFGD and the expansion of Corcovado Park made many residents particularly hostile toward the region's conservation agenda. In addition to the formal land expropriation process, small-scale gold miners were evicted from the park. These were the first of many such evictions. Most of the early efforts at protection of CNP concentrated surveillance and guards in the northern part of the park along the marine coast. The southern part of the park was left with little protection, which allowed the concentration of gold mining in this area.

Osa Conservation and Sustainable Development Area. In 1987, the government of Costa Rica created a National System of Conservation Areas, known as SINAC. The SINAC was to begin a process that would integrate protected-area management with surrounding areas and decentralize much of the decision-making power to regional levels. Nine Regional Conservation Areas (ARCs) were established, including one for the Osa. Each ARC comprises three land-use categories:

- core protected areas such as Corcovado National Park
- buffer zones and multiple-use areas such as the RFGD and the Guaymi Reserve
- lands for agriculture and other uses.

Each of the ARCs is headed by a director, who is responsible not only for park management but also for intergovernmental coordination of activities within his region. This interagency organization is composed of representatives from the park service, wildlife service, forestry department, mining sector, and energy department. It also has oversight of committees composed of local leaders and government officials designed to coordinate policies and programs. In November 1989, Executive Decree 19363, regarding the Ministry of Natural Resources, Energy and Mines (MIRENEM), established the Osa Peninsula Regional Conservation Unit (UNIOSA), which later (September 1991) became the Osa Conservation and Sustainable Development Area (ACOSA). The objectives of ACOSA are twofold: to achieve optimal protection and management of the peninsula's protected areas and surrounding areas through actions directed at the communities that live inside and near these areas; and to integrate, under one management unit, the protected areas and buffer zones under the perspective of sustainable development (ACOSA 1993). The model of how each of the regional conservation areas would operate is visionary, linking protected areas with buffer zones. However, in reality, it has been difficult to establish such links either within ACOSA itself or in the eyes of local people, who "see conservation as an imposition, and something for which they have been sacrificed" (Cuello 1997).

At present ACOSA includes Corcovado National Park, Golfo Dulce Forest Reserve, Golfito Wildlife Refuge, the Biological Reserve of Isla del Caño, Sierpe-Terraba Mangrove Forest Reserve, and Piedras Blancas National Park. The Guaymi Indigenous Reserve was part of ACOSA until April 1994, when it was excluded. Decision-making authority within ACOSA includes the director, a technical committee, and a local committee. ACOSA is structured to have eight management programs: protection, ecotourism, research, environmental education, agroforestry, coastal zone management, land surveying, and mining oversight (García 1994). The technical committee is composed of the eight program coordinators, ACOSA's director and deputy director, the heads of the different protected areas within ACOSA, the director of the BOSCOSA project (see section entitled "Linkages between Park and Buffer Areas," p. 172), and the local representative of the Institute of Agrarian Development (see "Land and Resource Tenure and Resettlement," p. 154). The technical committee is responsible for the operations of ACOSA and for evaluating the technical, social, and environmental feasibility of the different projects undertaken within the conservation area (García 1994). In 1994, ACOSA had sixty-four staff members, of which 40 percent were from the Osa Peninsula. For 1994, ACOSA had a budget of about $900,000, 60 percent generated internationally, 37 percent nationally, and 3 percent locally through ACOSA's collection of ecotourism revenues (García 1994). The sources of the funding for the Osa reflect a clear dependence by ACOSA on extranational funding.

The Golfo Dulce Forest Reserve

In 1978, the Golfo Dulce Forest Reserve was created, placing lands under the legal control of the Forestry Department. The lands for the RFGD were purchased both from the OPF in 1979 and from the government's official land reform agency, the Institute of Land and Colonization (ITCO). ITCO was directed to relocate and resettle squatters from the approximately 84,538 ha of newly created reserve—much of it land purchased from the OPF. The size of the reserve was scaled back to 61,350 ha to remove lands suitable for agriculture. Unfortunately, the government failed to follow its stated policies. Agricultural lands should have been passed to ITCO for land-use assessments, forest management plans, and organized resettlement programs. Such a process did not begin until the early 1990s. Tenure for those living within the RFGD has been unclear for the past eighteen years.

Golfito Wildlife Refuge (GWR)

The Golfito Wildlife Refuge was created in July 1985. It is located west of the city of Golfito and is a total of 2,810 ha. It was created to maintain wild plant and animal populations in a region nearly completely without forest cover. Until recently it was administered jointly by the University of Costa Rica and the government's Wildlife Department; it is currently managed by the Wildlife Department. The GWR is primarily composed of disturbed forest and abandoned pasturelands in various stages of regrowth. Due to its location, the refuge has a great ecotourism potential, and it also lends itself to the development of a biological corridor that would connect the RFGD with Piedras Blancas National Park.

Isla del Caño Biological Reserve

This reserve was established in 1976 and is located 20 km off of the Pacific coast of Agujitas. The island has a land area of 300 ha and a maritime band of 3 km surrounding the island. Its vegetation is relatively uniform and the wildlife is not abundant, but its importance lies in the coral reefs that surround it (Alvarez and Márquez 1992). Artifacts such as utensils and rock spheres and a burial ground indicate that the island served both residential and ceremonial purposes in pre-Columbian times, placing a value on its cultural history.

Sierpe-Terraba Mangrove Forest Reserve

This mangrove forest reserve is located at the north of the RFGD on the coasts of Coronado Bay, between the Sierpe and Terraba Rivers. It is 22,688 ha, which makes it the largest continuous block of mangroves in the country. The mangroves are between marine and terrestrial areas in the flatlands, which are low, wet, and in a state of pioneer successional vegetation. They are composed of diverse species, including five species of mangroves (Alvarez and Márquez 1992).

The use of wood from the mangrove forest is under the control of the

General Forestry Directorate (DGF), which has maintained the reserve's forest cover. Approximately 23 percent of the forest is virgin, 39 percent was altered but has recovered, and the rest is highly altered; 16 percent has dense thickets and 22 percent has been deforested and converted to other uses (Alvarez and Márquez 1992). The mangroves in the reserve have great importance for the migratory habits of some species that reside in the RFGD. The reserve is an important source of fuelwood, tannins, wood, and charcoal, as well as fisheries and the cultivation of the *piangua,* a large, edible mollusk, and other crustaceans.

Piedras Blancas National Park (PBNP)

Corcovado National Park was expanded once again on June 5, 1991, to include a portion of the Río Esquinas forest, which was called Corcovado National Park—Esquinas Sector. However, on April 19, 1994, a new government decree (20790-MIRENEM) turned this portion of CNP into a new park, called Piedras Blancas National Park, with a total area of 15,000 ha. The park borders on the Golfo Dulce, extending from an area north of Golfito to the entry of the Osa. One important element in the creation of PBNP is that the majority of the land is privately owned because the government does not have the money to purchase it. Deforestation in the area is taking place legally, by means of so-called management plans. This occurs because the law allows the owners of land in a declared protected area to extract timber until the government formally purchases the property.

Guaymi Indigenous Reserve

The Costa Rican government established the Guaymi Indigenous Reserve in 1981 with land officially taken from the RFGD and from the lands ITCO had to return to establish agricultural settlements. In fact, much of the land was formerly owned by OPF and had squatters on it. When the reserve was established, a resettlement process was initiated to "buy out" the land from the squatters who claimed it. The Guaymi, however, are not indigenous to the Osa and had settled there some time in the 1960s (see "Indigenous Peoples and Social Change," p. 184). Fundación Neotrópica (FN) has played an important role in supporting the Guaymi reserve by channeling international funding to the Guaymi for land purchases, training, a school, an aqueduct, and a bridge over the Rincón River, which provides access to the reserve.

Land and Resource Tenure and Resettlement

Costa Rican law has traditionally allowed peasants to claim land that was uninhabited or unimproved. By the mid 1950s, land clearing on the Osa had become the primary way to claim land. Land invasions increased dramatically in the 1960s, on both public and OPF lands. Although the management of OPF had anticipated a forty-year rotation schedule for logging, "embodying the prevailing contemporary land use ethic, untitled settlers and government functionaries soon formed common cause and took Wright's (the OPF managers) failure to rapidly clear and develop OPF land as grounds for censure and opposition to his company's hegemony" (Christen 1994). In order to prevent future claims for titles, OPF decided to rent land at a nominal rate to squatters. However, in the early 1960s, the residents of one community, Drake, refused to rent the land and insisted it was theirs. Protracted negotiations followed for the next three years.

This historical background is instructive because it shows the strength of threats to the resource base. This was one of the few areas where peasants had power: they threatened to act violently and to clear forests by burning. ITCO was responsible for negotiating between OPF and the settlers. Various schemes and land swap deals were proposed, but ITCO made few credible offers and moved extremely slowly, claiming for several years that they couldn't make an offer because of the difficulty in getting to the Osa. Nevertheless, ITCO ended up owning much of the land on the peninsula.

The legislation that created Corcovado Park assumed that the three hundred people who resided in CNP would be resettled. However, the legislation did not provide funds for this purpose. Corcovado Park was declared a disaster area in 1976 in order to provide the government with a mechanism to allow settlers to be paid with national funds for any improvements (*mejoras*) or for their rights of possession if they could prove they had been there for ten years. ITCO became the lead agency responsible for identifying who lived in the park and what mejoras they had made. Intensive negotiations finally led to a May 1976 agreement between settlers and ITCO for settlers to be compensated for their mejoras. Christen says, "By the time all the settlers were removed in mid-1977, payment of mejoras undoubtedly exceeded the actual value of settler activities, because the settler process was expedited by carrying out only oral surveys, with no site inspections" (1994). Resettlement was extremely complex because of the uncertainty of land tenure throughout most of the peninsula.

The lack of tenure security and the need to undertake resettlement have been a major theme of the creation of conservation areas on the Osa. The situation by the early 1980s can be summarized as follows:

- The government had to worry about compensation and resettlement of different groups of settlers from four areas: the original Corcovado Park, the extension of Corcovado Park, the Golfo Dulce Forest Reserve, and the Guaymi Reserve.
- While squatters felt that they had legal claims to the land, few in fact did. Many were squatters on land that was either state owned (which could not be claimed after the 1950s) or was owned by OPF.
- The government was in a difficult position—legally it was required to provide compensation for resettlement and the improvements, even when peasants lacked title.
- A serious mistake was made when ITCO and the Forestry Department failed to coordinate land and forest use in the RFGD.
- The construction of the first road (although a dirt one) and evidence of gold led to rapid rates of in-migration to the Osa Peninsula.

Another significant wave of migration to the Osa took place in 1984. Workers on the Golfito and Palmar banana plantations went on strike. Faced with the strike, declining yields, and opportunities for expansion in countries without labor problems, the Banana Company of Costa Rica, a United Brands (formerly United Fruit) subsidiary that had operated in the region since 1937, shut down its operations. Many unemployed workers migrated to the Osa to try gold prospecting as a way of earning a living. When the company withdrew, President Luis Alberto Monge declared a state of emergency in the southern zone, which was expanded to include the Osa Peninsula by February 1986. The state of emergency was declared, in large measure, due to the thousands of people who were left unemployed and "the grave socio-economic impacts of the exit of the Compañía Bananera de Costa Rica." Expansion of the state of emergency to cover the Osa suggests the substantial impact that the migration was having there.

Present Situation in the Golfo Dulce Forest Reserve

The land tenure situation is particularly complicated within the Golfo Dulce Forest Reserve. Table 6-2 summarizes the tenure issues from the early 1980s to 1995.

Lands that were granted in the early 1980s to ITCO, renamed the Institute of Agrarian Development (IDA), constitute twenty thousand hectares, or nearly one-third of the reserve. These lands are occupied by 450 households. The remainder of the reserve, about forty-one thousand hectares, is claimed by peasants. However, only about 5 percent of these families have title (Alfaro et al. 1995).

Much of this land is inadequate for either logging or agriculture, due to the high rainfall and rugged topography, which results in rapid erosion once trees are cut. However, the lack of clear title has also made it nearly impossible, legally, for those claiming the land to have access to any of the sources that might help them use the lands sustainably such as incentives for reforestation, legal permission for logging permits, or bank credits. This situation started to change in 1993 when IDA promised to grant titles within the RFGD. IDA's decision dates from its claim that it was never compensated by MIRENEM for lands it held (acquired as part of the Corcovado swap with OPF) inside the RFGD. Since IDA was not compensated, it did not have the financial resources to grant titles elsewhere—and, since it wasn't compensated, the transfer was not legal. Therefore, IDA believes that it is all right for it to grant titles inside the reserve. Since IDA receives a great deal of public pressure to title land, this political maneuvering is an expedient way to gain local support. A 1995 review (Cuello and Piedra-Santa) indicates that there is no legal agreement between MIRENEM and IDA for granting of titles within the RFGD. Fundación Neotrópica mediated between MIRENEM and IDA to obtain a legal agreement; but despite promising prospects, the document was never signed by the respective officials. However, an oral agreement between the two agencies opened the way for IDA to begin to grant titles to those who could demonstrate that they were in possession of the land before the declaration of the Forest Reserve.

In 1994, IDA granted 109 titles within the RFGD, for a total of 1,475 ha. IDA planned to grant another 50 titles in early 1995. However, it reversed itself at the last minute when its own legal office said that forest reserves are state lands, and therefore public, and no titles could be granted

Table 6-2. Tenure Issues in the Golfo Dulce Forest Reserve

Early 1980s	20,000 ha granted to IDA (ITCO) soon occupied by 450 households
	Remaining 41,000 ha claimed by peasants—only 5% have legal title. This lack of title makes access to bank credit, reforestation incentives, and logging permits nearly impossible for them to obtain.
1993	IDA promises to grant titles within RFGD and gains local support.
	IDA claims that it was never compensated by MIRENEM for land acquired as part of Corcovado swap with OPF and that the land transfer was therefore illegal. Because of this IDA claims it lacks the resources to grant titles elsewhere.
1994	Fundación Neotrópica tries to mediate between MIRENEM and IDA for granting titles within RFGD; while a verbal agreement is reached, no legal one is. IDA begins granting titles to those with proof of possession prior to declaration of reserve.
	IDA grants 109 titles within RFGD for a total of 1,475 ha.
1995	IDA plans to grant another 50 titles.
	IDA legal office claims forest reserves are state lands and therefore public and no titles can be granted within them.

Sources: Alfaro et al. 1995; Cuello and Piedra-Santa 1995.

within them. More specifically, such lands could not be "expropriated, altered, granted, transferred, delivered, rented under any circumstance, without having been previously classified by the DGF. If this institution considers them appropriate for forestry, such lands automatically will be incorporated into the State's forestry patrimony" (IDA 1995).

IDA then concluded that people occupying these areas are doing so illegally and should be resettled and compensated with respect to improvements they made. The exception are those residents who can prove that they were there ten years before the declaration of the Forest Reserve (i.e., 1968, when it would have been OPF or state land). A recent FN report says that if this new ruling is applied, titles would be withheld or rescinded and rental agreements would be canceled. The report predicts that this could have the following effects:

- increased deforestation from illegal logging
- increased illegal hunting in protected areas, including Corcovado National Park
- conflicts with government authorities
- a negative attitude change toward forest conservation (Cuello and Piedra-Santa 1995).

If the traditional understanding of mejoras is applied, it could lead to extremely rapid clearing as "proof" of settlement and for indemnization. At the present time it is unclear whether the government of Costa Rica will apply the traditional laws about mejoras, which implies clearcutting, to evaluate these lands or whether it will introduce a new system that regards forested land as the improved land—a new kind of mejora.

In 1995 a high-level ACOSA official told us that legislation was being prepared to be submitted to the Legislative Assembly for titling inside the RFGD. The purpose of the legislation will be to resolve the problem of land tenancy on the Osa. Such legislation, if it resolved conflicts between all interested parties and established adequate controls for forest management, could be an important step in promoting conservation in the zone. However, some conservationists are concerned that such titling could lead to real estate speculation and allow the land to be sold, especially to foreigners, as has happened elsewhere in Costa Rica.

A third major and potential source of rapid change is the influx of foreigners buying property on the Osa Peninsula. We were unable to obtain any concrete figures on the magnitude of this change, but anecdotal evidence suggests that it could be significant. For example, one informant told us that over fifteen lots in El Tigre had recently been purchased by foreigners. It was also mentioned that there were plans for more extensive developments in the area south and to the west of Puerto Jiménez, extending to Carate. We were told that there were plans for large-scale "planned communities" catering to wealthy foreigners; upon our return we reviewed real estate packets that featured a variety of Osa properties for sale at prices ranging from $2.5 million for a 650-hectare farm to $35,000 for 1.8-acre residential lots. These brochures indicated that of sixty-four available properties, twenty-nine had been sold. Plans for the properties include a proposed private forest reserve near Carate. It is clear that in the past five years numerous real estate offices and property management agencies that cater to foreigners have sprung up in Puerto Jiménez. Improved infrastructure, paving of roads, and electrification of the peninsula could lead to very rapid land-use changes, as they have elsewhere in Costa Rica. In Talamanca, for example, there was extremely rapid clearing of primary forest as peasants without title clear-cut land in order to prove ownership and sell it off to real estate speculators (Wells and Brandon 1992). This constitutes a significant potential threat to the Osa.

Present Situation in Piedras Blancas National Park

Piedras Blancas National Park (PBNP) is currently a paper park, as it has only been decreed but not consolidated or managed. The lands making up PBNP are still in private hands, and the owners are waiting to be compensated. We were told that many rich people have lands within Piedras Blancas Park. People were upset that there is strong inequality in the resolution of land compensation, that compensation has been quickly dealt with when it is a matter of rich elites. Such local perceptions, even if untrue, highlight the resentment and suspicion that some long-term residents have; they perceive they are subject to unfair treatment when new parks are established. Deforestation in the area designated as Piedras Blancas Park is reportedly high, especially near Chacarita, as smallholders seeking to demonstrate their claims become eligible for compensation through mejoras and sell off valuable wood while they are able. There are many conflicts in terms of the land tenancy in the area designated for PBNP. However, a local NGO has submitted a financing proposal for the purchase of park lands that are in private hands, with the aim of consolidating the area.

Resource Use

Little empirical information is available on resource use within Corcovado Park. All extractive or consumptive use of resources within the park is illegal. The primary types of resource use that affect CNP are illegal hunting and wildlife trade, gold mining, forest clearing, and tourism. In June 1995, the number of park guards was cut in half. Although ACOSA claims that twenty-five park guards are needed to guarantee protection of CNP, only ten are officially assigned there. By July 1995, ACOSA staff were saying that they couldn't guarantee the protection of even one sector.

Hunting and Wildlife Trade

We were unable to find any data on the levels of hunting within CNP. However, both ACOSA representatives and numerous residents of the peninsula said that there is hunting within the park for both subsistence and commercial trade. Hunting is a problem especially prevalent in the dry season, from December to March, when animals concentrate around water sources. Hunting was named as the second most important threat to CNP in three separate interviews with park staff. Most hunting is seasonal subsistence, although there are increasing isolated examples of hunting or wildlife collection for trade by outsiders. None of the following anecdotal information could be verified.

Commercial. Species most commonly hunted for commercial use are wild pigs and pacas for meat and macaws, monkeys, snakes, and ornamental plants for trade and sale. There were reports of local sales of pacas. Several people interviewed indicated that they thought there was an increasing capture of wildlife in CNP for sale in Panama, although no one knew the magnitude of such collections. One source indicated that the macaw population had declined dramatically due to trade. There was also mention of the number of foreigners who enter and collect species, particularly reptiles and amphibians, saying they are scientists. ACOSA authorities also allow INBio (National Biodiversity Institute) parataxonomists to extract specimens from CNP. Such collections are subject to an agreement between the government and INBio that allows INBio to extract samples from national parks and other protected areas in exchange for giving the government a percentage of revenues from the development of any products from protected-area biodiversity.

Subsistence. One source told us that wild pigs are easy to get because they are concentrated in Corcovado. Several people claimed that hunting has increased with the country's worsening economic problems. Particular communities said to put hunting pressures on the park were Los Planes, a community near Drake that is not receiving benefits from tourism, and Cerro Rincón. Several people commented that animals leave the park and are then killed in the RFGD. Although hunting outside CNP is illegal as well, there is no control of hunting by the Wildlife Department anywhere on the peninsula. Everyone interviewed commented that there is hunting within the park. The impact is not known, but one ex-miner said that Corcovado was becoming a "forest without animals" and that large-size pigs were now very hard to find. Who is doing the hunting is disputed—ACOSA staff blamed the gold miners for much of it. Studies in 1986 said gold miners don't hunt much; ex-miners interviewed said they hunt for subsistence and only if something is easy to get. ACOSA staff disputed this and said they find lots of animal remains near mining areas.

Marine. Several people mentioned the exploitation of marine resources, particularly of the marine area outside of the park, as problematic. Collection of turtle eggs on CNP beaches is largely a subsistence-level activity as is collection of some species of mollusk. Commercial off-shore fishing by Panamanians was mentioned as another type of resource use that affects CNP.

Gold Mining

Miners claim that there are gold deposits in numerous locations within Corcovado National Park and the Forest Reserve, but particularly in the Río Claro, Río Corcovado, and southern portions of the park. Artisanal mining has been one of the most significant sources of long-term resource use within CNP, while both artisanal and commercial gold mining has been extensive in the buffer areas surrounding the park.

Artisanal. There is perhaps no element about Corcovado National Park that is as well known as the conflicts with gold miners. Dr. Alvaro Umaña, the minister of MIRENEM during the Arias administration (1986–90), once remarked that he felt that most of his tenure was spent dealing with the gold-miner crises. When Corcovado Park was established, there were a few miners located within CNP. The damage from their activities was limited, as was the scale. The withdrawal of a large banana company near Golfito in 1985 led to increased unemployment, as mentioned earlier. Few were farmers, and offers of agricultural land were unattractive. Dreaming of fast and easy money, several hundred people invaded CNP and began panning for gold. For several months, both the National Park Service and the central government were unable to control CNP. The government's inability to deter such actions diminished the public's perceptions of the government's ability to manage the parks elsewhere in the country, which led to increased encroachment both in CNP and elsewhere.

In 1985, an international team of scientists reviewed the impact of miners in CNP and recommended that gold mining be halted and that CNP be managed so as to benefit local people and increase revenues from tourism. Finally, a court order forced the police to evict the miners, after compensatory benefits were negotiated between the outgoing government and nearly eight hundred "miners" and their families.

Unfortunately, even though the court decree said that miners should be compensated, there were many problems. People who hadn't been mining suddenly claimed they had and just hadn't been "found." The government did not deliver the promised compensation; miners, with few assets or economic alternatives, started mining again for income. A march from the Osa to San José, with miners camping out downtown, became a highly visible public symbol of the conflict on the Osa. The gold miners threatened to re-invade CNP if they were not compensated immediately. In an unprecedented action, the Costa Rican Congress approved compensatory changes in a single day, and in-kind (mostly food) and cash payments. These eventually cost the Costa Rican government nearly $3 million. This nearly tripled the yearly budget of the entire National Park System and took badly needed resources away from park management nationwide.

ACOSA staff still find miners within CNP. Two common types of gold mining are tunneling and panning. In fact, different communities are associated with different types of mining: El Tigre with tunnels, and La Palma with panning. We were told that 60–80 percent of the people in El Tigre survive by mining within CNP. There is strong disagreement over the level of ecosystem damage that miners do. Miners we interviewed downplayed the environmental impacts. They also claimed that the only hunting they do is opportunistic hunting—they will shoot something if it walks in front of them. However, gold miners have recently been found working a stream only two miles from La Sirena Biological Station in the heart of CNP. ACOSA staff claim that besides causing damage to vegetation and water, miners leave garbage.

Miners still have a tremendous amount of hostility toward CNP; many claim that their compensation was pitiful and their livelihood has been stolen. They also ask why they shouldn't benefit when the rich transnational mining companies did much more ecological damage and weren't even Costa Rican. Miners interviewed said that most of the miners are part-time panners who use gold as a safety net. They would like to see some system set up by which they could mine on a part-time basis, in an environmentally sound way, that was legal and recognized. They think that mining could help lure ecotourists, who could spend time in CNP with miners and pan for gold. They could stay in the communities where miners live and help provide another source of income.

The leader of COOPEUNIORO (a former gold mining cooperative that now has an ecotourism project) who is also the current president of ACOSA's Local Committee (CLACOSA) believes that the relationship between the miners and the government is less tense now than it was five years ago. Many miners have left because they have found little gold or steady employment to support them. Because the majority of the gold mining activity is artisanal, he believes the attention given to the gold miners as one of CNP's main threats is unwarranted. He claims that the National Park Service sees miners as the primary threat to CNP but does little about illegal hunting, which occurs near Los Patos Station inside CNP and in other places. Additionally, he believes that if one compares the damage done by the miners in the last fifty years with the damage done by logging in the last ten years, one will see what the true threat is to Corcovado. He claims that there are only thirty artisanal miners, who work the river with a shovel, a bar, and a pan and, therefore, cause little environmental damage. However, prior research indicates that even two miners intensively working a "medium-size stream" could destroy it for two to ten kilometers downstream (Janzen, cited in Naughton 1993).

To summarize, it seems that artisanal mining per se does not represent a substantial threat to CNP at the present time. However, the levels of mining and encroachment into CNP that existed in the past were sufficiently high to have caused problems. Care must be taken to insure this does not recur. The more significant impact at the present time is that of miners on CNP's wildlife. This could probably be controlled, especially if zoning was used to determine areas where mining would be permitted. However, the hostility against CNP due to the conflicts with gold miners has created a more substantial threat—that of a planned invasion of CNP. Another potential threat is the new crisis with Costa Rican banana companies, raising the possibility that yet again many people will lose their jobs. Concern has been raised that this could renew pressure on protected areas, especially CNP.

Commercial. Commercial gold-mining operations began in 1974 on the rivers in the southern part of the peninsula. These mining operations were all foreign-run ventures that relied on heavy equipment, which severely degraded the river ecosystems. For example, artisanal miners using a pick and shovel can remove approximately one cubic meter of material daily, while a mining company with machinery and other modern technologies can process up to two thousand cubic meters of material each day (González 1992). There were thirty-nine active mining concessions registered and a waiting list of over 100 on the Osa Peninsula. In 1995 we were told by officials in MIRENEM that there are no commercial mining concessions actively worked at the present time. However, others disputed this and said that there were, and that companies used some of the permits intended for exploration for actual mining. Specifically, it was claimed that in the dry season, big machinery capable of moving five hundred cubic meters of riverbed each day is brought in adjacent to CNP. We were told that this machinery is brought in without permits with funds from international investors. Such machinery is highly destructive to the environment. Ecologists have estimated that "90 percent of the reef complexes in the Golfo Dulce have been destroyed by sediment" from mining (Donovan 1994).

We were unable to confirm the assertions of current large-scale, illegal mining. But in Cerro de Oro, close to the border of the park and COOPE-UNIORO's ecotourism project, old mining machines have been discarded. Previous mining has, over the years, altered the course of the river.

Forest Clearing for Agriculture and Timber

There has been limited forest clearing within CNP since it was established, although prior to its creation, some parts belonging to OPF were cleared, mostly by settlers. There is an old logging road, now used as a trail, from Patos to Sirena that has been closed. One informant told us that a local delegate, backed by loggers, was proposing to open this road and log sustainably on either side. ACOSA staff indicated that this would be illegal under current law, which was unlikely to change. Significant levels of forest clearance are taking place in the RFGD, which surrounds CNP. Substantial clearance would turn CNP into an island and cause irreversible damage, given the peninsula's watershed and drainage. Strong tropical rains will increase soil erosion and runoff, and flooding will become more common in the flatter areas used for agriculture and pasture, especially on the Gulf side. This is already happening with increasing frequency in the flat area between Rincón River and Puerto Jiménez.

The Osa Peninsula contains one of the last three significant tracts of valuable hardwoods left in Costa Rica. Until about 1990, most of the deforestation was in the flat areas in the peninsula. Between 1990 and 1995, however, deforestation has increased on steeper slopes. One 1990 estimate placed the total amount of productive forest outside of CNP in the Osa at twenty-seven thousand hectares and the annual rate of clearing for lumber at twenty-four hundred hectares, or 9 percent annually. The deforestation rate on the peninsula is estimated to be 5.7 percent annually. One of the most common practices on the Osa is known as "high grading" or "creaming," which is extracting trees of high value. Over time, once the high-value trees have been cut, other trees are felled and the land is used for pasture. This was legal until 1992, when the DGF stopped issuing permits for creaming. The former head of ACOSA told us that even if the forest cover looks all right, he has witnessed substantial declines in overall forest health.

Logging is allowed in the RFGD with a permit and subject to regulations. In practice, these regulations are rarely enforced due to the extreme weakness of the General Forestry Directorate as an institution. For example, a thirty-three-year rotation would allow for the annual extraction of 16,279 cubic meters—however, between 1991 and 1994, an average of 23,500 cubic meters was legally approved (Cuello and Piedra-Santa 1995). Of course, all legal logging represents additional revenues for the Costa Rican government, which may act as a short-term incentive for unsustainable logging. Forest clearing as a requirement to prove ownership has also exacerbated deforestation during specific periods when settlers have felt their claims were particularly vulnerable. Lack of tenure and resource security has also meant that people log when they feel that they legally can, or when insecurity is so great that they want to extract as much as possible while they can. The granting and possible rescinding of titles in the 1990s in the RFGD is creating such a period of insecurity.

The DGF has been unable to control logging and forest clearing on the Osa. Although there is only one access road in and out of the peninsula, there are no DGF checkpoints on that road to control logging. Similarly, logging permits are legally granted without any site review or follow-up to see if the logging actually met legal requirements. Loggers are able to use peasants to get logging permits for limited cutting; they then use the permits on that land and on surrounding lands. Numerous people interviewed for this and other studies acknowledge that Osa residents do not benefit from logging; all agree that the prices received from loggers are way below market value. But the lack of mechanisms for peasants to log sustainably, sell wood at market rates, or get any value added has led to timber sales for whatever anyone can get.

There has been some reforestation on the Osa; a large corporation called Ston Forestal, a subsidiary of U.S.-based Stone Container corporation, has been contracting with farmers to plant *gmelina* (*gmelina arborea*) for pulp chips. The BOSCOSA project (see p. 172) has also initiated some small-scale efforts to promote reforestation: from 1989 to 1993 approximately 526 ha were reforested (Hitz, cited in Cuello and Piedra-Santa 1995). Another project, the Osa-Golfito project (see p. 168), has plans to reforest 800 ha of privately owned plantations and 350 ha of degraded land that is supposed to be protected.

Tourism

Tourism has become Costa Rica's second-greatest source of foreign income. In 1991, it generated $336 million; by 1993 the figure had climbed to $506 million (Burnie 1994). Yet the Osa has received few of the benefits of this tourism boom. While tourism to Corcovado National Park has increased significantly, from 4,390 visitors in 1990 to 19,164 in 1994, most tourists do not come in contact with the communities on the Osa. Instead, they go to a series of lodges on the north and western side of the peninsula and enter the park only for a day. Those who do come to Puerto Jiménez tend to be on a more limited budget, which means that they spend little in the local communities. While ecotourism has been promoted as one of the most effective ways to link parks with local communities, this has not happened on the Osa. Instead, tourism may pose a threat to the biological integrity of CNP. ACOSA staff voiced concerns about the dramatic increases in the levels of tourism coupled with a decline in revenue for protection. They felt that in some parts of CNP the levels of tourism were too high, but that there was nothing they could do about it given their limited resources. When a labor dispute took place in 1995, ACOSA staff planned to close CNP. Tour operators resisted this, and one section of CNP remained open for tourism. Yet, this reinforces the view of local communities that CNP is for rich gringos while they are excluded.

Organizational Roles

While numerous NGOs and development assistance projects have been involved in the Osa, not all of these groups have had a long-term commitment to the area or a high level of intensity. Fundación Neotrópica has had the most enduring effects on the Osa, through the BOSCOSA project, described on page 172. Few other groups have had any long-term presence on the peninsula.

Many of the people we consulted believe that although the Osa receives high levels of funding through NGOs and various projects, very little of this benefits local communities. Local-level frustrations are fed by the apparently large number of organizations associated with even simple undertakings. The complexity of who finances and provides what services on the Osa is demonstrated by a local training program provided by BOSCOSA, an FN project. Financing was provided by the German Cooperation Agency (GTZ), which supported a German NGO, which in turn provided the financing to FN. Local people are aware only that it took three institutions to get one course underway; they assume that most of the support went to these different organizations rather than to them. This example illustrates three difficulties that NGOs based outside the Osa have when working with local groups:

- the lack of understanding of local groups regarding the cost and complexity of projects designed to assist them
- the difficulty of identifying funding sources for activities desired by local groups
- the problem of communicating to local groups the nuances of national and international funding arenas, in order to avoid misunderstandings.

One of the most significant problems on the Osa is the tremendous lack of coordination or of a common strategy among the institutions and projects working there. This long-standing problem has never been rectified, despite occasional attempts. It is ironic that most of the projects have conservation as either a primary or a secondary objective, yet the lack of a coherent or strategic way of coordinating activities and channeling financial resources has left residents feeling cheated, donors frustrated that programs haven't worked, and, in reality, little concrete progress for all the money and effort expended. At the root of much of this lies the lack of strong political will to eliminate bureaucratic infighting and truly get all institutions cooperating to coordinate efforts in a way that will turn the rhetoric of sustainable development into a reality.

Several conclusions about the roles of different organizations on the Osa are evident. First, the initial impetus for the establishment of CNP was principally based outside the peninsula—the origins were in San José through government agencies, such as the National Parks Service, national elites, and the Tropical Science Center and from international organizations, such as The Nature Conservancy and the World Wildlife Fund. A second and related finding is that there has been a high level of sustained involvement from a select few external groups—especially the FN through the BOSCOSA project. A third finding is that there are few well-established local organizations on the Osa Peninsula in either conservation, local development, or cooperatives with the capacity to carry out a significant level of activity. Finally, there has been virtually no donor coordination on activities on the Osa. There have been a relatively high number of groups, adequate financial resources, and lots of interest in saving CNP and improving the quality of lives for people on the Osa. But the lack of a coherent and well-articulated donor strategy has neither saved CNP nor measurably improved people's livelihoods. Numerous opportunities to strategically target funding and expertise have been missed.

The following NGOs or projects have been, or are currently, active on the Osa Peninsula:

Fundación Neotrópica. Founded in 1985, FN's general objective is to promote the appropriate policies for sustainable forest management. FN works primarily near protected areas in Costa Rica, with a special emphasis on the Osa Peninsula, and the Tortuguero and Tempisque regions. FN has been a partner of The Nature Conservancy for numerous conservation initiatives in Costa Rica and for the PiP/Costa Rica program. It has a formal agreement with MIRENEM to provide support and technical assistance in sustainable development and conservation activities throughout Costa Rica, including the Osa Peninsula. FN's activity in the Osa Peninsula is made through its Forest Conservation and Management project (BOSCOSA), described on page 172.

The Osa-Golfito Project. This two-and-a-half-year, integrated, rural development project is funded by the European Economic Community (EEC) to promote agricultural production and increase living standards, thereby reducing pressure on regional forest resources. The funding, of about $22 million, is to be provided one-third each by the EEC, the Costa Rican government, and the Costa Rican national bank in the form of credit. The project intends to work through local organizations that want to become self-financing businesses. It has been active in the Golfito region for a while and in 1995 was beginning to expand to the Osa. A second phase will channel funding to other groups through loan guarantees. The project also intends to train local producers to produce crops for sale and to use credit from the national banking system. The major concern of this project is to develop productive activities that provide campesinos on the Osa Peninsula with alternatives to improve their living conditions, rather than to address environmental or conservation issues directly.

Proyecto REFORMA. The Regulation for Forest Management (REFORMA) project was started by the U.S. Agency for International Development in 1993, jointly with MIRENEM and the DGF, although implementation only began in 1995. The project was completed in 1998. The southern part of the Osa Peninsula is one of three critical forest zones on which the project focuses. The project is promoting the sustainable use of natural forest, with a substantial emphasis on promoting sound forest sector policies and changing the legal environment in which forest activities operate (USAID 1993). Specific elements include:

- simplifying regulations and legal codes governing forestry and sustainable forestry
- training DGF staff better and concentrating them in key areas of the country
- improving value-added processing capacity and lifting the log export ban to improve the value of wood
- reducing the cost of preparing logging permits and improving the monitoring and verification of permits granted
- improving the public's understanding of the importance of primary forest outside of parks and the problems caused by deforestation
- introducing a certification system for wood exports.

High levels of theft of project materials have been reported, such as of completely outfitted trailers that were to be used as headquarters and residences for forest service employees. No policy reforms have yet been reported.

TUBA Foundation. This organization, which is dedicated to the use and processing of residual woods that loggers discard, has had involvement in the Osa.

INBio. The National Biodiversity Institute was created by presidential decree as a nonprofit private organization in the fall of 1989. It has been training national and local parataxonomists to collect and identify the species found on the peninsula as well as to collect information on their distribution, abundance, habits, and habitats. At the national level, the goal of INBio is to collect, catalogue, and inventory all of the plant and animal species in Costa Rica.

FINCA. This nongovernment organization (NGO), which provides credit to local organizations, has provided a loan in El Tigre for the development of a small ecotourism hotel complex, which subsequently defaulted due largely to lack of careful financial analysis, insufficient training of local residents in hotel management, and overly optimistic assumptions about the arrival of ecotourists to El Tigre.

Organization for Tropical Studies. OTS represents a consortium of over forty U.S. and Costa Rican universities on tropical research and education; they sponsor research and have held several courses on the Osa.

Tropical Science Center. TSC was highly involved in the early stages of development of Corcovado Park and is still involved in research on the Osa.

University of Costa Rica. UCR has conducted scientific studies and evaluations and rapid ecological assessments, and monitored plans on the Osa.

CEDARENA. The Center for Environmental and Natural Resource Law, a San José–based NGO, has collaborated with FN on land tenure research on the Osa Peninsula.

Catholic Relief Service. CRS financed FIPROSA. It also financed BOSCOSA to provide an entrepreneurial training program, focusing on organization, administration, and marketing for small producers on the Osa.

World Wildlife Fund. Since providing some of the initial funding to establish CNP, WWF-U.S. has provided long-term support to CNP and the Osa, primarily through the BOSCOSA project. WWF also financed the formation of a group of local paraforesters, the majority of which are successfully working in the region.

The Nature Conservancy. The Conservancy has had a long-standing and supportive role in insuring the existence of Corcovado Park. More recently, through the PiP program, support has been provided for infrastructure and guide training and research. This support is channeled either directly to ACOSA or to FN. PiP support for Corcovado ended in 1994.

PRODERE. A program of the U.N. High Commission on Refugees, PRODERE had a project to deal with refugees until this year.

Some of the groups have had long-standing involvement in the Osa—for example, OTS and TSC—while other groups have undertaken specific activities and moved on when those activities have ended. Numerous international government agencies have also had a strong interest in the Osa, and USAID has been the strongest supporter of the BOSCOSA project in the past six years. GTZ, with financing from the German government via LUSO Consult, a German NGO, has financed the training for naturalist guides. The Danish International Development Agency (DANIDA) has supported the Tropical Youth Center with Nepenthes, a Danish NGO. The Development Fund, supported by the Norwegian Agency for Development Cooperation (NORAD), has supported the training. The International Tree Fund (Holland) has provided support. Groups such as the Massachusetts Audubon Society have also been involved in supporting local NGOs and research on the peninsula.

During the years 1994–96, ACOSA received a grant of about $4 million from the U.N. Environment Program (UNEP), to be shared between the Osa and La Amistad through the Global Environmental Facility. GEF funding does not include staff salaries, however, which are the responsibility of the Costa Rican government. The GEF funds are geared to strengthen four main areas:

- operational and institutional strengthening of ACOSA
- development of research programs in core areas and in buffer zones
- consolidation of sustainable economic practices in ACOSA's buffer zone
- generation of financial resources (for ACOSA's economic self-sufficiency).

There are also numerous Osa-based NGOs and cooperatives, although many of these are new institutions and, as yet, relatively inexperienced.

Linkages between Park and Buffer Areas

The Osa's population is approximately 11,922 residents, virtually all of whom arrived there in the past fifty years. The economically active population is 31.6 percent, distributed in the following productive sectors: 69 percent in agriculture, hunting, fishing, forestry, and ranching; 25 percent in services; and 6 percent in other activities, such as mining, construction, transportation, and others (Cuello and Piedra-Santa 1995).

In the late 1980s and early 1990s, one of the most widely discussed projects linking protected-area management with local development in Latin America was the BOSCOSA project, begun in 1987. After its first two years, the project was being hailed for its accomplishments. The achievements of this "pilot" project and the approaches developed to date are of considerable importance to the future design of ICDPs, or integrated conservation and development projects (Wells and Brandon 1992). Because of its reputation, its location on the Osa Peninsula, and its progress to date, BOSCOSA has been made the main subject of this section.

The BOSCOSA Project

The BOSCOSA project (Proyecto de Manejo y Conservación de Bosque de la Península de OSA)—the acronym combines *bosque*, or "forest," with OSA—was designed first by the Conservation Foundation (CF) in conjunction with FN and was one of the best-known conservation and development projects, both within and outside of Costa Rica, during its first five years. CF then merged into the World Wildlife Fund–U.S., and the program was put under the auspices of WWF's newly created Tropical Forest Program. In its early stages, the BOSCOSA project was conceived largely as a natural forest management project that would "complement a separate, unrelated protection program by fomenting grass-roots-level sustainable economic alternatives for people in Corcovado's buffer zone. The assumption was that deforestation could be slowed by providing rural campesinos with economic alternatives" (Donovan 1994). Since its inception in 1987, the project has moved through several phases. These will be briefly described below.

The First Phase

In the first phase of the project, the key elements of the BOSCOSA design were

- working with local grassroots community organizations
- emphasizing local decision making
- supplying only technical assistance not financing or materials
- using partnerships between BOSCOSA and community groups for resource leveraging
- helping link local initiatives with regional and national efforts and with government policies and programs (Donovan 1994).

By 1992, it was actively working with twelve local organizations with a total of seven hundred members, providing technical assistance on organizational strengthening, agriculture, reforestation, and natural forest management. While the linkages between the project activities and CNP were indirect and diffuse, the overall strategy of rapid start-up, with strong participatory components, was admirable. The project was also aware of the regional change context and had made early advances on establishing many horizontal (among different groups) and vertical (from local to national levels) linkages.

The BOSCOSA project also organized the Osa Inter-Institutional Committee, which brought together local people, BOSCOSA staff, and government staff from the wide range of governmental institutions represented on the Osa. The meetings were an attempt at dealing with the lack of coordination among government agencies and the lack of any local participation in planning for the peninsula; they resulted in a planning document: Osa 2000. This was the result of a well-coordinated effort at local participation from different groups to generate an overall plan for the development of the region. No one we interviewed in 1995 had been involved with or had heard of this document or the process behind it from the late 1980s. This highlights the rapid social change and mobility on the Osa. In 1995, FN undertook a survey of one of the communities where BOSCOSA had focused, Rancho Quemado, and found that about half of the population who were involved in the project in the late 1980s no longer lived there.

Other innovative elements that were initiated include the establishment of Community Rainforests and the FIPROSA Fund. Community Rainforests were initiated by the BOSCOSA project as part of its technical assistance in forestry as a way of helping farmers without title undertake natural forest management. Twenty management plans were established in conjunction with two different groups covering a total of 863 ha. The principal characteristic of these community management plans was that the groups were provided with technical assistance in forest management, including sustainable logging, even though they did not have title to the properties. Eleven of the areas were within the RFGD and the remaining nine were outside it. The FIPROSA fund was established to provide loan guarantees through two programs: the Program of Forest Incentives (PROINFOR) and the Collateral Program (PROAVAL). Through PROINFOR, the BOSCOSA project provided cash incentives to participants to maintain forest cover and as a substitute for the sale of trees to loggers while families developed a reliance on other sources of income. The PROAVAL component was to allow local groups to obtain start-up loans for projects that promote forest conservation, such as ecotourism, or development of nontimber forest products, although in practice, the majority of loans were granted for agricultural projects. Funding for FIPROSA (Commission for Small and Medium Producers on the OSA Peninsula) was initially provided by WWF and Catholic Relief Services. Additional support was received from a Norwegian NGO. Funds in 1995 were approximately $221,000.

Consolidation

In its early phase, the project had been primarily conceived of and run through FN but without direct involvement. The first director of the project was a U.S. national who was able to get many things up and running in a short time. By virtue of not being a Costa Rican, he was also able to push the government and other institutions in ways that would not have been viewed as acceptable for a Costa Rican. However, at the local level his nationality led to ironic jokes against BOSCOSA and its director, with the pejorative title of *gringo* applied (Camacho 1993). Rumors swept through certain areas that the project was sponsored by Osa Forests Products in an attempt to reclaim its lands—this confusion was due in part to the nationality of the director, as well as to WWF's panda symbol on project vehicles. People unfamiliar with a panda associated it with a bear (*oso* in Spanish) and with the OPF (Christen, pers. comm., 1995). The first Costa Rican director took over BOSCOSA in 1990; that signaled the beginning of the assimilation of BOSCOSA into FN. Subsequent directors of BOSCOSA have been responsible for making the project more integral to FN and for its evolution into a more Costa Rican project.

Few of the basic elements of project design that were apparent in 1989 and 1990 are in place now. Few of the project staff in the field or at FN are aware of the early history of the project and its accomplishments. One apparent difficulty is that the vision the word BOSCOSA conjures up to many is of the first phase, when it was a WWF-run project. This early vision of the BOSCOSA project, which in reality ended in 1991, has been promoted as recently as 1994 (Donovan 1994), even though most of the elements from 1991 were no longer functioning. FN still receives queries from all over the world about elements of the BOSCOSA project that had ended by 1991. Similarly, a recent USAID evaluation of the project used the first phase of the project as the benchmark for the evaluation. What is lost is the process and rationale for scaling back the project, internalizing it within FN's vision, and making it something that was manageable given the capacities of FN and the changing situation on the Osa. FN has acknowledged that the history and evolution of the project have not been adequately documented; it is currently undertaking such an effort.

Elements of the various transitions are nonetheless problematic. Few of the twelve community groups that were functioning in the early days of the project exist today. This has no doubt led to a certain discontinuity in the project when viewed from a local perspective. For example, in terms of agriculture alone, the project has gone from focusing on improved production for subsistence, to production for regional sales, to production for national markets, and finally to production for international markets. However, these changes in the agricultural program came about due to more analysis of the changing local, national, and international markets for products, as well as the rapid integration of the Osa resulting from improved access. On the positive side, some elements of the project have been absorbed by other institutions; for example, the Costa Rican government has attempted to institutionalize the process of interorganizational coordination through ACOSA, so the Inter-Institutional Committee has ceased to function.

Current Activities

In recent years, the BOSCOSA project has emphasized forestry, agriculture, environmental education, and training. In forestry, the project has resulted in two hundred hectares under a well-run natural forest management regime and three hundred hectares reforested. It has also been able to attract and implement a rotating fund that has provided financial incentives for forest protection. The training component has worked with different local organizations on institution building, administration, and marketing. The fundamental objective of this training has been to strengthen the business capacity of local groups. Within the agricultural component, production of roots and tubers has been successful. However, FN has decided to work on activities that are directly tied to forestry, halting the agricultural component of project activities and letting the Osa-Golfito project take over that component. The environmental education component has shifted from outreach to individual communities, to bringing communities to the Tropical Youth Center, which is now the main BOSCOSA headquarters. The Tropical Youth Center was largely the result of DANIDA promotion and funding for a center for environmental education. In the past few years, the project has hosted international, national, and local groups to learn about the rainforest. For national and international groups, the center makes money by renting rooms and providing environmental education services.

Since 1987, financing for BOSCOSA projects has amounted to over $7 million. A recent evaluation notes that the impact of the project was greatly reduced because of its attempt to deal with the whole buffer zone surrounding CNP (Hitz, Alpizar, and Montoya 1995). It has also been unclear—both to local people and to BOSCOSA's own staff—whether BOSCOSA was an institution or a project.

Other Initiatives

Ecotourism is often cited as one of the most clear and direct links between protected areas and local communities. In the case of Corcovado, it has had mixed results. The park has good relations with people around Drake, where there is a relatively high level of employment in ecotourism. Yet a community near Drake, Los Planes, which receives few benefits from tourism, has put substantial pressure on CNP through hunting. One person interviewed commented that BOSCOSA's Tropical Youth Center (TYC) would take away tourists from the local communities and keep the profits for the project, rather than widely distributing them. While FN has repeatedly told the local population that it does not intend to bring tourists to the TYC, such comments in the course of field interviews reveal either a lack of trust in what is being said, or the need for greater information dissemination. Yet field interviews also singled out training for local guides sponsored by FN as valuable. One local informant said that the financial support provided by the PiP program had provided important local benefits, such as the construction of the ex-miner–operated ecotourist lodge in Cerro de Oro.

The elements of buffer zone management that are the most clear and direct, such as local employment linked to the park, have been thwarted by government policy. For example, park guards may be hired through the civil service or locally. Recent government cutbacks forced ACOSA to fire half of the guards, all of whom were locally hired staff. Employment for local people is an obvious and visible benefit from the park that extends beyond the employees themselves into their communities. Firing them, while retaining the civil servants from outside the area who have much higher wages and guaranteed jobs, has created hostility. Even worse is the attitude of many of the civil servant park guards, who are from elsewhere in the country, that the Osa is a poor posting. The lack of employment opportunities for spouses, good schools for their children, and electricity meant that many left their families in the Central Valley and commuted to the Osa, returning home when possible. This transient attitude toward the Osa created feelings of resentment among residents and some of the local guards.

The key problems on the Osa are:

- lack of tenure security amid rapid land-use changes
- regional change.

None of the projects are tackling these issues in any concerted way. The CEDARENA project was addressing tenure issues, but it was cut short as the BOSCOSA project scaled back activities. Nor is there any correspondence between key threats to CNP—hunting, deforestation, and mining—and buffer zone activities. The best opportunities for clear and direct linkages, such as the community rainforest or those offered through the fund for keeping lands forested, have not been expanded—despite the fact that they are small but successful elements of the BOSCOSA program. Expanding such activities would require clarifying land and resource tenure components.

The Osa-Golfito project is the largest new initiative on the Osa. It is incorporating the agricultural component from BOSCOSA and introducing farmers to new crops and getting them linked into national credit systems.

Yet there is no link between the Osa-Golfito Project and maintenance of CNP or the forest reserve. There is a high level of physical proximity of certain projects such as Osa-Golfito, REFORMA, and PRODERE to CNP. Yet these projects incorporate a portfolio of different activities that are scattered widely around but without the focus that will lead to any real change. This is a common problem of the ICDP approach (Wells and Brandon 1992). The Osa has received an extraordinary amount of money in the last decade, but the money has not been clearly targeted—it has been widely dispersed to address a variety of problems. It is, therefore, impossible to see specific results, especially given the rapid land-use changes underway. One of the early elements of the BOSCOSA project, the Inter-Institutional Committee, offered one of the best mechanisms for participation of all groups in sharing what they felt was happening, as well as working out some common vision for the peninsula. Such mechanisms for participation have been transferred to CLACOSA.

Conflict Management and Resolution

The conflicts over land use on the Osa, including those over rights within CNP, have been intense. These conflicts started in the days of OPF, when the first manager did a survey and claimed rent or harassed those who would not pay. It intensified when the second manager established a policy of burning houses as a way of forcing squatters to move. The gold miners and former banana plantation workers who settled there were largely independent and had what is often referred to in reports by foreigners as a "wild west" mentality—they saw the Osa as the last frontier in Costa Rica. Few people there had long-standing presence in the area. Even among settlers, there was competition over land claims and resource uses. The lack of any tenure security, the lack of basic services, and the creation of the protected areas exacerbated the level of insecurity and conflict. There is a long history on the Osa of arrests of miners, people being resettled, park and RFGD reserve guards being threatened or shot at, and various people being jailed. Given all of this hostility, it is surprising that the actual level of violence, as opposed to threats, has been as low as it has.

The problems over who has what rights have been complicated; even within groups there is conflict. For example, gold miners interviewed claimed that there were really two groups of gold miners—those who were in CNP before it was established, and more recent opportunistic migrants who hoped for a settlement from the government and claimed to be miners. Many of this latter group are in fact miners, but the "old-timers"—those who were actively mining before the expansion of the park—feel that while they have legitimate rights that have not been reasonably recognized, the latter group should not get compensation. However, they also acknowledge the strength in numbers and, depending on the kind of settlement that could be reached, acknowledge that they are not ready to dissociate themselves from the other miners unless it is useful.

Several of the old-time miners told us that some miners were planning to invade CNP. During the period of the site visit, the national teachers union was striking against the government and other unions were striking in sympathy. Since the government was perceived to be weak at the time, there was, according to several informants, a discussion about whether it would be a good time to take over part of CNP by force. We were told that all of the people who intended to invade CNP were armed, and that they also had explosives available to blow up sections of CNP, making the gold more accessible. We were unable to confirm these statements.

One element that galvanized local NGOs on the Osa Peninsula was a recent fight with Ston Forestal about building a port facility in the Golfo Dulce. Local groups, together with organizations in San José and outside the country, were able to stop the project. A recent effort to build a processing facility and a small port/loading complex in the Golfo Dulce was attacked and stopped by the local communities and by the national and international environmental movement because of the pollution and interference with the biodiversity of the region it would cause.

Large-Scale Threats

The threats to CNP can be classified as direct and indirect—as well as by the scale of the threat. The most direct threats to CNP in 1995 are hunting and mining. Indirect, but with potentially greater impact, are:

- deforestation and hunting in the RFGD that are impelled by socioeconomic problems, particularly poverty
- regional land-use change and development on the peninsula

- lack of political will to manage CNP and surrounding protected areas

Illegal activities within CNP do not appear to be large scale; the threats are from the aggregate of many small-scale activities, particularly mining and hunting. However, both ACOSA staff and local residents indicated that given a declining economic situation in Costa Rica, pressures on CNP, particularly hunting and mining, will intensify. As mentioned previously, the present (1995) number of gold miners does not appear to currently pose a serious threat to CNP. But if the number of small-scale miners and panners even begins to approach past levels of two thousand people, then mining would again become a serious threat. If external factors, such as the economy, or internal factors, such as someone finding a large gold nugget, were to occur, the number could again rapidly escalate, becoming a large-scale threat.

It is difficult to differentiate tough talk from a serious potential threat. The threats to take over CNP by force have apparently persisted for years; nevertheless, it seems that such threats should not be entirely dismissed. However, the fact that no conflict took place at a time when the park and the government were perceived to be weak is a positive sign.

The more immediate threat to CNP is due to its small size; the survival of many of CNP's species relies upon an intact RFGD as a functioning buffer zone. If logging and clearing continue in the RFGD, CNP will become an ecological island. Similarly, the RFGD extends the range of species but is more accessible for hunting. One ex-miner told us that the RFGD was becoming a "forest without animals." As discussed in the section "Resource Use" (p. 159), 1995 estimates indicate that at present rates of clearing, the RFGD will be substantially cleared by the year 2000 unless immediate and strong action is taken.

The second potential source of indirect threat is the rapid land-use changes caused by the recent paving of roads and extension of electricity, although the magnitude of such changes is unknown. Costa Rican policy allows foreigners to freely buy land. Several people interviewed said that there was a huge influx of foreigners buying land on the peninsula—one report indicated that prices had increased five times in as many years. It was reported that land is being sold for residential estate developments and hotels. What is evident is that there is no overall development plan for the region, nor is it clear who would do one or how it would be implemented. The Osa 2000 plan, prepared through a participatory process in 1990, is virtually unknown on the peninsula. In addition, many of the key actors and groups who participated in 1990 have changed.

We heard that the Costa Rican Tourism Institute (ICT), the government's tourism agency, has plans to develop major resorts in the Osa. However, we were unable to confirm this, and the minister of tourism recently told FN officials that Osa was not a priority in ICT's immediate plans. MIDEPLAN, the government planning agency, was not undertaking planning at regional levels. ACOSA was unaware of the plans of government agencies in ministries outside of MIDEPLAN and did not feel any responsibility to be aware of the impending actions of other agencies. Other land-use changes underway may be positive—for example, converting cattle pastures into gmelina plantations. However, it is unclear whether such monocultures will provide substantial benefits for biodiversity, local employment, or the local economy. What is obvious is that there are rapid changes underway, but their magnitude, sources, and potential impact are poorly identified and understood. For a peninsula where approximately 66 percent of the land is protected, there are strong linkages between land uses.

The most direct threat to it is the lack of political will to manage it and surrounding protected areas. This is described in the following section.

National Policy Framework

Donovan, (1994) says "Inconsistent policies in government agencies and the desires of interest groups dominated by people from outside the Osa often have dictated the conditions under which the Osa's campesinos live."

Lack of government policy, conflicting policies, and their inefficient application have been at the root of many of the problems on the Osa Peninsula. The effects of a lack of political will for both park and forest management, as well as policy conflicts, are described in this section.

Lack of Political Will to Support Park Management

An important weakness within the SINAC is its legal status; it operates under a presidential decree (Umaña and Brandon 1992). This means that the legal processes to make the SINAC official have not been recognized by the Legislative Assembly. Therefore, all actions taken are provisional in nature and could be changed at any time by the Legislative Assembly. All of the protected areas in the country are faced with uncertainty in their management.

ACOSA's lack of ability to, at least, be aware of the actions of other governmental agencies on the Osa, if not coordinate them, is unfortunate. The Osa Inter-institutional Committee did serve as a clearinghouse for governmental agencies. CLACOSA, the local committee of ACOSA, is intended to provide a formal mechanism to link protected-area management with important local leaders. Unfortunately, this local committee has only an advisory role in providing opinions to the ACOSA leadership, and the lack of decision-making power has led to some frustration on the part of its members.

Another significant problem is the lack of funding to support CNP. This was graphically illustrated in June 1995 when CNP was closed because the government cut back half of the guard positions. The guards who remained said that they were unable to manage CNP adequately. An outcry from tour operators on the northwest portion of CNP led to a policy whereby CNP was open in that section but tourists would enter at their own risk. Staff felt this was unacceptable but had no options. To make matters worse, local guards from the peninsula were fired, while those from the National Park Service were retained. Since employment can be a strong and direct link between parks and local communities, this was extremely unfortunate and reinforced the notion of "outsiders" controlling the Osa. Discontent has been further fueled by the NPS guards receiving much higher wages than those who are locally contracted.

ACOSA staff told us that the level of protection at the time was inadequate for either the northern section of CNP, where the threats are largely from hunting and fishing, or the southern portion, which is threatened by encroachment and mining. PiP funding has been essential in the purchase of equipment, but much of this will need to be repaired or replaced in the near future, and there are no funds available.

Lack of Political Will for Forest Management

Despite all the attention directed at the Osa, there is little evidence of meaningful improvements in forest management there. The USAID-sponsored REFORMA project has begun on the Osa, but has not yet had any impact. The levels of DGF staffing to ACOSA are low; there are only two people for the entire Osa. Staff are unable to conduct inspections of sites either before or after issue of logging permits, and levels of deforestation on the peninsula are unknown. There is a lack of technical information for the permit process, such as information on the suitability of a given site for logging and the acceptable scale. Also, there is no monitoring or policing of logging trucks that leave the Osa, the existing checkpoints having been closed due to lack of funds. Many groups, such as the World Bank, USAID, and Inter-American Development Bank, have conducted studies of Costa Rica's forest policies and the perverse incentives that promote deforestation. Despite these studies, there have been no significant changes in forest policy leading to improved resource use. In short, the legal system is unable to respond to deforestation, even though this is one of the principal threats to Corcovado National Park. Further, unchecked deforestation represents a huge financial loss, both short and long term, at local and national levels.

Conflicting Policies and Lack of Inter-governmental Coordination

Lack of coherent and consistent policies, especially regarding land and resource tenure and deforestation, are at much of the root of the poor resource use on the Osa. Each government agency has historically appeared to be more interested in asserting its own regulations than in trying to bring about meaningful reforms on the peninsula. In the face of such conflictive policies and programs, it has been impossible for peasants on the Osa to adopt any long-term perspective about resource use. It also means that there is no local land-use planning, not even coordination or phasing of infrastructure development, such as roads and electricity. If national-level plans exist for big tourism development on the Osa (as some people told us they do), no one in ACOSA is aware of, or has even thought of, the importance of such plans. Even the disparate elements housed within ACOSA appear not to deal with one another. For example, no one at ACOSA except the mining expert knew anything about mining. All of this detracts from any shared vision about the key threats facing CNP.

Indigenous Peoples and Social Change

The Osa's indigenous population had disappeared by the latter half of the twentieth century. In the 1960s the Guaymi moved to the lands that are now part of the Guaymi Indigenous Reserve. In the early 1970s they began having conflicts with OPF over ownership and began lobbying to get the land declared a reservation. The reserve was legally established in 1985 with 1,700 ha and expanded to 2,713 ha in 1991. Some of the reserve was created using land from the RFGD, and other lands were IDA's, which it had received in the land swap with OPF. All of the land used to create the reserve had numerous nonindigenous families living on it. The establishment of the Guaymi Reserve meant that non-Guaymi had to be bought out. However, the lack of funds meant that few people received compensation until at least 1992, when money from a U.S.-based group was channeled through the BOSCOSA project.

There was hostility between the Guaymi and their nonindigenous neighbors, since both groups had colonized the area at the same time. Nonindigenous groups saw no reason the Guaymi should get special treatment in an area that was not their traditional homeland. One significant difference between the Guaymi and the colonists on the peninsula is that the Guaymi have maintained most forest cover on their lands and have been interested in natural forest management (Donovan 1994). Hostility by many campesinos over the creation of the Guaymi Reserve is still high. There have been threats made against the Guaymi, and even in the early 1990s, campesinos would shoot at the Guaymi if they traveled out of their reserve on pathways owned by campesinos (Wessels 1992). More recently, ACOSA staff have helped the Guaymi keep campesino hunters off of their lands.

Initially, the Osa Guaymi Reserve was included as part of the Osa Conservation and Sustainable Development Area (ACOSA). However, according to the latest modification of ACOSA, which took place in April 1994, the Guaymi Indigenous Reserve was left out of the conservation area. This resolution could have serious implications for Corcovado National Park's protection due to the fact that the Guaymi Reserve is along the border of CNP and, therefore, has to be considered as part of the buffer zone.

There are only about 120 Guaymi within the reserve. Currently, they do not appear to be putting any substantial pressure on the resources either within their own reserve or in CNP. The Guaymi reserve appears to be a stable and biologically significant buffer to CNP at the present time. However, there are many small children; substantial forest clearing and increased hunting could occur in the future if the children, when grown, decide to stay on the reserve.

Conclusion

The history of Corcovado National Park demonstrates the inherent difficulties of developing many protected areas. While the government of Costa Rica has shown strong political will in establishing the CNP, it is clear that similar political will is needed to maintain it in the future. The ecological fate of CNP is likely to be decided by the year 2000. In ecological terms, CNP is too small to be viable without the maintenance of the RFGD as an intact and functioning buffer zone. The analysis demonstrates that the existence of the RFGD is threatened. Strong political will is needed to:

- end the squabbling among governmental agencies
- resolve land tenure problems
- provide effective forest management to maintain the forest cover.

Existing programs are not fully or thoroughly addressing either the direct or the indirect threats to CNP. There is a need to effectively stabilize rapidly changing land and resource use patterns on the Osa. Historically, there have been high levels of commitment from the international community for CNP and the Osa—it appears that both funding and commitment for the Osa are strong. Yet a lack of strategic targeting and strong donor and governmental coordination may continue the trend that has existed—of high-cost programs with a low-level payoff. What is urgently needed is a catalyst to bring about the needed changes. Otherwise, within a decade, Corcovado National Park could lose its substantial biodiversity and its claim as "the crown jewel in Costa Rica's park system."

Postscript

The situation in CNP and the rest of the protected areas that constitute ACOSA (the protected areas on the Osa) has significantly worsened between 1995 and 1998. Lack of protection, financial resources, and clear and coherent policies have made the future of all protected areas on the Osa uncertain. Illegal hunting of wildlife and small-scale gold mining have increased in all areas. Rapid social changes are underway with the arrival of electricity to Puerto Jiménez and other communities.

The situation on the Osa has become so critical that Alvaro Ugalde, one of the best-known figures in Costa Rican conservation, wrote an open letter to President José María Figueres. In the letter, he warned that the park was in the gravest danger and stated that it only had twenty employees, ten of whom focused their attention on tourism and care of facilities, while the other ten were responsible for park protection (Ugalde 1997). Ugalde claimed that the situation in the other protected areas comprising ACOSA is equally precarious. Piedras Blancas National Park has two employees who devote their entire time to attending to the public. The Golfito Wildlife Refuge was recently invaded by squatters, and the Sierpe-Terraba Mangroves and the other areas completely lack protection (Ugalde 1997). In early 1997, ACOSA had fifty-two employees, most of whom were employed in a variety of nonprotection-related activities. Ugalde noted that the personnel assigned to ACOSA are fewer than the total number that Corcovado used to have. Ugalde declared that all of ACOSA's areas were "protected areas only on paper" (1997). Concerning Corcovado, he said:

> In only three years, this national park returned to its situation at the beginning of the 1980s, with imminent danger of returning to the situation of the 1970s, when it wasn't a national park, and it was invaded by occupants who were rapidly transforming it to pastures. From my point of view, Corcovado is a park that is rapidly dying, a park in extreme peril.

The director of ACOSA shares Ugalde's concerns, and in a letter to the director of the SINAC at the beginning of 1997, he stated that it was impossible, given the limited human resources available, to establish the number of people who enter the park and violate its rules. However, based on prior experience and sporadic visits to sections of the park, it appears that there are sections where permanent settlements have been established. He claimed that hunting was probably the most significant problem within the park, resulting from gold miners, commercial hunters, and sport hunters from other parts of Costa Rica and Panama (Mora, cited in Ugalde 1997).

In the GDFR, considered to be one of the last remaining remnants of natural forests, timber extraction is approximately thirty-five thousand cubic meters per year, likely to increase to fifty-one thousand cubic meters per year (Jiménez 1996). This level is vastly above the sixteen thousand cubic meters recommended in the forest reserve's management plan in 1992 and is the optimum amount for sustainable management (Alvarez and Márquez 1992).

A recent study by Fundación Neotrópica found evidence of increasing forest fragmentation. Aerial photographs showed that the forests along the southern part of the Rincón River were more significantly altered than those along the northern part of the river. On the Osa twenty-five thousand hectares are overutilized; this is to say that those lands, which should be forested, have been converted to pasture. A significant number of those pastures are found within the GDFR. It is worth noting, however, that the overall distribution of land is highly unequal on the Osa. For example, 132 parcels of forestland smaller than ten hectares constitute 1.2 percent of the Osa's remaining forested lands, while ten blocks of forested land greater than five hundred hectares make up over 85 percent of the forested land. If protection of these forested lands can be secured, a great deal of forest cover can be maintained. However, the fragmentation of blocks held by small-holders could make it extremely difficult to maintain cover on sufficient contiguous lands to support the range of species that ACOSA seeks to protect.

One of the reasons for increased fragmentation of lands has been the construction and improvement of roads on the Osa. For example, the road from Rancho Quemado to Drake has been improved, directly increasing logging on both sides of the road and opening the whole area to pressures for logging and colonization. All of this area is supposed to be protected.

A recent study analyzed the diversity and density of trees and plants in one area of the Osa (Thomsen 1997). Its conclusions indicated that:

- The Osa's forests are less dense and contain trees much bigger than other neotropical rainforests: the basal area per hectare was 28 percent higher than the average, the basal area per tree was 57 percent higher than the average, and the maximum height was 42 percent higher than the average.
- Mature forests ranked third in species diversity compared with 89 Neotropical sites analyzed.

Despite the high levels of deforestation, timber markets are restricted, and the species of wood that are most in demand are the woods that are least abundant. The State Forest Administration has given the market value of the Osa's woods to be: very high value (0.3 percent), high value (1.6 percent), medium value (4.4 percent), low value (25.4 percent), other commercial use (51.5 percent), and no value (16.8 percent). There are over seven hundred species of loggable trees on the Osa, although quality woods make up only 2.7 percent of the total volume (Jiménez 1996). It is worth noting that in January and February of 1996 alone, state authorities authorized the logging of 11,405 cubic meters of wood, of which 67 percent was represented by only twelve species of trees. Significantly, three of those twelve species are in danger of extinction (Jiménez 1996). The rest of the wood was distributed among sixty-seven species of which at least six species were endangered. Of the 11,405 cubic meters authorized, 2,434 cubic meters, or 21.3 percent, were of endangered species. Given that these are levels permitted by authorities, one can only imagine what happens in the vast majority of areas that are logged without permits.

There are almost no permanent, established logging companies in the zone. Instead, loggers from other areas come in to the Osa. There are only three loggers with land on the Osa, and none reside there. There is only one example of any integration of logging with local industry, on the Hacienda Cópano in Tamales de Puerto Jiménez. This implies that virtually all economic benefits from logging leave the Osa.

The fragmentation of the GDFR diminishes the prospect for a biological corridor between Piedras Blancas and Corcovado National Park. What was once a promising biological corridor is now a fragmented series of parcels being converted from forests to other uses (Maldonado 1997).

The availability of electricity has led to new patterns of consumption and the establishment of new agro-industries on the peninsula. Between 1995 and 1997, numerous factories have been established for processing chips made from tubers and roots; milk products; and palm heart (*palmito de pejibaye*). The presence of these industries has started to lead to changes in cropping patterns. There is talk of paving the main road from the Pan American Highway all the way to Puerto Jiménez, but no plan to do that could be confirmed. Other new roads, such as the one between Rancho Quemado and Drake, while unpaved, have opened access to loggers, allowing them to penetrate into the GDFR.

A new forestry law, passed in February 1997, has created a favorable environment for logging the natural forests left on the peninsula and elsewhere in Costa Rica. Local communities on the Osa are alarmed, blaming the incompetence of state representatives responsible for controlling the situation.

Because the situation at CNP and the other areas within ACOSA is similar to that of other conservation areas throughout the country, there is talk among conservationists, and even within the government, of the need for radical change in the park system, particularly in administration and finance. Some have suggested that the parks system be privatized, and that individual parks be turned over to private NGOs, such as Monteverde in Costa Rica and Rio Bravo in Belize (described in chapter 8).

Glossary

ACOSA—Area de Conservación y Desarrollo Sostenible de Osa (Osa Conservation and Sustainable Development Area)

ARC—Areas Regionales de Conservación (Regional Conservation Areas)

BOSCOSA—Proyecto de Manejo y Conservación de Bosque de la Península de Osa (Osa Peninsula Forest Management and Conservation Project)

CEDARENA—Centro de Derecho Ambiental y de los Recursos Naturales (Center for Environmental and Natural Resource Law)

CLACOSA—Comité Local de ACOSA (ACOSA's Local Committee)

CNP—Corcovado National Park (Parque Nacional Corcovado)

coligalleros—small-scale gold panners

CONAI—Comisión Nacional de Asuntos Indígenas (National Commission of Indigenous Affairs)

COOPEUNIORO—a former gold-mining cooperative that now has an ecotourism project

Costa Rican Banana Company—Compañía Bananera de Costa Rica

DGF—Dirección General Forestal (General Forestry Directorate)

FIPROSA—Fideicomiso para Pequeños y Medianos Productores de la Península de Osa (Comission for Small and Medium Producers on the Osa Peninsula)

FN—Fundación Neotrópica

gringo—pejorative name for non-Latinos, usually U.S. citizens

GWR—Golfito Wildlife Refuge (Refugio de Fauna Silvestre de Golfito)

ICT—Instituto Costarricense de Turismo (Costa Rican Tourism Institute)

IDA—Instituto de Desarrollo Agrario (Institute of Agrarian Development)

INBio—Instituto Nacional de Biodiversidad (National Biodiversity Institute)

ITCO—Instituto de Tierras y Colonización (Institute of Land and Colonization)

mejoras—improvements

MIDEPLAN—Ministerio de Planificación (Government Planning Ministry)

MIRENEM—Ministerio de Recursos Naturales, Energía y Minas (Ministry of Natural Resources, Energy, and Mines)

OPF—Osa Productos Forestales (Osa Forest Products)

OTS—Organization for Tropical Studies (Organización de Estudios Tropicales)

PBNP—Piedras Blancas National Park

PROAVAL—Programa de Avales (Collateral Program)

PRODERE—Programa para Refugiados del Alto Comisionado de las Naciones Unidas (U.N. High Commission on Refugees)

PROINFOR—Programa de Incentivos Forestales (Program of Forest Incentives)

REFORMA—Regulation for Forest Management Project (Programa de Regulación para el Manejo del Bosque)

RFGD—Reserva Forestal Golfo Dulce (Golfo Dulce Forest Reserve)

SINAC—Sistema Nacional de Areas de Conservación (National System of Conservation Areas)

TSC—Tropical Science Center
TYC—BOSCOSA's Tropical Youth Center
UCR—University of Costa Rica
UNEP—United Nations Environment Program
UNIOSA—Osa Peninsula Regional Conservation Unit

Del Este National Park

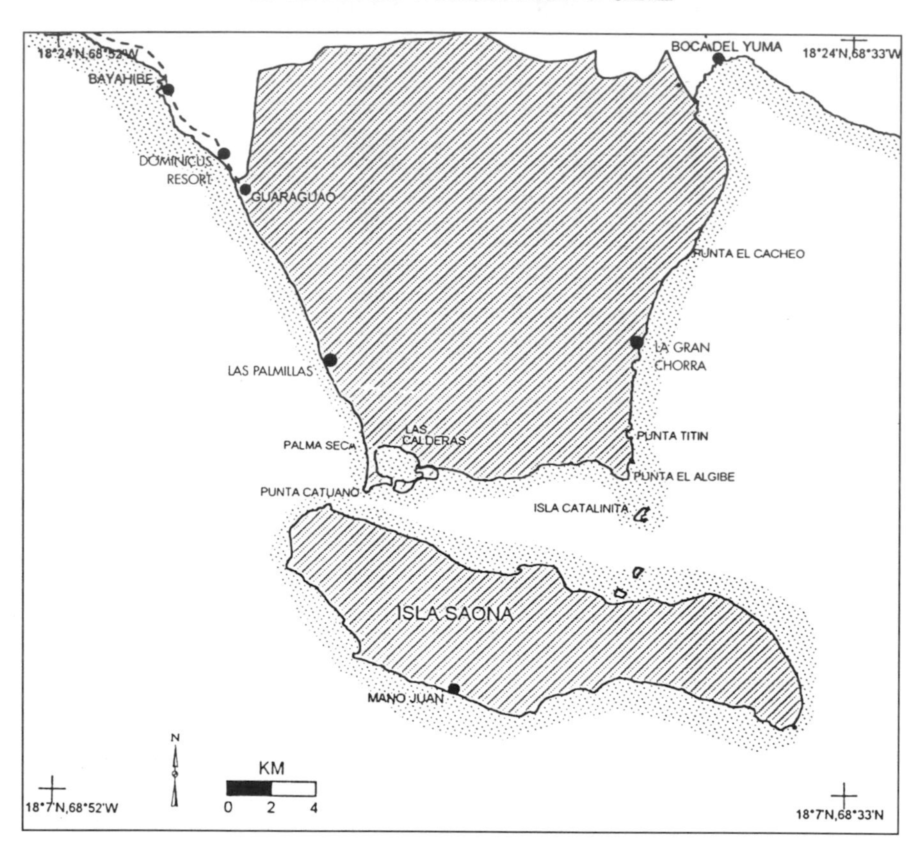

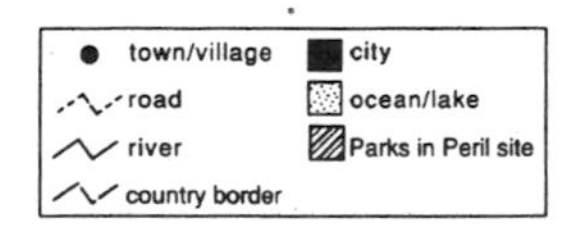

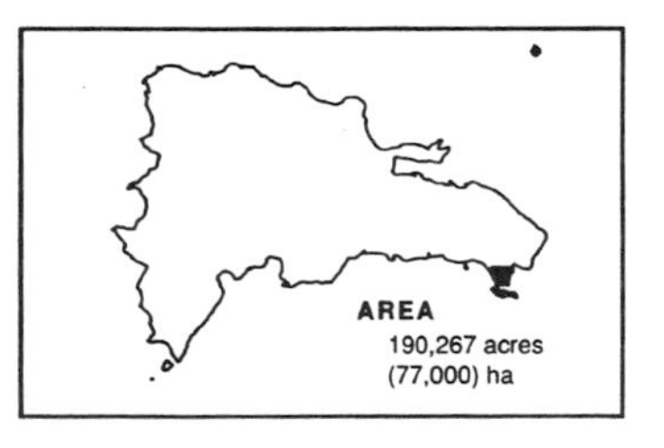

7. Dominican Republic: Del Este National Park

Kelvin Guerrero and Debra A. Rose

Del Este National Park (DENP) is located in the extreme southeastern coast of the Dominican Republic. The park represents some of the complexity often found in the Caribbean in protected areas that combine biodiversity conservation with land that can also be viewed as prime beachfront real estate. The park includes representation of the majority of Hispaniola's mammals as well as numerous endangered species, such as Caribbean manatees and three species of endangered turtle. A rich cultural history, with pre-Columbian pictographs and petroglyphs in underground caves, can be found within the park. The park also contains spectacular beaches, a major drawing point to many of the tourists located in nearby complexes. Mangroves and offshore coral reefs are important elements in or near park boundaries. Finally, DENP includes a number of local communities, which add to the complexity of managing for a variety of ecosystems, tourism, and sustainable livelihood activities.

Park Establishment and Management

Del Este National Park was created by presidential decree (No. 1311) on September 16, 1975. The park covers a total area of forty-two thousand hectares in the province of Altagracia in the extreme southeast of the republic; thirty-one thousand hectares extend from Boca de Yuma and Bayahibe south to the mainland coast, and eleven thousand hectares encompass Saona Island, separated from the mainland by the Catuano Channel. The park boundaries were amended in 1976 to exclude the resort complex Dominicus, located two kilometers from the community of Bayahibe on the northwest border of the park.

The terrain of the park is generally flat and formed of porous calcareous rock, which reaches a maximum of sixty feet above sea level and severely limits the availability of fresh water. Interior vegetation is composed primarily of dry subtropical forest, humid subtropical forest, and transition forest. Several large caverns are located within the park, some of considerable archaeological and anthropological value due to the presence of pictographs and petroglyphs, shell deposits, and burial sites. These have long been subject to exploration by organizations such as the Dominican Society of Speleology (caving), Espeleogrupo, and the Dominican Museum of Man.

The eastern and western coasts of the park are lined by rocky shores interspersed with sandy beaches dominated by coconut palms. On the western coast, beaches serve as important tourist attractions. Mangroves and coastal lagoons are found on the southern coast north of the Catuano Channel. Barrier reefs dominated by the coral *Acrophora palmata* are located to the south of Saona Island, with additional coral formations located in the waters to the east and west of the park. The marine waters surrounding the park are not currently protected.

Studies to date have registered 148 species of birds, among them the endangered white-crowned pigeon *(Columba leucocephala)* and the Hispaniolan parrot *(Amazona ventralis)*; 17 mammal species, including the *solenodonte (Solenodon paradoxus)* and the *hutia (Plagiodontia aedium)*, both endangered and endemic to Hispaniola; 27 reptile species, including green, leatherback, hawksbill, and loggerhead sea turtles (*Chelonia mydas, Dermochelys coriacea, Eretmochelys imbricata,* and *Caretta caretta*), and the endangered rhinoceros iguana (*Cyclura cornuta*); and 8 amphibians. Also present in the park are an abundance of vascular plant species, representing 106 families and 572 species, of which 53 are endemic to Hispaniola, 484 native, and 35 introduced (Abreu and Guerrero 1997).

A management plan for Del Este National Park was published in 1980, stating that park objectives are: the protection of native flora and fauna and their habitats; protection of areas of scenic, geological, and historic or archaeological interest; recovery of areas altered by human activity; conservation of marine ecosystems; provision of facilities for research, interpretation, education, and recreation; development of the community of Mano Juan, in harmony with its environment and resources; and integration and development of the park within the general development of the province of Altagracia (DNP 1980).

The management plan proposed modification of the park's original boundaries in order to exclude an area of agricultural cultivation in the northwest area of the park and include the Catuano Channel (Bahía Catalinita) and a marine zone extending five hundred meters from the coasts of the peninsula and of Saona Island. However, these proposed modifications were never implemented. The plan also proposed zoning within the park to provide for several categories of use: Intangible Zones (mangroves in the area of La Caldera and the nesting area of the Paloma Coronita, to be used only for research), Primitive Zones (forest areas to be managed for moderate public use), Extensive Use Zones (buffer zones between communities and more strictly protected areas), Intensive Use Zones (beaches for tourism use), Historic-Cultural Zones (primarily caverns), Recovery Zones, and Special Use Zones (Saona, park administration, and services). Zoning guidelines have been observed, although regulations governing human use have in some instances been only weakly enforced.

Authority for the management and protection of all parks and protected areas in the Dominican Republic is vested in the National Parks Directorate (Dirección Nacional de Parques, or DNP), an autonomous agency. However, by decrees of the president and/or agreements with DNP, several nongovernment organizations (NGOs) have been granted management and protection responsibilities in specific protected areas. To implement the PiP in Del Este National Park, the Parks Directorate has since 1993 signed yearly agreements with a Dominican NGO, PRONATURA (The Integrated Fund for Nature), and with The Nature Conservancy, for cooperation in management and fund-raising for the park. In late 1994, a board of trustees, or *patronato,* was created by executive decree to serve as an advisory body to the park. Operating rules for the patronato were approved in October 1995.

Land and Resource Tenure

Most of the land area now included within DENP was donated to the Dominican government by the Gulf and Western Corporation, with the condition that a national park be created on the donated lands. However, the declaration of the park was accompanied by the expropriation without compensation of several additional land parcels, including three communal landholdings, twenty-two individual landholdings, and eight commercial landholdings. In 1976, two parcels and 293 hectares lying within a third parcel were excluded from the expropriation. Among the lands exempted were those belonging to the resort Club Dominicus.

The owners of the remaining landholdings were given the right to remain provided they did not expand the area under agricultural cultivation or engage in cattle raising. The 1980 management plan for the park estimated the human population within the peninsular portion of the park at twelve families, or a total of sixty individuals, located in Gran Chorra in the eastern zone of the park, with an additional eight families, or a total of thirty-five individuals, residing in Las Palmillas in the western zone. Land use in the peninsular area of the park included thirteen agricultural plots with a total area of 200 *tareas* (1 tarea = 625 m^2); seventeen areas dedicated to livestock production, with a total area of 2,200 tareas; eleven small coconut plantations, or *cocotales;* and thirty-seven honey producers with a total of fifty apiaries (DNP 1980). Resident families, formerly concentrated in the settlements of Guaraguao, Gran Chorra, and Las Palmillas, gradually abandoned their lands as a result of land-use restrictions and lack of infrastructure, and guard stations are now located at these sites.

The human population of Saona Island was estimated in the 1980 management plan at a total of 370 individuals concentrated in the community of Mano Juan, with 5 families located in Punta Catuano at the post of the Marina de Guerra; the 1994 population was estimated at 324 during the course of a rapid ecological assessment conducted by several organizations with the support of PRONATURA, the Conservancy, and DNP (Abreu and Guerrero 1997). Residents of Saona were permitted to remain on the island but were similarly faced with severe restrictions on land use. The construction of permanent buildings and infrastructure on the island has been prohibited since the park's creation, leading to a shortage of housing to accommodate intrinsic population growth or tourism development. Expansion of lands dedicated to agricultural and livestock production is also prohibited. In the 1980 management plan, land use on the island was estimated to include forty-two agricultural plots with an area of 861 tareas; 450 head of free-ranging cattle; an unknown number of areas dedicated to coconut production; and five apiaries (DNP 1980).

Communities adjoining the park include Bayahibe, located northwest of the park with a population of 582, and Boca de Yuma, northeast of the park with a population of 1,210. Padre Nuestro, northwest of the park, is a small settlement of some 30 individuals employed primarily in charcoal production. Most of the agricultural lands bordering the northern limit of the park are owned by a large private agroindustrial complex, the Central

La Romana, for the production of sugar cane and cattle. The small landholdings of surrounding communities are used for subsistence agriculture and domestic livestock, and to a lesser degree the cultivation of fruit trees. In the 1994 socioeconomic survey, 83.61 percent of the population in Bayahibe and 57.40 percent in Boca de Yuma reported that they owned neither land nor livestock. Only 6.01 percent of the population surveyed in Bayahibe and 4.69 percent of those surveyed in Boca de Yuma owned land, while 10.38 percent and 37.91 percent, respectively, owned livestock (Abreu and Guerrero 1997).

Coconut and honey production on areas cleared before the creation of the park were authorized under the 1980 management plan, and continue, although the producers now reside outside the park. In addition, several tourism concessions have been granted to tour organizers located outside the region of the park to bring visitors to beach areas. Concessionaires are not permitted to construct permanent structures, but in some instances they have developed shelters and other temporary infrastructures to serve visitors. The Parks Directorate has recently begun to charge concession fees of US$5.00 per visitor; previously, tour operators paid fixed annual fees based on an estimate of visitation.

The presence of a resident community on Saona Island, and land and resource shortages in regions north of the park, imply that resource use within the boundaries of the park will be ongoing and significant. However, with the exception of coconut plantations, apiaries, and tourism concessions, there are no true buffer zones established for Del Este National Park in which local communities are granted rights to resource extraction or other low-impact activities. Furthermore, official attitudes toward authorized productive activities by residents of Saona and nearby communities have been ambiguous, leading to uncertainty regarding future rights of tenure and resource use. For example, following administrative and personnel changes within the Parks Directorate, the park administration began in 1995 to remove ovens and other infrastructure located within the coconut plantations, storage sheds located within apiaries, and domestic livestock on Saona Island as well as the mainland as a means of limiting their impacts within the park. The recently formed patronato subsequently indicated that it would reverse that action. Such ambiguity in land and resource tenure hinders efforts to engage local communities in dialogue regarding sustainable resource use or to propose alternative conservation and development projects.

Resource Use

Agriculture continues to be one of the most important economic activities in the region of the park, involving approximately 10 percent of the economically active population. In Bayahibe and Boca de Yuma, 34.4 percent and 24.5 percent, respectively, of the populations surveyed in a 1994 socioeconomic analysis reported agriculture as their primary economic activity. In the community of Mano Juan on the island of Saona, crop and livestock production have declined sharply as a result of restrictions imposed on land use within the park's boundaries; in 1994, 7 percent of the economically active population reported agriculture as their primary source of income, with an additional 5 percent reporting livestock raising (Abreu and Guerrero 1997). Many residents have taken to fishing as the only available alternative or chosen to migrate to La Romana to seek greater economic opportunities.

Illegal land clearing within the park for small agricultural plots, or *conucos,* and agricultural and livestock production by resident and neighboring communities continue on a small scale. Estimates of the area of park lands cleared for this purpose increased from 2.10 square kilometers in 1984 to 5.52 square kilometers in 1994 (Cano 1993; Abreu and Guerrero 1997). The rural poor and landless also make limited use of the park's resources for collection of firewood and charcoal production. Firewood and charcoal are used for cooking by 15 percent of the households in Bayahibe, 23 percent in Saona, and 56 percent in Boca de Yuma (Abreu and Guerrero 1997). In addition, wood from the park is used for the construction of fences, fish traps, and other small-scale uses, and several local plant species are collected for medicinal use. Residents of Boca de Yuma and San Rafael de Yuma frequently engage in subsistence hunting and the collection of crabs, wild pigs, white-crowned pigeons, and sea turtle eggs. They also hunt commercially and collect parrots, iguanas, sea turtle eggs, and orchids. Hunting and collecting are accompanied by negative human impacts such as fires, small-scale forest extraction, and vandalism (Abreu and Guerrero 1997).

When Del Este National Park was created, several coconut plantations and ovens for the production of copra and balanced feed existed within the park boundaries. In most cases, their owners were granted concessions to continue working the coconut plantations, again with the conditions that production be limited to existing groves and that plantings not be expanded. In addition, several residents of Boca de Yuma derive a significant portion of their income from apiaries. The majority of these apiaries are located within the park boundaries with the permission of the parks administration, under the condition that areas cleared for beekeeping not be expanded. Producers each own an average of some thirty to forty boxes and produce honey, wax, and, to a lesser extent, pollen, for sale to large-scale dealers in Santo Domingo and for export. In late 1994, the spraying of pesticides for mosquito control by a local resort resulted in a massive bee die-off, severely affecting production in 1994 and 1995.

Fishing continues to follow agriculture as the second-most-important economic activity in the region, with 40 percent of the economically active population surveyed in Saona, 30 percent in Bayahibe, and 22 percent in Boca de Yuma reporting fishing as their primary economic activity (Abreu and Guerrero 1997). The most important fisheries species in the area include snapper, grunt, file fish, grouper, tuna, lobster, queen conch, and crabs. Fishing techniques in order of importance are hook-and-line, trap, harpoon, scuba and free diving, net, and longline. Most fishing is conducted from small wooden boats, or *yola,* equipped with sails, oars, and/or outboard motors of 5–6 hp. Because of equipment limitations, fishing trips are restricted to a single day (Pugibet 1995; Abreu and Guerrero 1997).

There are no cooperatives or fishermen's associations in any of the communities within or bordering DENP. Fishermen operate either independently with their own boat, or are contracted by local fisheries seafood dealers. In a 1995 survey of sixty-one fishermen in the area of the park (twenty-seven of whom were from Saona), 53 percent reported that they owned their own boats, while 26 percent rented and the remaining 21 percent used boats under contract with dealers (Pugibet 1995). There are four seafood dealers located in Boca de Yuma, two in Saona, and none in Bayahibe. These vendors equip a number of local fishermen in return for exclusive rights to purchase the catch at a reduced price, or buy directly from fishermen within the community. Those in La Romana also equip fishermen, visit the area to purchase the catch of independent fishermen, or purchase products transported by local fishers. There are no refrigeration or warehouse facilities available to local fishers; the catch may be stored on ice for short periods but is transported almost immediately to La Romana or Santo Domingo.

Fishermen in Bayahibe, Boca de Yuma, and Saona report that fisheries resources are abundant, but that they have declined significantly. Of the sixty-one fishermen surveyed in 1995, 85 percent reported that capture had declined in recent years. Of those, 64 percent believed that the cause of that decline was an increase in the number of fishermen (Pugibet 1995). Although many residents of Bayahibe have stopped fishing, a growing proportion of the residents of Boca de Yuma and Saona pursue fishing as their primary economic activity. In addition, overexploitation of fisheries in other areas has prompted an influx of fishers from La Romana and other nearby ports to waters surrounding the park, although no conflicts have been reported to date. Reports of overfishing are supported by observations of biologists who report the virtual absence of large fish in the region of the park (Abreu and Guerrero 1997).

Dominican legislation (Law No. 67) prohibits sport and commercial fishing within natural protected areas, except as authorized by the Parks Directorate on the basis of prior research demonstrating that the activity will not cause any ecological alteration. However, the marine waters surrounding Del Este National Park have not formally been included within the park's boundaries. Thus, although land-use restrictions within the park appear to be pushing an increasing proportion of resident and neighboring communities into fishing, there have been no government-sponsored studies to evaluate fisheries activities, stocks, or impacts. Although fishers active in the area are required to conform to nationwide fishing regulations, such as a closed season for lobster, there are no marine patrols to ensure compliance, and there are no entry limits or special restrictions for fishers within the area of DENP.

It is noteworthy that yola are also used for commercial transport, to conduct foreign visitors into the park, and for the transportation of illegal emigrants to Puerto Rico, particularly from the community of Boca de Yuma. In some cases, illegal emigrants take refuge within the park while awaiting transportation. Thus, in addition to fishing, emigration is an important economic alternative for human populations in the area of the park, where unemployment is currently 28 percent of the economically active population (Abreu and Guerrero 1997).

Tourism to Del Este National Park has grown rapidly, from virtually none until the mid 1980s to 10,333 visitors in 1989 and 85,062 visitors in 1993, according to official figures. Evaluations conducted in 1994–95 as part of a rapid ecological assessment, however, estimate the number of visitors to Saona at approximately 300,000 per year (Troncoso 1995). The majority of the visitors entering the park do so as part of package tours to the park's sandy beaches that are offered by hotels and resorts outside the area. Visitation is typically for a single day.

Most tourists are attracted by the recreational potential of the area's beaches and waters without necessarily being aware of the broader ecological values of the area or even of the park's existence. Visitors enter the park directly by boat from Bayahibe, Club Dominicus, or the nearby port of La Romana, or are transported to Bayahibe by bus from surrounding tourist zones in order to be transported by boats owned primarily by tourism concessionaires. The park's visitor center, located at Guaraguao, offers a collection of the park's common flora and fauna and topographical maps but is seldom visited due to difficult access and its location outside the tourist zone. In addition, three to four tour buses pass through the community of Boca de Yuma daily in order to visit the nearby Cueva de Berna, situated on private lands rather than within the park.

Tourism concessionaires charge some US$30 to $35 per person for package tours and until recently were required to pay a concession fee based on an estimate provided by the concessionaire of the number of visitors entering the park. In 1993, assuming total visitation of 85,062, these concession fees generated some RD$1.3 million, of a nationwide total of some RD$1.7 million, or US$136,000, in revenues for the Parks Directorate (Lladó 1994). However, the revenues generated by tourism concession fees have not been used to develop additional park infrastructure, as revenues obtained from DENP were returned to the Parks Directorate and distributed among protected areas nationwide. The patronato has declared that future revenues will be reinvested in the park.

Residents in the communities of Boca de Yuma, Saona, and Bayahibe complain that tourism offers few economic benefits for residents. Not only do visitors typically return to their point of origin at the end of the day, but tourism packages for visitors include transportation, food, and beverages, severely limiting the benefits for local hotels and cabins (*cabañas*), restaurants, and businesses. On the island of Saona, there are no hotel facilities, as the construction of additional personal structures on the island has been prohibited since the park's creation; with internal population growth, current population already greatly exceeds available housing capacity. One hotel was recently constructed by an Italian investor next to the park guard station at Mano Juan, but it remains closed due to park restrictions. Three restaurants have been established in Mano Juan, offering seafood products purchased from local fishers.

In addition, the community of Mano Juan constructed a row of beachfront stalls for the sale of handicrafts by local women, but tour groups stay too briefly in the community to permit more than sporadic sales. Moreover, the handicrafts offered for sale are brought from outside the community, increasing their price without adding to locally generated benefits. Additional stalls have been constructed at Punta Catuano. At present, the only locally collected or produced items offered for sale are shells, mainly of queen conch.

The community of Bayahibe has benefited more than any other from the increase in tourist visitation. According to the 1994 socioeconomic survey, 14.5 percent of Bayahibe residents reported tourism as their primary economic activity, compared to 2.0 percent in Boca de Yuma. If the number of persons owning tourism facilities or providing cleaning services in local cabins are added to this percentage, it rises to 30 percent in Bayahibe and 17 percent in Mano Juan (Abreu and Guerrero 1997). In Bayahibe, recent years have witnessed the construction of new cabins and improvements to local restaurant facilities. Some of these offer their own package trips to the national park and to other nearby attractions, such as the Altos de Chavón in nearby La Romana, which hosts the Regional Archaeology Museum (Museo Regional de Arqueología). In addition, most local fishermen have left fishing and now operate small outboard-engine motorboats owned by tourism concessionaires. A few local fishermen have acquired their own motorboats with which to transport visitors into the park; these may also offer their services to tourism concessions. A handful of foreign investors have purchased or constructed new properties within the area.

Tourism activity in Bayahibe has, however, been limited by the lack of infrastructure; there are currently no paved roads in the community, no electrical service, and no water service. Electricity is provided by solar panels or small electric generators, and fresh water is obtained from a nearby spring. Possibilities for future infrastructure development are limited because the community is situated on lands belonging to the Central La Romana, which also owns tourism facilities in La Romana.

The rapid increase in park visitation has already been accompanied by a number of observed or suspected ecological impacts, many of them associated with the heavy traffic of motorized craft and tourists along the western coastal zone of the park. Concentrated visitation has occasioned chemical and noise pollution, disturbance to benthic communities, disturbance of mangroves and nesting areas, accumulation of solid waste on concession beaches, and inadequate treatment and disposal of human wastes. Enforcement even within this relatively small area of the park has been inadequate due to limited personnel and resources.

Organizational Roles

The National Parks Directorate has signed agreements with a number of conservation NGOs for collaborative management or the provision of resources to national parks and other protected areas (UNIDOS 1995). The Integrated Fund for Nature, a consortium of seventeen conservation and sustainable development NGOs in the Dominican Republic, has signed agreements relating to two national parks, Jaragua National Park and DENP. Both agreements allow for joint execution of the PiP by DNP, the PRONATURA, and the Conservancy, and for NGO support to some conservation and management activities. Ongoing coordination of PiP projects is implemented through trimestral meetings. Among the objectives of the agreement is the commitment of each of the parties to work to obtain additional funding and to secure long-term funding for the project. In both cases, the agreements have incorporated other local NGOs. In Del Este National Park, two local NGOs have been created recently and have received assistance from PRONATURA and the PiP but to date have not begun to play a major role in project implementation. Beginning in 1995–96, greater participation by the Conservationist Society of La Romana and by the marine NGO the Dominican Foundation for the Study and Conservation of Marine Resources (MAMMA), which will implement marine projects in the park, has been planned. The recent emergence of the patronato as an active force in park management will continue to encourage this trend (D. Marte, pers. comm., 1995).

Since the initiation of the PiP in Del Este National Park in 1993, PRONATURA has begun to address several of the most critical limitations to park protection, including the shortage of park infrastructure and personnel, lack of baseline and socioeconomic data, and rapid growth in fishing and tourism activity. Infrastructure development funded by PiP includes the construction of three observation towers with stations underneath and solar energy units to power radios and lighting, installation of a radio communications system in the three observation towers and five preexisting guard stations, acquisition of a motorcycle for use by park guards, the construction of twenty-seven boundary markers along the northern border of the park, and placement of advisory signs in the four main entrances to the parks. Funding from the government of Japan to the PiP was provided for the purchase of two motorboats for marine patrols, while the Spanish Agency for International Cooperation donated a third. PiP provides funding for the maintenance and operation of vehicles, equipment, and infrastructure and for training of park guards and volunteers. PRONATURA also supports the maintenance of infrastructure, which was provided to the park through assistance from the Spanish Agency for International Cooperation in 1991–93, including the visitor center in Guaraguao, ten reflecting signs indicating the entrance to the park, an inventory of flora and fauna found within the park, and land-use and topographical maps (Cano 1993).

Several organizations supported by PRONATURA, the Conservancy, and the National Aquarium, with funding from PiP, also conducted a rapid ecological assessment in 1994, with separate components for socioeconomics and tourism as well as terrestrial and marine ecology (Lladó 1994, Pugibet 1995, and Troncoso 1995). The assessment not only contributed to the availability of ecological information, but also updated background information on human communities and resource use needed for future project development. Following the findings and recommendations of the assessment, a fisheries survey was completed in early 1995 by the National Aquarium in Santo Domingo, and a number of activities were initiated to address the importance of the park's marine areas and activities. Additional funding was obtained to establish an ongoing marine monitoring program. The PiP is currently covering the salary of a marine biologist within PRONATURA to assist with the program, and cooperation with the John G. Shedd Aquarium in Chicago, the National Aquarium in Baltimore, and the University of Miami will provide additional research support in future years. PRONATURA has also recently reached an agreement with MAMMA, a member of the PRONATURA consortium, which will assume responsibility for ongoing marine research, regulation, and community education and outreach.

The Nature Conservancy has provided support to build upon the ecotourism evaluation to develop an ecotourism plan and has suggested regulatory framework for the park. A draft ecotourism management plan elaborated in 1995 with funding from the Conservancy is currently under review by the Parks Directorate. The draft plan proposes the zoning of the current area of the park into Buffer Zones; Public Use Zones, composed of areas with guard posts, beach concessions, and trails; Zones of Special Protection, for protection of nesting sites and other critical habitat areas; and Zones of Complete Protection, covering more than 90 percent of the area of the park, which, despite their name, would permit fishing, apiculture, collection of coconuts, controlled ecotourism, and research. The plan also proposes a number of improvements needed to develop ecotourism activities, including training of park guards in order for them to serve as guides, further development of trails and camping areas, development of interpretative materials, regulation of tour boat velocity and entry into mangroves, construction of dry latrines at public visitation sites, and development of a waste recycling system (Troncoso 1995).

PRONATURA and PiP have thus contributed significantly to park protection by providing the infrastructure and most of the resources needed for the operations of existing park personnel. However, decision-making authority continues to be centralized in the Parks Directorate, and the patronato is too recent in origin to allow predictions of its success. Consequently, the success of NGO efforts is ultimately dependent on approval, support, and implementation by other bodies. An official commitment either to this process in the present, or to the continuation of project work in the long term, was not evident prior to the emergence of the patronato. For example, although implementation of measures to ensure park protection depends in large part on available personnel, the number of park guards employed by the Parks Directorate actually declined during the project period, from twenty-two in 1993 to fifteen in 1995. Low salaries for park staff have prevented the hiring of administrators and guards with adequate training and experience, contributed to poor morale, and led to frequent staff turnover that dilutes the effectiveness of NGO-supported training programs. Operation of infrastructure, vehicles, and boats acquired during the project period formerly depended on funding provided through PiP, although this responsibility has now been assumed by the patronato. Similar problems have been experienced in developing relationships with human communities, as discussed in the following sections.

Given this political context, one of the most important roles of PRONATURA and PiP may be to facilitate projects designed to provide the types of data and information likely to alter official attitudes toward park management. For example, information made available through the rapid ecological assessment suggests that more in-depth studies of the livelihood strategies and diversified resource use of human communities within and bordering the park could help to illustrate the need for local participation and administrative flexibility, suggest the likely consequences of prohibiting or limiting a given activity, and provide the information necessary for the development of ecologically and economically viable alternatives. Similarly, information gained from studies of tourism within the park has pointed to the inadequacy of DNP visitor registration and revenue collection. Follow-up analysis could examine the potential for permitting regulated use of the park's resources by communities within and adjacent to the park, explore options to create meaningful buffer zones in order to encourage local participation and control the impacts of resource use, or develop suggestions for improving the capture of revenues.

Linkages between Park and Buffer Areas

Pressures arising from the presence of human communities both within and bordering Del Este National Park make the establishment of linkages an issue of great importance in the administration of the park. However, community participation and environmental education have received little attention to date. There has been no local involvement in park creation or in the development of management and operational plans, and no effort to provide educational materials to schools in and around the park or to adults. The creation of the patronato raises the possibility that this situation may be rectified, but the future direction of park management remains an open question.

Many local residents benefit directly and indirectly from the resources protected by the existence of the park. These people include beekeepers, coconut producers, fishermen, and those employed in tourism. Park guards are also hired from within local communities. Some locals have served as guards for a decade or more. However, in general, low salaries and inadequate resources and training have led to problems of staff turnover and low morale. Moreover, many residents have experienced or are aware of conflicts with the park's administration over resource use. Some of the individuals removed from or forced to leave the area now included within the park's boundaries still reside in nearby communities, particularly Boca de Yuma. Recent actions by the Parks Directorate to further restrict resource use within the park have worsened local tensions.

Furthermore, although residents of Saona and nearby communities are aware of the benefits offered by tourism to the park, their attitudes toward the park are often negative because the benefits of tourism are captured by outside interests. When asked for suggestions regarding measures to increase the local benefits from tourism, residents respond that the government should act to ensure that more tourism revenues remain within the communities. The benefits of park tourism thus appear to be outweighed not only by the attitude that the park exists for the benefit of others, but also by perceptions that the Parks Directorate could and should correct this situation but refuses to do so.

Residents of Saona Island feel particularly prejudiced by the park's creation. In the 1994 survey, for example, 88 percent of the respondents indicated that the prohibition of housing construction was the principal social problem facing residents. Not only is this issue directly linked to the existence of the park, but residents believe that this policy indicates the intention of the Parks Directorate to relocate residents (Abreu and Guerrero 1997), engendering further distrust and resentment. Limitations in education and health services and the shortage of land available for agricultural production were also commonly identified by residents as important social problems; again, both the lack of government-provided infrastructure and limitations on productive activities are linked to the existence of the park. In some cases, lack of infrastructure development may directly hinder park conservation; for example, sanitation and solid-waste disposal facilities are greatly needed on the island to protect water quality. In other cases, benefits gained by restricting development may be offset by tenure insecurity created among local residents, which undermines efforts to develop conservation and development projects on the island.

Attitudes toward the park vary widely among and within local communities. Among respondents to the 1994 socioeconomic survey, 65 percent were indifferent, while only 26 percent believed that the existence of the park benefited them, and 9 percent believed that the park harmed them. Interestingly, 93 percent believed that it was possible to live in harmony with the park (Abreu and Guerrero 1997). The latter finding suggests that many residents may be receptive to initiatives designed to develop more constructive linkages between the park and buffer areas, particularly if residents feel that efforts are being made to involve them more directly in the park's benefits.

Community participation is among the objectives of PiP, and some efforts by PRONATURA to involve local residents in decision making and outreach programs and to develop community-based projects have been planned. However, these efforts have been hindered by negative attitudes toward the park among local communities, particularly on Saona Island, and by the lack of local organization. Although PRONATURA had planned to promote the formation of local ecological societies in Bayahibe and Saona in 1994–95, relations with these communities and indifferent and negative attitudes toward the park in Bayahibe and Saona, respectively, led the NGO to postpone such activities. In the absence of social or producers' organizations on Saona that would facilitate interaction with communities, NGO projects and work plans have been developed without input from local residents. Furthermore, the potential for promising conservation and development projects in apiaries and coconut groves was diminished by recent and drastic actions by the DNP. Given the present political climate, PRONATURA's ability to overcome the effects of past DNP policies toward local communities is limited.

Nonetheless, PRONATURA plans to begin working with the residents of Saona in environmental education and small-scale productive activities. In cooperation with the Conservationist Society of La Romana (CSLR), the organization plans to foster the establishment of an Ecological Committee for Saona Island, design and implement training courses for the members of the committee, and identify and implement two sustainable development activities using coconut by-products for Saona inhabitants.

Efforts by PRONATURA to strengthen constituencies outside the immediate area of the park have been more successful. The NGO has supported the creation and activities of two local ecological groups, the Yumera Ecological Society of San Rafael de Yuma and the CSLR. The activities of these groups are oriented primarily toward environmental education, the Sociedad Ecológica Yumera conducting talks on environmental themes in San Rafael de Yuma and neighboring communities, and the CSLR having recently developed a ten-minute television program for local channels. In addition, the two groups have collaborated in the formation and training of a volunteer corps of park guards for Del Este National Park, with funding from the UNDP GEF Small Grants Program through Espeleogrupo of Santo Domingo, Inc. The corps consists of thirty volunteers who initiated work in late 1994 and who will be available to assist park guards in beach cleanups and other activities requiring additional labor. The organization of these constituencies is likely to contribute significantly to long-term park protection by expanding regional capacity to monitor and contribute to the effectiveness of park management.

Conflict Management and Resolution

Local organization in each of the communities within and bordering the park is extremely weak and lacking in associations of producers or providers of services. To date, the formation of organizations or networks that could assist in conflict management or resolution has not been promoted. Therefore, despite the existence of several conflicts over resource use between local residents and the parks administration, there is presently no forum for their resolution.

The patronato for Del Este National Park is currently the most promising vehicle available for local participation in decision making and for the management of conflicts among competing interests. Provision for local participation could include representation by mayors of neighboring communities and by local ecological societies as well as leading research and conservation interests. Reformulation of the composition of the patronato could also serve as an incentive for the organization of additional stakeholder associations, for example by providing for future participation by producer organizations, tour guide associations, and other local interests.

Large-Scale Threats

Important threats to park conservation include land clearing for agriculture, grazing, and firewood and charcoal production, particularly on Saona Island and on the southeastern portion of the mainland area of the park. The absence of physical demarcation of park boundaries and insufficient enforcement infrastructure have facilitated intrusions into the park, although PiP efforts to demarcate the northern border and improve the patrolling capacity of park guards appear to have improved the situation. In some instances, lack of regulation of apicultural and copra production activities has permitted unauthorized clearing, construction, and fires. In addition, exotic species introduced for agricultural and livestock use and as personal pets compete with native species and in some instances have altered or destroyed natural plant communities. Hunting has contributed to population declines of the endangered white-crowned pigeon, Hispaniola parrot, and rhinoceros iguana, while overfishing has contributed to the diminishing abundance of fish, lobster, queen conch, and other mollusks (Abreu and Guerrero 1997).

Although tourism may offer the greatest possibilities for generating alternative employment opportunities and resources necessary for effective long-term park protection, it may also pose the most severe long-term threat to the park. Rapidly increasing visitation within Del Este has occurred as part of a nationwide tourism boom that accelerated sharply after 1992 and now contributes some 15 percent of the gross domestic product (Lladó 1994). Del Este is the most frequently visited protected area in the Dominican Republic, generating more revenues than all other parks and reserves combined, so that there are few incentives for government agencies to limit visitation to the park. Furthermore, past growth has occurred with little planning and has emphasized large-scale resort development. An Ecotourism Program exists within the Parks Directorate, but the program appears to be oriented more toward attracting greater numbers of visitors than toward limiting and redirecting existing levels of visitation, despite indications that the park's carrying capacity has already been reached and enforcement of park regulations is already clearly inadequate. The existence of private lands on the coastline adjacent to the park, rapid development in Bayahibe, and new construction currently underway at Dominicus suggest likely future threats to resources within the park from continued tourism development outside the park boundaries.

National Policy Framework

Both the national system of protected natural areas and the agency that administers them were created by Law No. 67 of 1974, which remains in effect (Congreso Nacional 1974). The law created three types of natural protected areas: national parks, natural scientific reserves, and botanical gardens. To date, sixteen national parks, eight scientific reserves, one marine sanctuary, and one scenic route have been included in the system. The law establishes strict controls on human activities within these areas, prohibiting, for example, extraction of animal and plant specimens and products except for purposes of scientific research; agriculture, grazing, apiculture, and commercial activity; collection or extraction of marine, geological, or archaeological objects; and sport and commercial fishing, except under permit from the directorate. The law also prohibits the cession of lands to or establishment of installations by persons, groups, or organizations.

Although Law No. 67 clearly intends the strict protection of these areas except for scientific, cultural, and recreational use, in practice nearly all protected areas have been established with resident human communities, are affected by resource use within their boundaries by neighboring communities, or both. The development of means to address this reality has progressed on an ad hoc basis, hindered by political and economic limitations such as frequent administrative and staff turnover, lack of adequate funding, and irregular disbursement of allocated revenues.

With regard to Del Este National Park, the lack of a clear and consistent approach to dealing with human communities in and around the park has been one of the principal problems facing park administration. Incoming administrators have been forced to rely on the conflicting mandates provided under Law No. 67, the 1980 management plan, or established practice as the basis for their actions, creating uncertainty regarding future resource tenure, use rights, and even residence. Within this context, the provision of funding and other support by PRONATURA and external donors has resulted in significant improvements in park administration, maintenance, and protection. However, staffing levels remain insufficient and park guard salaries remain low, prompting frequent turnover, and vehicle operation and equipment maintenance continue to be funded by the PiP. Nor have the necessary steps been taken to reduce dependence on external funding for park protection. Del Este now provides more tourism revenue than all other parks combined, but concession fees generated in the park have been dispersed throughout the protected-areas system, thus reducing their impact on this park.

In 1995, by Presidential Decree No. 309-95, the six protected-areas categories of the World Conservation Union were adopted as guidelines for the Dominican National System of Natural Protected Areas. The decree also charged the National Parks Directorate with evaluating and recommending any needed changes in the classification of existing protected areas. This development may allow for the future development of more consistent criteria for protected-areas creation and regulation.

The success of NGO efforts to improve park management, conservation, and linkages with local communities is highly dependent on approval, support, and implementation by other bodies. To date, NGO efforts to influence protected-areas policy and encourage its institutionalization have achieved limited success. However, the creation of the patronato and the passage of new legislation establishing guidelines for protected areas and their management open new opportunities for NGO participation in decision making affecting the long-term direction of protected-areas management.

Conclusion

The way in which DENP was established has undoubtedly complicated park management, although it is noteworthy that the boundaries were redrawn to exclude a resort. Yet the prohibitions placed on residents within the park, both on livelihood opportunities and infrastructure development, such as new house construction, have led to some hostility toward park management. It also appears that many local residents may undervalue the benefits they derive from the park. However, it is significant that despite these restrictions and the tensions that have existed between communities and park managers, 93 percent of residents believe that it is possible to live in harmony with the park. While the magnitude of tourism is such that it could provide substantial benefits for park management and local communities, it has not been organized in such a way as to substantially benefit either. Tourism to DENP does appear to offer substantial economic benefits to those outside the area. First, substantial benefits accrue to the tour operators and their employees who bring tourists from nearby complexes. Second, other national parks throughout the Dominican Republic are subsidized by the revenues from DENP. Tourism to the park is both the park's greatest potential ally and its greatest potential threat. Policy changes will be required to allow for park management to better capture and manage revenues. The information and experience provided by PRONATURA and the Conservancy through PiP may therefore prove extremely valuable to efforts to redirect official policies toward human communities, tourism development, and infrastructure and resource planning in Del Este National Park.

Postscript

The board of trustees, or patronato, was created in late 1994 to serve as an advisory board to the park and was given management authority over the park in 1996. This change in management was initially supported by the Dominican conservation community, in order to strengthen the park's integration with regional development plans and long-term protection. However, a great deal of controversy and conflict stemmed from the patronato's centralized management style and lack of collaboration. Eventually, the National Parks Directorate relinquished its role in Del Este's management. The NGO community became wary of the patronato and cynical of its intentions as chief administrator of the park that generates the greatest tourism revenues in the country. Due to the turmoil that developed around the patronato's role in Del Este's management, a split occurred among staff and board members of EcoRomana (a conservation NGO based in La Romana). The split resulted in the creation of EcoParque (Friends of Del Este Park, or Amigos de Parque del Este), the current primary local NGO partner of USAID and The Nature Conservancy.

Over time, all conservation NGOs opposed the patronato's management of the park or wanted its role and function to be modified. In May of 1997, the newly elected president of the Dominican Republic signed a decree that returned the park to the National Parks Directorate. Although the patronato was not dissolved officially, it was supposed to serve as an advisory body and relinquish park administration. It no longer controls any of the park's finances. EcoParque is seeking methods for involving local participation in park management as well.

In April of 1997, the National Parks Directorate, PRONATURA, and the Unión Dominicana de Voluntarios Incorporados (UNIDOS), a Dominican voluntary organization, held a workshop at which they presented and reviewed the newly published Integrated Ecological Assessment of Del Este National Park. This study investigates biological, socioeconomic, and tourism issues for both marine and terrestrial areas of DENP and provides recommendations for future resource protection. The workshop brought together many stakeholders, including the PiP implementation team, government representatives, and NGO members of the Dominican conservation movement, as well as residents of towns and cities around the park's periphery. The primary threats to the park were reviewed, and proposed management recommendations were discussed. It was widely agreed that no other Dominican park has the benefit of such comprehensive information and that the National Parks Directorate should take advantage of this information and work diligently to improve the situation. A great deal of hope was placed in the new leadership that has recently arrived throughout the government's environmental sector. Specifically, it is hoped that there will be improvements in park management throughout the country and increased collaboration with the NGO community.

Efforts continue in the development of a five-year management and financial plan for the park. Fiscal year 1998 will be dedicated largely to increasing the self-sufficiency of Ecoparque and positioning it as a strong liaison between the National Parks Directorate and local communities. Discussions are still being led by UNIDOS to modify the primary laws governing the establishment, management, and use of Dominican protected areas. These laws are in great need of modernization in order to handle the challenges and opportunities presented by the next century. Parque del Este still has poorly defined marine boundaries and no form of government–NGO comanagement exists. The physical aspect of the park within its boundaries remains the same, and changes within the communities in the Park have been minimal.

Glossary

cocotales—small coconut plantations

conucos—small agricultural plots

DENP—Del Este National Park.

DNP—Dirección Nacional de Parques (National Parks Directorate)

Espeleogrupo—Sociedad Dominicana de Espeleología (Dominican Society of Speleology)

MAMMA—Fundación Dominicana Pro-Investigación y Conservación de los Recursos Marinos, Inc. (Dominican Foundation for the Study and Conservation of Marine Resources)

patronato—board of trustees

PRONATURA—Fondo Integrado Pro Naturaleza (Integrated Fund for Nature)

Sistema Nacional de Areas Naturales Protegidas—National System of Natural Protected Areas

Sociedad Conservacionista de La Romana—Conservationist Society of La Romana
Sociedad Ecológica Yumera—Yumera Ecological Society
tarea—625 m^2 or 6.25 percent of a hectare
UNDP GEF—United Nations Development Program Global Environment Facility
UNIDOS—Unión Dominicana de Voluntarios Incorporados (Union of Dominican Volunteers, Incorporated)
yola—small, wooden sailing boat

Rio Bravo Conservation and Management Area

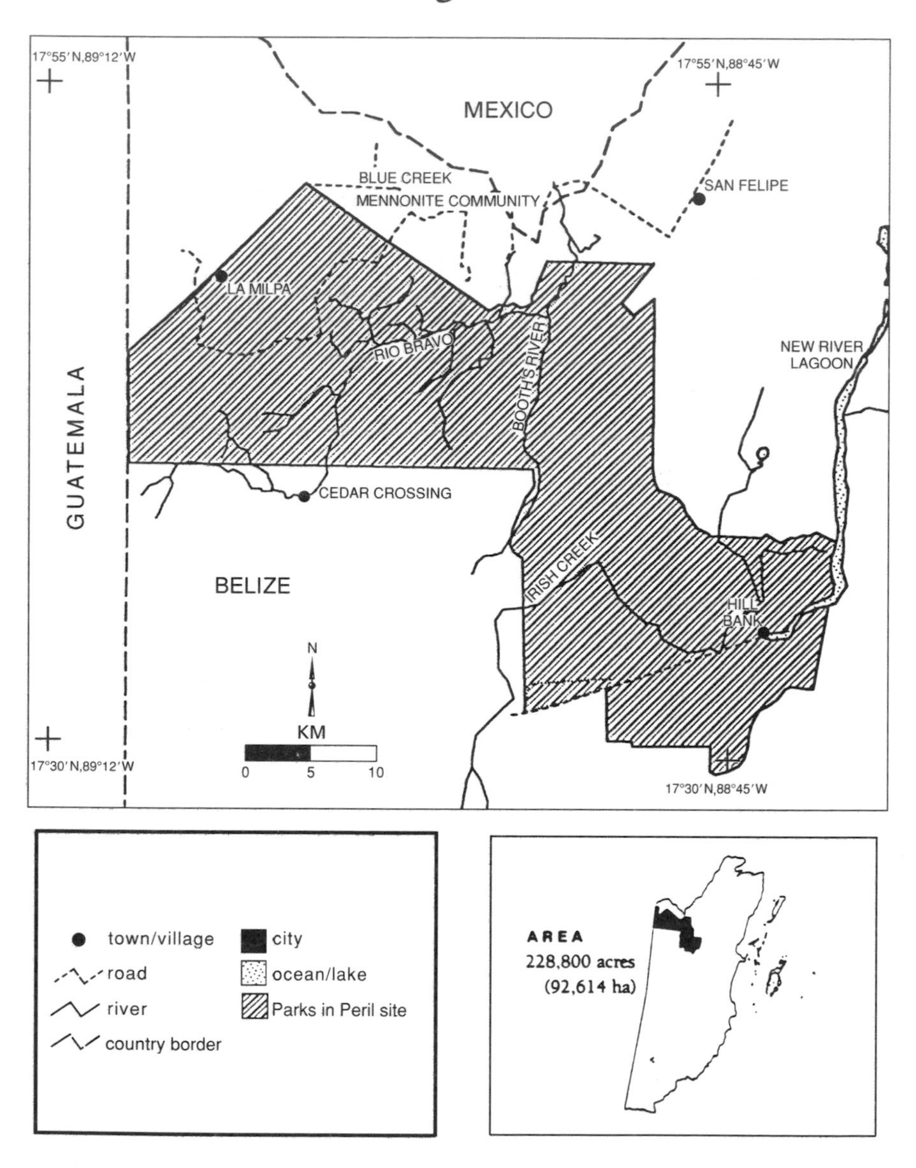

8. Belize: Rio Bravo Conservation and Management Area

Audrey Wallace and Lisa Naughton-Treves

The number and variety of nongovernment organizations (NGOs) working to conserve biological diversity in the Neotropics is without parallel. A novel institutional arrangement is illustrated by the Rio Bravo Conservation and Management Area (RBCMA) in Belize, where the NGO Programme for Belize (PfB) legally owns and manages a large (228,800 acres, or 92,614 ha) and biologically rich area. The government of Belize formally authorized PfB to manage this area (equivalent to 4 percent of Belize's terrestrial area) for the public good in perpetuity. By describing how PfB achieved stewardship of this land, and their progress to date in protecting the site, we hope to illustrate the conditions necessary for NGOs to successfully assume responsibility for protecting parks elsewhere.

Programme for Belize defines the central management goals for RBCMA as follows (PfB 1995):

- preservation of the natural heritage and biological diversity of Belize
- production of sufficient cash return from sensible utilization of the area to pay for its perpetual care
- participation in the proper economic development of the greater Rio Bravo area, including surrounding population centers, to further the national economic interest.

RBCMA contains an exceptional variety of vegetation types including hardwood forest, savanna, and wetlands, as well as diverse aquatic habitats. Within RBCMA one also finds land suitable for agriculture. This is a rare feature for a protected area; most parks are located on marginal or unproductive land. The forest of RBCMA is recognized as having the richest stock of mahogany in Belize. One of the most extraordinary features of RBCMA is that it contains an intact wildlife community with long-term viability, including populations of large and rare species.

Relative to other protected areas in Central America, RBCMA currently faces only minor human pressure on its resources. Forest cover is still extensive in the region, and human population density is low. Relations with surrounding communities are by and large neutral or positive. Recognizing the promising conditions of the site, several international organizations have contributed an impressive level of support to PfB. This support includes comprehensive biological field inventories, surveys, and monitoring. These physical, social, and financial conditions create a favorable conservation context for RBCMA, one that should allow PfB to rigorously test popular but vague concepts in the sustainable use of biodiversity. Indeed, PfB is initiating projects in natural forest management, sustainable harvest of nontimber products, and ecotourism. Having secured the protection of RBCMA (at least in the short term), PfB is now challenged to plan for future population growth and land shortages by developing land-use practices that reconcile economic development and biodiversity conservation.

Park Establishment and Management

The Rio Bravo Conservation and Management Area was established in 1989 as a result of "the land deal of the century" involving a Belizean entrepreneur, international corporations, and Belizean and international conservation activists (Spearhead 1987). RBCMA was once part of an enormous forested landholding (approximately 700,000 acres, 280,000 ha, or 12 percent of terrestrial Belize) belonging to the British company Belize Estate and Produce Company (BEC) (Spearhead 1987). BEC exploited the forest primarily for mahogany and cedar for over one hundred years, until it abandoned extraction activities in the early 1980s and sold its entire property to a private individual who rapidly sold it again to a Belizean entrepreneur named Barry Bowen. Faced with a heavy debt burden, Bowen subdivided the land into three sections, retaining one section (Gallon Jug) and selling two others (232,000 acres each) to Yalbac Ranch and Cattle Company of Houston, and Coca-Cola Foods Ltd. (also known as Refreshment Product Services). At that time, Coca-Cola Foods Ltd. divested some of its surplus lands. One parcel was donated to the government. Another parcel was sold to New River Enterprises, which already held timber rights to the land. A third parcel was sold to a member of the Blue Creek Mennonite Community.

In 1988, the Massachusetts Audubon Society (MAS) began searching northern Belize for land to purchase in order to protect Nearctic migrant birds. Coca-Cola donated 42,000 acres of marsh, forest, and savanna habitat for protection. This initiative prompted MAS to undertake the Programme for Belize to manage the land, entitled Rio Bravo Conservation and Management Area. PfB subsequently began working with The Nature Conservancy and other conservation organizations to raise funds to purchase a further 110,000 acres from Bowen. The 42,000 acres were not actually signed over to PfB until 1990. PfB became independent from MAS in 1988.

In 1989, Bowen further subdivided his land, selling 110,000 acres to PfB for US$3.5 million. Meanwhile, Coca-Cola Foods abandoned its plan to grow citrus in the area, succumbing to pressures from international and national conservationists. In addition to those pressures, Belizean citrus producers lobbied their government to constrain Coca-Cola's activities, as they threatened to overwhelm national interests (Spearhead 1987). Coca-Cola eventually donated 50,000 acres and $50,000 to PfB in 1992. The Conservancy acted as a facilitator and banker for this transaction for tax purposes. Shortly thereafter, RBCMA was consolidated into a single block of land with the purchase of an additional 28,800 acres from the logging company New River Enterprises.

The creation of a large, private reserve in less than five years was possible due in part to the previous regime of large-scale land speculation by foreign companies in the area. What turned this colonial climate into positive conservation achievement was the activism and leadership of Belizean and international conservationists, the substantial legal and technical support of the Conservancy and others, the financial donations of individuals and corporations, and the favorable attitude of the Belizean government toward conservation action by NGOs.

Current PfB documents state that "the phase of opportunistic land purchase has . . . been passed" and that RBCMA is an adequate size for the prescribed management objectives [PfB 1996]. However, additional, adjacent properties are being considered for acquisition or comanagement due to their strategic location. Aguas Turbias National Park (3,540 ha) lies between PfB land and the Mexican border and is highly vulnerable to timber theft, as is Sunnyside Farms (15,390 acres, also located along the RBCMA border). To the south of RBCMA lies land owned by New River Enterprises (26,890 acres), which contains a central-western Belizean forest variant of seasonally moist, lowland, evergreen broadleaf forest, which is poorly represented in national protected areas, and which also forms a corridor to protected areas within Belize Valley wetlands. Purchasing this land would increase the debt burden of PfB and might provoke charges that PfB is constraining economic activity over an excessive area of land. PfB favors the alternative strategy of creating management arrangements or associations with neighboring landowners that enhance biodiversity protection while allowing economic activities.

Land and Resource Tenure

The land tenure system of Belize was shaped by British colonialism. Some historians have concluded that the imposition of British extractive interests (focused primarily on timber) led to the formation of a "monopolistic structure of land ownership and distribution." Even as recently as 1977, 3 percent of landowners owned 95 percent of freehold land, and 90 percent of freehold land was owned by foreigners (Bolland and Shoman 1977). Other analysts emphasize that Belizean history has produced a society that respects private property and an individual's freedom to decide how to manage his or her land (G. Likes, pers. comm., 1995). Setting aside political interpretation, a paramount feature of land tenure in Belize is that it evolved under conditions of low population density and minimal competition for land.

Land and Resource Tenure within RBCMA

RBCMA is privately owned by PfB as a freehold property in trust for the Belizean people. Management of this trust property is subject to the terms of a formal agreement signed by PfB and the Belizean government in 1988 and amended in 1991. These terms are very broad in their definition, authorizing PfB to carry out projects for "the proper development and conservation of natural resources . . . including forest, fisheries and wildlife" in conformity with government policy (Trust Deed 1988, p. 1). Within the formal agreement, PfB states that economic development is to be incorporated into conservation activities for the site (Trust Deed 1988). (PfB is exempt from duties due to a development concession and from land taxes due to "trust" status. All incomes derived from economic activity on the property—i.e., fees., product sales, etc.—must be reinvested in the management of the area. Donations received by PfB are similarly used. There is also a mechanism for PfB to contribute part of its revenue to the government of Belize to support conservation throughout the country.)

Within the trust deed of 1988, PfB was obligated to transfer its freehold property "at the expiration of a period of ten years from the date hereof [1988]" to the government of Belize (Trust Deed 1988, p. 7, Second Schedule). This obligation has been lifted in recent agreements, which indicate that the trust is held by PfB indefinitely and cannot be alienated unless PfB were to misuse its trust (PfB 1996).

PfB adheres to certain restrictions regarding extractive activities within RBCMA that correspond to how a certain area was obtained (see previous section, "Park Establishment," p. 218). The majority of the 110,000 acres acquired via donations from Adopt-an-Acre, United Kingdom, USAID, and others are being managed in the spirit of the donation—pure protection. However, the outer perimeter of the 110,000-acre tract does allow for controlled nontimber harvest as a buffer. Meanwhile, certain lands donated by Coca-Cola (92,000 acres) are designated for natural forest management and other extractive activities because the terms of the donation expressly state alternative economic development and experimentation as management goals.

While PfB owns the land within RBCMA, it does not hold property rights to any water body, or to Mayan antiquities or artifacts on site. PfB does not own rights to subsurface deposits or mineral rights for RBCMA (see "Large-Scale Threats," p. 242). PfB holds hunting rights but has not granted permission for any hunting; however, its authority does not extend to rivers.

Land and Resource Tenure around RBCMA

PfB defines five patterns of land ownership around RBCMA, which correspond to unique socioeconomic groupings and land-use practices: large private holdings, government reserves, Mennonite communities, Mestizo communities, and Creole (Afro-Caribbean) communities. The discussion that follows is organized by the same categories.

Large private holdings. At one time, all of the large properties neighboring RBCMA were part of the Belize Estate and Produce Company (BEC; see "Park Establishment," p. 218). Today they share the characteristics of extensive land use and small resident populations. To the south of RBCMA lies the 130,000-acre parcel owned by Gallon Jug Agroindustry, Ltd., which operates an exclusive ecotourism lodge and leaves most of its land under forest. Further south still is Yalbac Ranch and

Cattle Company (200,000 acres), a mixed management area of citrus and logging. Sunny Side Farms is owned by a trust and has been selectively logged. To the east, New River Enterprises owns 24,300 acres, which it uses for logging.

Government reserves. To the north lies Aguas Turbias National Park, which is subject to frequent timber theft and illegal hunting. Lamanai is an archaeological reserve. As one of the three primary Mayan artifact sites in the country, Lamanai attracts a considerable number of tourists. Crabcatcher Lagoon is used by the Belize Defense Force as a bombing range.

Mennonite communities. Immediately to the north of RBCMA lies Blue Creek, a community of approximately fifteen hundred Mennonites who practice mechanized agriculture. Each family holds formal title to a large farm. Recently, they have begun posting No Trespassing and No Hunting signs on their property boundaries.

Mestizo communities. Beyond Blue Creek and extending approximately thirty miles to the north lies a population of 9,450 Spanish-speaking agriculturalists, 12 percent of whom are recent immigrants. The closest neighbors of RBCMA from these communities are those living in San Felipe, which lies five miles from the RBCMA boundary at its nearest point and eighteen miles by road to the North Gate. Most agriculturalists lease the land they farm, and many work seasonally or regularly as wage laborers. In a separate area near New River Lagoon are three villages with 384 inhabitants, over half of whom are recent immigrants. Here residents are primarily subsistence farmers. Their living conditions are worse than those of other mestizo communities in the area.

Creole (Afro-Caribbean) communities. To the southeast of RBCMA in the Belize Valley area lie ten small, English-speaking Creole communities (total population less than 2,000). These villages date back to the first penetration up the Belize River, at least to the eighteenth century, if not earlier, when what are now villages were logging camps and way stations. These were originally established during the logging and chicle (natural gum) boom in the early 1900s. Currently, people here practice small-scale, subsistence farming on leased or freehold land.

Resource Use

The ecological viability of RBCMA depends in large part on the condition of surrounding land. Fortunately, RBCMA is not an island of protected forest in a degraded landscape, but rather forms part of the largest remaining tract of Central American forest. Currently, no intensive resource extraction or agricultural activities are occurring in RBCMA or its buffer zone (defined as a three kilometer region lying inside the area's perimeter). This is an optimal arrangement for the maintenance of ecosystem processes (e.g., disturbance and recovery cycles) and the conservation of viable populations of large and rare species. In the future, population growth, the intensification of agriculture, and increased timber extraction threaten this ecosystem. Recognizing this threat, PfB hopes to identify and promote economic activities that are compatible with biodiversity protection. We will first discuss land-use practices in the neighboring areas before we describe PfB plans for sustainable resource extraction within portions of the reserve.

Resource Use around RBCMA

Large private landholdings. Gallon Jug Agroindustry devotes a portion of its land to logging, cattle raising, and agriculture, but the majority of the land is forested and is likely to remain so considering that ecotourism is a priority activity. Yalbac Ranch and Cattle Company engages in citrus production and selectively logs its land but leaves most of its property under forest.

New River Enterprises (NRE) has been logging near RBCMA since 1989. Currently, the majority of the timber comes from the 24,300 acres. For most of the past five years, NRE primarily extracted valuable hardwoods, such as mahogany and cedar. Recently, it has diversified its harvest to include numerous hardwood and softwood species and has expanded its processing techniques to include plywood, furniture, and doors. NRE is currently the largest wood-processing company in Belize. Seventy-five percent of its production is for internal consumption; the rest is exported, primarily to Mexico. The volume of timber currently processed averages one truckload a day (this is a rough figure, as production is seasonal). NRE also purchases timber from private individuals and processes it at a rate of approximately one truckload a week.

Of great concern is the imminent exhaustion of large valuable species within the NRE property. So far, the company does not replant, but it practices liberation thinning, directional felling, and careful use of skidders, while leaving forest intact along river courses. Managers claim their sawyers are technically skilled, thanks to their previous employment with BEC. NRE intends to put approximately twelve hundred hectares (three thousand acres) into a rotation trial of clear-cutting and replanting with softwoods, aiming at a harvest cycle of ten to fifteen years.

NRE currently employs 125 people but hopes to expand its labor force to 200 by the end of 1995. The company appears to have progressive policies concerning the hiring of women, promotion of household production, and technical training. Moreover, the NRE director sits on the board of Belize Audubon and participates in regional and international initiatives (e.g., Tropical Forestry Action Plan—TFAP) aimed at promoting ecologically and socially sound logging practices. NRE appears to be an environmentally aware logging company. Nevertheless, in a recent interview, a top-ranking NRE official revealed his questionable support for conservation when he stated that PfB could not "incarcerate" the valuable timber of the region forever.

NRE once attempted to participate in the Rainforest Alliance's Smartwood certification program, which guarantees consumers that their products are sourced from well managed forests. They later quit, purportedly because the restrictions were unacceptable to them. The fact that NRE buys logs from independent sources without concern for the source of the logs or the impact of the harvest is likely to preclude certification.

NRE recently acquired a logging concession for Chiquibul in southern Belize via a British Overseas Development Administration (ODA)/TFAP program. NRE worries that it will be difficult to make a profit given the long list ("over eighty") of environmental restrictions on the concession (J. Loskot, pers. comm., 1995) PfB will certainly be interested in how NRE manages the concession, as NRE will play an important role in forest extraction activities at RBCMA (see "Resource Use within RBCMA," p. 227).

Government reserves. The Aguas Turbias National Park is subject to frequent timber theft and illegal hunting. This is a problem characteristic of the border region between Guatemala, Belize, and Mexico. Each country tends to identify the citizens of its neighbors as being responsible for poaching timber or wildlife (B. Romero, pers. comm., 1995). In 1993, approximately four hundred trees were logged from forest within RBCMA property along the Guatemalan boundary. The armed forces of both Guatemala and Belize were called to the site to establish control. This intervention, followed by frequent patrols by PfB rangers, appears to have prevented further timber poaching (see "Transboundary Issues," p. 245).

Mennonites. To the north of RBCMA lies a band of forest approximately one kilometer wide belonging to individual Mennonite farmers at Blue

Creek. Beyond this narrow strip of forest, Mennonite farmers have completely transformed the landscape for intensive agriculture. Using bulldozers, pesticides, herbicides, fertilizer, and a fierce work ethic, they have left scarcely a tree standing. Mennonites produce over half of Belize's maize and rice, and they completely dominate poultry production. The increasing national demand for beans and rice will likely be met by Mennonites neighboring RBCMA.

A stark contrast is visible between Mennonite lands and the adjacent farms of Mexican slash-and-burn subsistence farmers, where forest and milpas (small-holder cornfields) coexist. This contrast presents a strong exception to the rule that poverty alone drives the overexploitation of resources in developing countries. A few Mennonites have also played an intermediary role (that of transport) in illegal timber extraction on the Guatemalan border within RBCMA.

Thus it is initially surprising to learn that the RBCMA manager views the Mennonites as a buffer against extractive activities. The explanation lies in the exceptional self-sufficiency of the Mennonites. As they demand nothing from the government, likewise they demand nothing from RBCMA. They purchase, delineate, and defend their large farms from trespassers. A hunter wishing to poach a peccary in northern RBCMA must first cross miles of private Mennonite farms, a difficult prospect.

PfB has managed to establish a positive dialogue with the Mennonites. It has also proposed organizing exchanges between Blue Creek Mennonites and Amish communities from land-scarce sites in Mbaracayú, Paraguay, and/or from the United States, in the hopes that lessons in wise land stewardship emerge. This effort has been stymied by logistic obstacles and the general reluctance of Mennonites to participate in public dialogue. In the long term, RBCMA would benefit if PfB could somehow influence the Mennonites to reduce the environmental impact of their land-use practices (e.g., use less pesticides, maintain forest corridors, etc.). PfB, through the Carbon Sequestration Pilot Project, is identifying the economic rationale for landowners to retain forest on land unsuitable for clearance—such as much of the land near the RBCMA. In a recent document outlining steps toward improving national protected-area management, PfB and Inter-American Development Bank (IDB) formally recommended the creation of tax incentives for good land stewardship (PfB/IDB, in prep.).

Mestizo communities. To the north of RBCMA, mestizo farmers practice subsistence agriculture and produce sugar cane. Others are employed on Mennonite farms or work seasonally on large sugar cane plantations. Many youth seek employment in urban centers, returning home for weekends. Recent immigrants from Guatemala augment the number of subsistence farmers in the area. If immigration from Guatemala is not controlled, land shortages will become serious, and RBCMA will face increased pressure for access to its arable land (L. Nicolait, pers. comm., 1995).

Creole communities. The residents of the Belize Valley area traditionally earned their income as loggers and later as wage laborers in mechanized rice production. During the 1950s and 1960s two or three villages in Belize Valley were known for their honey production; however, beekeeping in the area declined in the 1970s with the arrival of Africanized bees. In the absence of these activities, residents have turned to subsistence agriculture and livestock production. As in the communities to the north of RBCMA, most youth seek employment in nearby urban centers. Seven of the ten communities in the valley have created a wildlife reserve, the Community Baboon Sanctuary ("baboon" is the local name for the howler monkey). The sanctuary was established using private forested lands that are managed on a voluntary basis according to a set of principles that advocate using the forest in a way that attracts howlers and other species while helping landowners reduce riverbank erosion and reduce cultivation fallow time. However, farmers in the area complain that tourism revenue from the sanctuary is accrued by only a few families in one village (Bermudian Landing).

Other than the case of the sanctuary, natural resources are managed by individuals on lands privately owned or rented. Communal systems of production or marketing are absent. Elders of the valley complain that the youth of the area no longer show a sense of community, and that it is difficult to motivate them to work on community projects.

Some lands near the RBCMA are extensive landholdings of foreign and often absentee landowners, many of whom are from the United States or Taiwan. These lands are often poorly managed; many were purchased in land speculation deals. Inasmuch as these lands have largely remained in more natural states, PfB has not concerned itself. However, as there is increasing land scarcity in Belize, if extensive, foreign-owned lands are not utilized, there could be mounting pressure to improve their utilization.

PfB is aware of the potential positive (e.g., greater local control over land use) and negative (e.g., forest clearing and high-intensity agriculture) impacts that a sudden push to promote use could bring.

Resource Use within RBCMA

RBCMA has been subject to various regimes of resource exploitation for several centuries. Mayan populations settled in the area as early as 800 B.C. By the Classic period (A.D. 600), population growth was rapid (2–3 percent per year) and resource use was both intensive and extensive. Some historians believe that the area was almost entirely cleared of forest, and that Mayan agriculturists used terracing, raised fields, and irrigation systems on a regular basis. The reason for the sudden collapse of the Mayan population in the early ninth century remains a mystery. Mayan communities persisted along waterways, but inland sites were abruptly depopulated.

British companies and Baymen were next to exploit the resources of RBCMA (mid seventeenth to early twentieth century). Their principal activity was cutting logwood, primarily mahogany (*Swietenia macrophylla*) and Mexican cedar (*Cedrela mexicana*). The forests of RBCMA contained the richest supply of mahogany in the country and sustained logging operations for over two hundred years. The average extraction rate was approximately 4,850 trees a year during 1975–81 and reached a peak of 7,950 in 1982, the final year of industrial logging by the British Estate and Produce Company (PfB 1996). During the 1970s and 1980s, RBCMA constituted an open-access site. Looting of archaeological sites was rampant and marijuana production reached industrial scale (PfB 1996). Widespread hunting, chicle and timber theft, smuggling, and squatting occurred throughout the area. The official creation of RBCMA greatly reduced these activities; however, illegal resource extraction still occurs in the area. PfB works to control these activities in two ways: (1) restricting resource access by patrolling the area and confiscating any illegally harvested materials, and (2) substituting illegal resource use with planned, low-impact resource uses.

Hunting and fishing. Hunting of peccary, deer, cracids, and other wildlife is a widespread tradition among the mestizo communities (B. Cruz, pers. comm., 1995). Over recent decades, subsistence hunting has been largely replaced by commercial and sport hunting. Citizens of San

Felipe frequently buy game meat (mostly venison) from commercial hunters, although the game does not necessarily come from RBCMA (G. Dominguez, pers. comm., 1995). PfB rangers report that 90 percent of the poachers they catch in RBCMA are sport hunters who illicitly enter RBCMA in their vehicles, often to hunt in a group (B. Cruz, pers. comm., 1995). Some of these individuals come from as far away as Belize City. Subsistence hunters are few (less than 5 percent), partly because RBCMA is too far from most villages for a hunter to reach by foot (B. Cruz, pers. comm., 1995). In 1995 a sport hunters' association was formed in Orange Walk. This association invited the head ranger of RBCMA to attend its first meeting. Officially, it plans to negotiate hunting access rights on private lands in the area.

As a group, creole communities are less avid hunters than the Maya/mestizo communities (G. Ku, pers. comm., 1995). However, several individuals from Creole and Mennonite communities poach wildlife and capture parrots and parakeets in the region of the east gate. On the other hand, Creole communities are more likely to be found fishing in RBCMA. The villagers from Lemonal, Rancho Dolores, and even as far away as Burrell Boom fish in the New River Lagoon (southeast RBCMA). PfB rangers cannot control fishing within the reserve; all they can do is control access to the water, which they prevent unless a fisherman comes exclusively by boat. The only illegal fishing is for hiccatee, an endangered species of river turtle (*Dermatemys mawei*).

Hunting is prohibited within RBCMA. Upon finding a poacher, PfB rangers confiscate the prey and ammunition and record the identity of the hunter, who is then released. Should an individual be caught a second time, fines and arrests are planned, but as yet no poachers have been encountered twice in RBCMA. Rangers report that after three years of patrolling, hunting has been reduced in the north (i.e., the 44,500-ha [110,000-acre] parcel). In September of 1995, a ranger station was set up on the southeast border of the Rio Bravo, and four new rangers were hired. This increase in the ranger force coupled with an established presence at the east gate is expected to increase control in that area.

Legalized fishing and hunting has not been formally considered in RBCMA. However, some consider it a promising alternative for generating revenue, once harvest rates and control systems are established (B. Cruz and G. Ku, pers. comm., 1995). Particularly for hunting, the difficulties in establishing harvest rates and control systems make authorization a distant prospect.

Chicle and other nontimber forest products. PfB has identified several nontimber forest products from RBCMA of potential commercial value, including chicle, sabal thatch and leaves, essential oils, honey, cohune (*Orbignya cohune*) oil and artifacts, and house plants. Ideally, the sustainable harvest of these products would provide a source of income to finance conservation activities. Achieving sustainable harvest is difficult from both an ecological and an economic perspective. The current biological information is inadequate for determining what harvest rates are ecologically sustainable. Experimental nontimber harvest (chicle and thatch) is most actively pursued in the buffer to the 110,000 acres. Experimental timber harvest is concentrated in the vicinity of Hill Bank. Much work also remains to be done in product development and marketing. PfB has initiated pilot projects to study these products with mixed results.

Chicle extraction was a major industry in Belize during the late 1800s through the mid 1900s, after which time production levels fell due to overexploitation, problems in production quality, and market collapse. Estimates of the best return times for chicle tapping from a typical sapodilla tree vary from five to fifteen years. Seven years is commonly accepted and is being used by PfB as the basis for tapping, or "bleeding," trees. During the period of peak production earlier this century, trees were bled with increasing frequency, reaching cycles of five years or less (P. Cowo, pers. comm., 1995). Overexploitation resulted in lower-quality chicle and tree mortality. Chicle quality also suffered from collectors mixing the sap of other trees with sapodilla sap during the processing. A decline in international demand due to alternative, synthetic sources of gum eventually reduced the industry to a minor activity.

PfB is attempting to resuscitate chicle extraction in RBCMA, where there is an abundant population of sapodilla trees (5.3 mature, productive trees per hectare). PfB entered into a trial business arrangement with a small company, Wildthings, Inc., by which PfB was to supervise the harvest of twenty-five tons of chicle per year and sell it to Wildthings, Inc., which would establish a gum factory on site. Included in the agreement was an option for PfB to purchase shares and eventually achieve a majority holding in the company. The plan ultimately broke down for a number of reasons:

- Controlling the rate and extent of collectors' extractive activities proved difficult.
- To ensure an adequate supply in financial terms, chicle would have to be harvested over a large portion of RBCMA, and this conflicted with other management goals.
- Wildthings, Inc., was not able to raise sufficient capital for the endeavor.

PfB has not abandoned chicle production. Rather, it has tested harvest rates on trial plots and discovered that the chicle is of high quality and that yields of 2½ to 5 tons of chicle per year are possible. Problems persist, however, including the tendency of collectors to overharvest, even when their working areas are defined. Supervision of collectors proved very difficult. Sustainable chicle production is also hampered by the limited capacity to predict yields under variable climatic conditions. Drought seriously reduces chicle production. Finally, marketing chicle has proven difficult; PfB now has a stockpile of unsold chicle.

PfB is currently initiating a European Union (EU)–funded project (Superplants) to produce ornamental plants and native tree seedlings for reforestation. A micropropragation laboratory, based on propagation from tissue culture rather than seeds, will be constructed at Hill Bank. The production goal is set at 750,000 plantlets per year, targeting an international market. The microprop will have purely commercial product lines—not necessarily even Belizean species. PfB hopes the project will be a means to a conservation end—revenue generation—rather than an end in itself. Apart from creating revenues for the reserve, other linkages to the RBCMA are:

- education
- research
- including lines of native RBCMA species for micropropagation, particularly for small, nonthreatening offtake on founder stock
- conservation by propagating species depleted through wild collecting for reestablishment in wild areas (not just RBCMA).

The above-mentioned activities are all oriented to commercial production involving formalized agreements, production goals, etc. PfB is also investigating possibilities for community-based harvest of sabal thatch, cohune oil and artifacts, and honey. An integrated approach is necessary for promoting these activities (e.g., environmental education, community development, etc.). (See "Linkages between Park and Buffer Areas," p. 237).

Forestry. Timber harvesting is an integral part of the history of RBCMA. PfB intends to continue to include timber harvesting in the area's future and is currently developing a major program designed to sustainably harvest timber, called the Rio Bravo Forestry Development Programme. In fact, the budget for the activities listed under this program currently outweigh any other single management activity at RBCMA (excluding payments on land purchases). PfB outlines an ambitious set of objectives for this program, including:

- developing a model for other sites in the region of sustainable forestry
- contributing to local industry and employment
- achieving economic viability
- minimizing impacts on the environment and biodiversity
- minimizing impacts on the global environment (e.g., carbon sequestration)
- reinforcing other management activities on site.

PfB documents vary in their degree of optimism for achieving these goals, which echo changes underway in the Belizean timber industry overall. One key trend is that only recently has the timber industry diversified from a dependency on mahogany. Earlier there was overoptimism and a belief (erroneous but included in official reports, including the Belize TFAP) that the RBCMA was well stocked with mahogany—it is, but only in small-size classes. One document (PfB 1995) describes problems including overly competitive market conditions and an already seriously depleted stock of hardwoods within RBCMA.

PfB's 1995 management plan highlights the opportunity to experiment with sustainable timber harvest. While studies of forest dynamics have been ongoing since 1989, forest inventory appears to be incomplete. Apparently, a visit from a team of consultants (forest industries specialists, natural forest management experts, business planners, and a silviculturalist) convinced PfB to proceed with promoting timber harvesting. Stock surveys began in January 1995 and intensified from November 1995 onward. A reconnaissance inventory conducted in 1975 is still valid and likely to be of substantial use. A three-year pilot project will test technical and operational aspects of forest management regimes in the timber extraction zones. PfB will assess the sustainability of timber extraction largely through the certification process of Smartwood or Woodmark. This experimental timber approach is meant to be a cautious one, designed to demonstrate intent to include timber harvest in RBCMA management and so protect against pressure to allow concessions, while testing the actual potential in the mid and long term. Since the myth of a timber-rich forest prevails, there will always be a belief that PfB is "incarcerating" a valuable resource. That perception is a threat, against which the forestry project is a preemptive action.

The Carbon Sequestration Pilot Project of the Rio Bravo Forestry Development Programme is financed by Wisconsin Electrical Power Company. It has two components:

- financial support to PfB for the purchase and protection of forested land under threat of clearance (power companies can purchase land at $300/ha and claim the carbon in on-site natural vegetation as an offset); and
- carbon sequestration, which is used as a means of financing experimental work to develop sustainable approaches to timber harvest in hardwood forests and to implement techniques for rehabilitating pine stands or degraded savanna.

The logic behind the carbon sequestration component is that simple purchase and set-aside over the extensive areas required to combat climate change is not a viable long-term strategy because social pressure will overwhelm the area unless it is supporting an alternative land use—such as forestry.

The number and diversity of players involved in the Forestry Development Programme is imposing: forest ecologists and field biologists from several research institutions, power companies, international development agencies, conservation NGOs, international certification schemes, and the Belize Forestry Department. Coordinating the activities of these institutions will be challenging. The role of NRE, the only wood-processing factory in the region, will be pivotal.

PfB has previously demonstrated that it is willing to cancel an agreement involving harvest regimes within RBCMA if there is overexploitation or a loss of control—for example, as there was with chicle harvesting during 1994. They also refused US$1.6 million at the last minute from a prestigious U.S.-based foundation for an earlier forestry development scheme when PfB identified conceptual flaws that could have led to unsustainable practice. This experience has informed much of the structure of the present forestry program, including:

- insistence on experimentation in the first phase
- cut-off points
- avoidance of actions that tie management into approaches that may prove unsustainable later
- validation and breadth of input from different quarters.

PfB's willpower may again be tested by one of the many parties involved in the natural forest management experiment. Maintaining ecologically sustainable timber harvest rates in the future is likely to become more difficult as timber resources are exhausted elsewhere and the wood becomes ever more valuable. Several factors indicate that PfB is committed to a cautious approach:

- It continues to obtain technical input from expert forest ecologists.
- It is carefully delineating preservation zones on fragile habitat.
- It has budgeted considerable investment into future research and monitoring of environmental impacts.
- It is planning low-level extraction rates (six trees/ha).

Tourism development. PfB identified ecotourism as a key sustainable development activity in its management plan of 1991. That same year, PfB made an agreement with the Save the Rainforest (STR) organization based in the United States to provide two-week field ecology courses to STR students. STR provided funding for the construction of a dormitory for accommodation, since the only facilities were a central multipurpose building and four staff cabins. This agreement with STR and the construction of the student dormitory were the first development of ecotourism on the Rio Bravo. The average size of a group is twenty-two students, and these courses have been the largest revenue-earning activity conducted by PfB on the Rio Bravo. At the end of fiscal year 1995, total revenues earned from tourism-related activities on the Rio Bravo including gift shop sales, covered 45 percent of the operating expenses of PfB.

PfB also established within the organization a Tourism Development Unit in November of 1993 to market tours in Belize, most of them involving visits to the Rio Bravo. From this source, approximately ten non-STR groups of ten to twelve people each are booked for three- to four-day stays on the Rio Bravo. An encouraging number of individual tourists are also staying one or two nights at the Rio Bravo as part of their travel itinerary in Belize.

In 1993, PfB hired a consulting firm that developed an Interpretive Planning Prospectus, which proposed plans for three areas (in three corresponding phases) of tourism development on the Rio Bravo. Phase I of the plan involves renovations to the existing research station to improve the accommodation and educational facilities. Two new activity areas comprising tourism and research facilities were also conceived. The Hill Bank site in the belly of the New River Lagoon was envisioned as a water resort to be developed in Phase II, with a more remote jungle lodge to be developed in Phase III.

By 1995, PfB had received funding from the National Fish & Wildlife Foundation and private donors for portions of the research station renovation. This money was used for the construction of a new "green" dormitory at the La Milpa Field Station, which features solar power energy and composting toilets. The old dormitory was transformed into an education center with meeting facilities and a reference collection section. Costs for Phase II are estimated at US$1,500,000, requiring an additional working capital of US$250,000, while Phase III would cost approximately US$1,000,000, and an additional US$250,000 of working capital. With borrowed funds from a foundation or other conservation-minded entity, a 3 percent loan over twelve years provides the necessary funding to allow a payback and to create a discretionary cash flow from tourism income of approximately 10–12 percent.

Organizational Roles

Programme for Belize (PfB) is a Belizean, nonprofit organization, established in 1988 "to promote the conservation of the natural heritage of Belize and to promote the wise use of its natural resources" (Trust Deed 1988). PfB focuses almost entirely on managing one site, RBCMA. Included in PfB's management activities are research, conservation education, professional training, community outreach, and traditional protection practices (e.g., patrols by rangers). PfB's abilities to raise funds and its receipt of high-level technical support are reflections of deliberate and highly focused PfB policy and development strategy maintained over many years. PfB devotes a considerable part of its resources to developing and maintaining its own agenda and then soliciting support to pursue it. It acts opportunistically only when the proposed activities mesh with PfB's own plans; as a result, it has occasionally refused involvement with particular projects and donors. Associated with this is a high level of awareness of the need to maintain its own institutional profile while working alongside powerful partner organizations. Another important element is that PfB has actively sought to maintain a range of types of support, including technical, financial, and institutional, with a range of foundations and other agencies. In general terms, PfB is a proactive organization, and that is probably the underlying reason for its relative success.

PfB has also had a policy of providing relatively high wages to attract relatively highly trained Belizean staff with previous experience. A comparison between the staffing and organization of the different government and NGO bodies in protected-area management demonstrates that PfB is different in several respects. First, it has a strong accountancy contingent and a heavy emphasis on business, financial, and administrative backgrounds among the seven staff in the main office. The deficiencies in natural resource management are made up in service. This makes for a strong backup to the field staff, whose numbers are due to double over the next three years, even further accentuating the field–office ratio in favor of the former.

There is widespread debate regarding the appropriate role of an NGO. Many of the activities of PfB are characteristic of conservation NGOs throughout the Americas. What makes PfB atypical is that it is the legal owner of the protected area it manages and defends. Many consider PfB's status as steward and owner of an area to be an ideal arrangement. Indeed, RBCMA is well managed and protected compared to many public parks and reserves in Central America. But before using PfB as a template for privatizing parks, it is essential to first consider how Belizean conditions compare to those of other Central American countries, as well as how PfB's institutional mission and resources compare to those of other NGOs. A list of key factors to address in such a comparison is suggested below.

Conditions unique to Belize:

- Belizean legal structures and policy conditions favor private property (some would say that local land alienation is a tradition [e.g., D. Vernon, pers. comm., 1995]; (see "Park Establishment," p. 218).
- Land is relatively abundant, and the population density is low.
- Ecotourism is a major source of international revenue (100,000 tourists visited Belize in 1994).
- The Belizean government supports NGO participation in protected-area management. (See "National Policy Framework," p. 243).

PfB goals and strategies:

- PfB manages only one site. (At the national level, PfB participates in policy analysis and reform, as well as technical training.)
- PfB maintains a consistent positive relationship with the government via formal and informal communication and working agreements.
- PfB has an outstanding ability to raise funds from international sources.
- PfB covers 33 percent of recurrent expenditures through self-generated income (mainly tourism).
- PfB places twenty-eight out of thirty-five staff in the field.
- PfB has strong internal leadership.
- PfB receives a high level of technical support from advanced field biologists, planners, natural forest management specialists, archaeologists, etc.
- PfB maintains program focus. In fact, it has rejected well-funded donations to work on projects outside RBCMA, e.g., coastal ecosystem protection.

It is also useful to briefly compare the situation of PfB with that of the Belizean Audubon Society (BAS), another well-respected Belizean NGO. BAS comanages seven national parks with the Belizean government. BAS must coordinate operations with the government on a much more detailed basis than PfB. The national parks it manages are vulnerable to future government decrees to reverse its decision to declare an area a national park (O. Salas, pers. comm., 1995). Meanwhile, RBCMA has greater land tenure security. Finally, RBCMA has a considerably bigger budget than any of the seven national parks, or indeed of all seven parks budgets combined.

The fact that PfB owns the land it manages allows it more autonomy than BAS and buffers it from the policy fluctuations currently affecting public lands. However, the land tenure arrangement alone does not guarantee the protection of RBCMA. Privatizing a protected area where funding is scarce, or where the policy environment undermines NGO actions, or where land is scarce and land pressure great would likely result in a less than ideal outcome. Moreover, in most countries there is an inadequate number of stable, competent NGOs to assign to each protected area. Privatizing protected areas is a promising strategy in some contexts but is certainly not a panacea for conservation.

Linkages between Park and Buffer Areas

Buffer zone and outreach activities at RBCMA are best described in two categories:

- extractive activities promoted within RBCMA boundaries aimed at testing models of sustainable resource use, generating employment opportunities, and producing sufficient cash return for PfB to pay for the perpetual care of RBCMA
- outreach activities beyond RBCMA's boundaries designed to educate the public regarding the importance of conservation and build positive relations between RBCMA and neighboring communities, as well as to promote local alternative sources of income, including tourism.

In describing the various activities below, it will become apparent that there is considerable overlap between the activities within and outside of RBCMA.

PfB distinguishes buffer zones as areas of sustainable resource use within RBCMA's boundaries. Even though RBCMA's 100,000 ha (229,000 acres) are contiguous, for land-use planning purposes they are conceptually divided into categories. The broadest categories of management (protection vs. sustainable use) reflect how each parcel of land within RBCMA was acquired. The 45,000 ha (110,00 acres) purchased with Adopt-an-

Acre, United Kingdom, USAID, and other donations are restricted to protection and research activities only (see "Park Establishment," p. 218). Within the 45,000 ha, however, a one-kilometer border area is accorded buffer zone status. (Note: In certain areas this is extended to three kilometers.) All other areas within RBCMA (total 48,000 ha, or 118,000 acres) are potential sites for buffer zone–type activities, such as experimental forestry, the harvest of nontimber forest products, silviculture, and agroforestry. Within the 48,000 ha available for sustainable-use activities, decisions regarding what project will be undertaken where are dictated mainly by habitat variation and management history. For example, within buffer areas, riparian forests or endangered plant communities are off limits for timber harvest.

Within PfB's management plans regarding resource use within RBCMA, the emphasis to date is on planned activities, formally negotiated with existing research and commercial institutions (e.g., natural forest management; see "Resource Use within RBCMA," p. 227). The participation of rural communities in these activities is discussed in terms of employment opportunities. Meanwhile, informal collection of nontimber products by community members themselves (e.g., thatch collection) has received less attention and investment. This is partly due to a low level of demand within neighboring communities for the nontimber products permitted for use (thatch, cohune, honey, etc.).

PfB oversees a diverse set of outreach activities, reflecting the diversity of socioeconomic conditions surrounding RBCMA. One strategy PfB has been employing to create linkages with neighboring communities is to establish a presence in those areas of RBCMA where illegal incursions occur frequently. Such a presence could be in the form of a ranger station, a full-fledged research station, or a project site. A natural extension of this strategy is employment opportunities for the villagers. In addition, however, under PiP, PfB is able to make a deliberate effort to build relationships with its neighbors through community outreach and environmental education programs.

Community outreach remains a critical component of PfB's management plan for the long-term protection and management of the RBCMA. Under PiP, PfB has initiated an outreach project entitled Friends of Rio Bravo with the main goals of educating the communities on the importance of conservation and exposing them to alternative, sustainable economic activities. To date most of the activities have centered on the

Maya/mestizo villages northeast of the reserve. The reason for this focus is quite logical. The first 45,000 ha (110,000 acres) of land acquired for conservation were in this region, and the research station built there was the first official presence of PfB. These communities were the first to be affected by RBCMA, and they were the most accessible.

PfB initially started out working with five villages in the northeast, namely San Felipe, August Pine Ridge, Trinidad, San Lazaro, and Yo Creek (ranging from approximately thirty to fifty kilometers from RBCMA). Later Yo Creek was dropped from outreach activities when PfB staff judged that its residents were not greatly affected by the reserve, and that its close proximity to Orange Walk (the second-largest town in Belize) allowed its residents considerable economic and employment opportunities.

Between 1992 and 1995 the Friends of Rio Bravo project has attempted to provide villagers with skills training to enable them to participate in PfB's economic development projects, as well as develop their own income-generating ventures. The specific activities embarked upon have been chosen by the communities that were consulted to identify their needs. PfB works steadily with various community groups, striving to equip them with the necessary tools for their continuity and independence. The four focus villages to the northeast of RBCMA exhibit various stages of development and levels of social organization. Outreach activities have been most visible in the village of August Pine Ridge, where there are two strong community groups dedicated to their projects and committed to their sustainability. Other villages have not been as promising. In San Lazaro, for example, the village is divided by politics, and community spirit is virtually nonexistent. As a result, outreach activities there are progressing very slowly. The level of interest and dedication from the community groups basically determines how the different projects will evolve.

For the most part, apart from training in sustainable agricultural practices, small farms management, and livestock management, alternative economic activities initially focused on the development of a handicraft business complete with skills training, equipment, and start-up supplies. In addition, a handicraft center was built in August Pine Ridge as a formal site to display and sell the products. Participating local crafts groups were expected to benefit primarily from the flow of tourists visiting Rio Bravo, Lamanai, and Chan Chich Lodge. Unfortunately, the traffic has proven insufficient to maintain a steady flow of income. When asked about their sales record, craft group members from both San Felipe and August Pine Ridge expressed their gratitude to the RBCMA station manager of La Milpa Field Station, because he brought them their only tourists. The craft groups have also sought markets in other parts of Belize with limited success. The challenge facing PfB now on this particular project is to salvage this investment by aggressively seeking markets for these crafts and offering more training to bring the level of craftsmanship to a competitive level.

The transformation of the Hillbank site into a fully operational field station in September 1995 has sparked community outreach activities in nine villages to the southeast of the RBCMA. A ranger station has been established at Hillbank, and three major forestry projects will be based there starting in September 1995. Employment opportunities for the neighboring villages are already evident, and a team of workers from these villages is already working at Hillbank. The new Hill Bank Station manager is from the village of Isabella Bank. The nine villages targeted for community outreach are Bermudian Landing, Rancho Dolores, Lemonal, Isabella Bank, St. Paul's, Willows Bank, Double Head Cabbage, Flowers Bank, and Scotland Half Moon. Outreach activities have already started with four villages in the areas of ecotourism, honey production, and the canning of fruits and jellies.

The history of these villages dates back over seventy years to when logging and rice production were booming businesses. Today, the picture is very different, and the lack of job opportunities in the villages has resulted in most of the employable population working outside the community. The few that reside permanently in the villages are usually subsistence farmers and hunters. The objectives of this community outreach extension are basically the same as in the northern villages:

- to establish community groups to participate in ongoing community projects
- to provide training in areas that will allow them to participate in sustainable development projects and small business management
- to promote the concepts of sustainable development and the interrelationship of conservation and economic development through education and community outreach.

Environmental education goes hand in hand with community outreach, and PfB employs a full-time education coordinator to work with adult groups and student bodies. The overall aim is to create an environmentally aware and educated Belizean public through various media. PfB's environmental education program has grown and now targets both the national school system and international student groups. The primary focus of environmental education, however, is with the communities neighboring RBCMA. At the root of trying to change attitudes and behavior is that one recurring ingredient: education. The main activities associated with environmental education are school presentations and lectures, teacher workshops, field trips to Rio Bravo, and collaborative efforts with other environmental NGOs.

Undoubtedly, the Friends of Rio Bravo outreach program has been successful in establishing solid relationships with the communities where they have come to respect the boundaries and restrictions of RBCMA, as well as to understand the basic concepts of protection of biodiversity and the linkages between conservation and development.

In the area of developing alternative economic activities in the communities, the outreach program has not yet been as successful. The actual number of families benefiting economically from these projects is minimal in relation to the total neighboring population. For example, the population of one of the targeted northeastern villages is roughly two thousand. Participation in craft or dance groups for each village averages less than twenty individuals. Again, these small groups may have a significant impact on local attitudes toward RBCMA; however, their impact on local economies is likely to be minor.

Valuable lessons emerge from PfB's experiences in outreach activities thus far, which can be applied to new sites targeted for community outreach. It is clear that greater research and market analysis have to be conducted before selecting these projects. The realization that some communities have limited capacity for (or interest in) communal production activities is another key factor that must be taken into consideration. Promoting opportunities to participate in communal credit or communal technical training may ultimately be more reasonable than expecting a community to enter into communal production activities (B. Cullerton, pers. comm., 1995).

Finally, outreach activities between 1992 and 1995 have revealed that a substantial investment of time, energy, and skilled personnel is necessary to significantly influence community development. PfB does not have the resources or the mandate to become a development NGO. At some sites, PfB has contracted another NGO, Belize Enterprise for Sustainable Technology (BEST), to offer training in community banking and small farm management. PfB will continue to face the challenge of identifying key sites for outreach work and the development activities most important to the long-term conservation of RBCMA.

Conflict Management and Resolution

The fact that PfB has exclusive property rights over its land simplifies conflict management regarding land tenure or resource use within RBCMA. Within its boundaries, PfB has so far been able to expropriate squatters (see "Resettlement," p. 245), remove poachers, reject any industrial exploitation, etc. However, conflict management and resolution are likely to become more problematic as resources and land become scarce outside RBCMA's boundaries.

The international conflict between Guatemala and Belize over their shared boundary makes portions of RBCMA particularly vulnerable to timber poaching. If the conflict between the two countries dramatically worsened, the vulnerability to illegal resource extraction would also increase (see "Transboundary Issues," p. 245).

Large-Scale Threats

Currently, RBCMA faces no single large-scale, external threat. Oil exploration was carried out within RBCMA in 1991, promising indications were found of fields extending under the RBCMA, and exploratory wells were proposed. Exploration has not been initiated because of the state of the oil market. PfB assumes that at some point in the future, it will be active in at least ensuring good practice in oil extraction in sensitive areas, probably close to but not in RBCMA. However, the government ultimately retains subsurface rights, so hypothetically there is risk to RBCMA if rich oil deposits are discovered in the future.

A more likely threat to RBCMA will be the increasing demand for its valuable hardwoods, although PfB is attempting to thwart this through its sustainable logging program. Maintaining biologically sustainable timber harvest rates in the future is likely to become more difficult as timber resources are exhausted elsewhere and the wood becomes ever more valuable. The forests adjacent to the Guatemalan and Mexican borders are likely to continue to be vulnerable to timber poaching.

Other analysts forecast problems for RBCMA if there are acute land shortages in the future, caused by population growth or uncontrolled immigration (L. Nicolait, pers. comm., 1995). This potential threat was a key justification for inclusion of RBCMA in the Conservancy's Parks in Peril program.

National Policy Framework

The current political climate in Belize favors reducing the size of the public sector. While this trend constrains the proper management of national parks in Belize (e.g., personnel shortages), it is unlikely to directly affect RBCMA. Parallel to the effort to shrink the public sector is the drive to privatize the Belizean economy and portions of the public service sector. Again, RBCMA is unlikely to be directly affected by this shift. However, some professionals working in social welfare and development describe a growing social and economic crisis. Small-scale farmers are losing access to infrastructure and materials and are unable to compete in the market (D. Vernon, pers. comm., 1995; B. Cullerton, pers. comm., 1995). Elsewhere in Central America, the creation of a large class of landless, marginalized peasants has led to invasions of forested lands by individuals forced to carry out spontaneous agrarian reform.

Biodiversity protection in Belize suffers from an absence of a unifying national conservation policy. Management priorities and regulations are variable between protected sites according to circumstance and the institutions involved. Moreover, the institutional framework for biodiversity conservation is diffuse. Three government bodies, each in a separate ministry, have statutory responsibility for protected areas: Ministry of Tourism and Environment, Ministry of Natural Resources, and Ministry of Agriculture (PfB/IDB 1995). The Ministry of Tourism and Environment is described as the most supportive of NGO activities in sustainable development and even allows NGOs access to government funds (B. Cullerton, pers. comm., 1995). Meanwhile, the Ministry of Natural Resources is less cooperative, mainly because it is presently suffering from a leadership crisis.

The Belizean government invites NGO participation in numerous governance activities—e.g., identifying policy priorities, analyzing policy and carrying out legal reform, mediating intersectoral conflicts, and implementing policies. PfB has recently completed an analysis toward defining a National Protected Area Systems Plan (PfB/IDB 1995). The government also solicits technical advice from NGOs. In the field setting, government extension workers often collaborate with NGOs in development projects. NGOs have been participating in these activities for over ten years. The reasons behind the government's pro-NGO approach are debated. Some explain that the government is obliged to assume this approach because it has a severely limited technical and financial capacity (present annual budget: approximately US$175 million). Others would add that the small size of Belize (200,000 citizens) allows for familiarity and cooperation between public and private institutions. Critics argue that in many cases, inviting NGO participation is only a sort of window dressing for public agencies, which ultimately establish policy in a black box (D. Vernon, pers. comm., 1995). Others would dispute this perspective and instead identify problems of implementation.

An area of conflict between conservation NGOs and the government has been the policy of not charging tourist entry fees to protected areas. In 1995 tourists were not charged entry fees in protected areas, which meant that, in effect, protected areas were subsidizing the tourism industry while suffering a severe shortage of funds (PfB/IDB 1995). When conservation NGOs (in consultation with international ecotourism experts) recommended levying a US$12.50 conservation tax on every tourist, the tourism industry was outraged and lobbied the government against the proposal. The debate continued for two or three years until finally, the conservation tax was dropped to $3.75. This tax revenue will be placed in a trust fund (Protected Areas Conservation Trust, or PACT) with investment income directed toward support of conservation activity. The fund will be controlled by a board of trustees composed of representatives of public institutions, NGOs, and private industry. The PACT is intended to augment, not replace, government subventions. Meanwhile, the government has agreed that entry fees to protected areas should be charged, and rates are currently being discussed (ranging from $2.50 to $5.00).

Relations between the government and conservation NGOs in Belize are gradually changing in two ways:

- Roles are becoming more formally defined; (e.g., where before the government approved the management plans of Belize Audubon Society with a handshake, now official contracts are signed (O. Salas, pers. comm., 1995).

- As the conservation movement increasingly emphasizes economic development, government interest and participation are increasing (I. Fabro, pers. comm., 1995).

Transboundary Issues

Timber theft is considered a serious and present-day threat. Much damage can occur within a short time period, and tracking down those responsible can be extremely time consuming. In February and March of 1992, theft was discovered, and by April it had been stopped. The following year the area was more thoroughly explored, and the level of cutting was found to be more extensive than had previously been believed. Along the western border of the park, 423 trees (mostly mahogany) were cut (see "Conflict Management and Resolution," p. 242).

Resettlement

Five families of farmers were present in RBCMA at the time it was established. While they had no title, they claimed to have resided in the area for over ten years. However, they had been asked to leave after five years of residence. Squatter's rights in Belize apply only if no action has been taken. PfB advised them that they would have to leave their farms but offered to pay for the farms and their labor investments (assessed by the Belizean Ministry of Natural Resources) as well as land elsewhere. PfB successfully lobbied to get them title to land of equivalent quality elsewhere. Four families left, and one refused. PfB reiterated its intent to give compensation and took civil action through the courts. The squatter was finally evicted, after several attempts by the local bailiff, whereupon he picked up his compensation check from PfB and moved to the new land allocated to him. Since then there have been no squatters or re-invasions.

Conclusion

The case of Rio Bravo demonstrates how a unique set of circumstances led to the establishment of a large protected area owned by a conservation NGO. RBCMA is well managed and protected compared to many public parks and reserves in Central America. PfB's ownership of the land it manages allows for greater autonomy and buffers PfB from the policy fluctuations currently affecting public parks elsewhere in Belize. However, using the case of Rio Bravo as a template for privatizing parks is inappropriate. One must first consider how Belizean conditions (e.g., low population density) compare to those of other countries, as well as how PfB's institutional mission and resources compare to those of other NGOs.

PfB is embarking on large-scale experimentation in sustainable resource use, with particular emphasis on natural forest management. Establishing ecologically and economically sustainable resource use is an imposing challenge, but several factors indicate that PfB is committed to a cautious approach:

- PfB recruits technical input from expert biologists and ecologists.
- PfB delineates preservation zones on fragile habitat.
- PfB budgets considerable investment into future research and monitoring of environmental impacts.
- PfB aims at low-level extraction rates.

Through PfB outreach activities at RBCMA, considerable achievements have been made in building positive public relations with neighboring communities. The economic impact of community development activities thus far is less substantial, although there is a steady growth of local employment opportunities at RBCMA.

Postscript

Many of the plans that PfB outlined at the time of the site review are now underway. The micropropagation laboratory, which will produce ornamental plants and native tree seedlings for reforestation, was constructed and put into operation in January of 1997. The 1995 PfB management plan, which proposed sustainable timber harvest, has moved well beyond the planning stage. The first extraction took place in April of 1997, and PfB is undergoing a certification assessment by a joint team from Woodmark and Smartwood.

The Carbon Sequestration Pilot Project of the Rio Bravo Forestry Development Programme was initially financed by Wisconsin Electrical Power Company. It has been joined by a number of other utilities: Pacificorp, Detroit Edison, Cinery, and Utilitree (a consortium of power companies). The first component of the land purchase, financial support to PfB for the purchase and protection of forested land under threat of clearance, was completed by December 1995. The second, to finance experimental work to develop sustainable approaches to timber harvest, was underway in 1997.

PfB has instituted an annual forest liaison meeting to allow the diverse stakeholders in the Forestry Development Program to meet. This group now includes forest ecologists, field biologists, power companies, international development agencies, conservation NGOs, international certification schemes, the Belize Forestry Department, and NRE (the only wood-processing factory), as well as representatives from local communities and other local interest groups. Low-level extraction rates, such as 6 trees per hectare, were envisioned in 1995 as necessary to maintain ecologically sustainable timber harvest rates. This planned low level has dropped even further, to 1.5–2 trees per hectare.

The Protected Areas Conservation Trust, with investment income directed toward support of conservation activity, is now fully operational. The leadership crisis at the Ministry of Natural Resources has been resolved.

Glossary

BAS—Belize Audubon Society
BEC—Belize Estate and Produce Company
BEST—Belize Enterprise for Sustainable Technology
chicle—natural gum
MAS—Massachusetts Audubon Society
milpa—small-holder cornfields
NRE—New River Enterprises, Inc. (a logging company)
ODA/TFAP—British Overseas Development Agency/Tropical Forestry Action Plan
PACT—Protected Area Conservation Trust
PfB—Programme for Belize
RBCMA—Rio Bravo Conservation and Management Area
STR—Save the Rainforest

Machalilla National Park

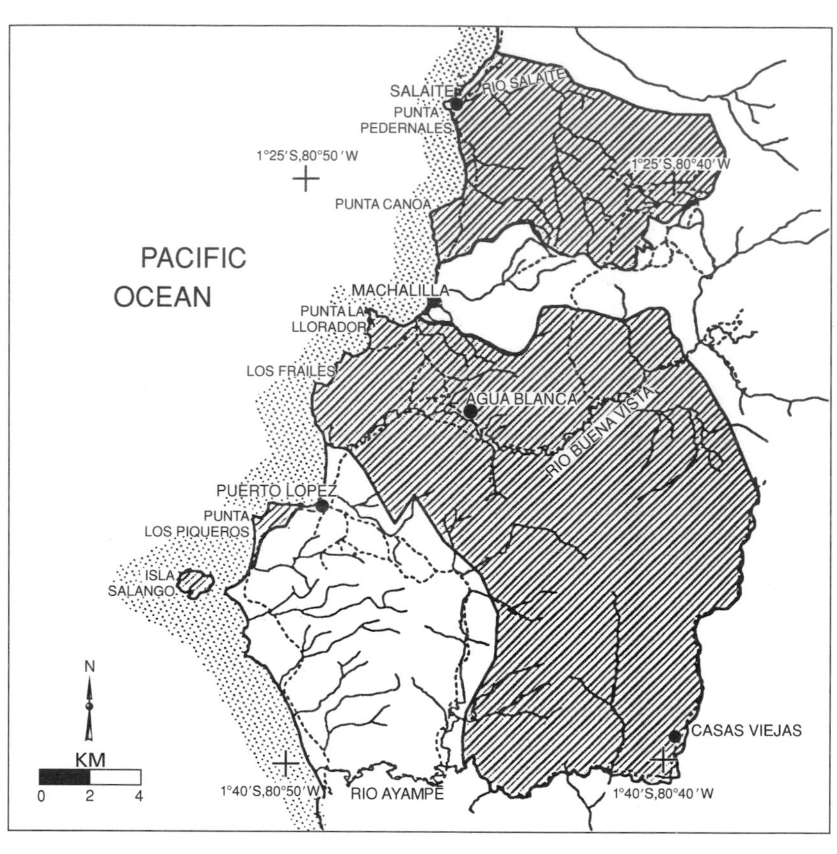

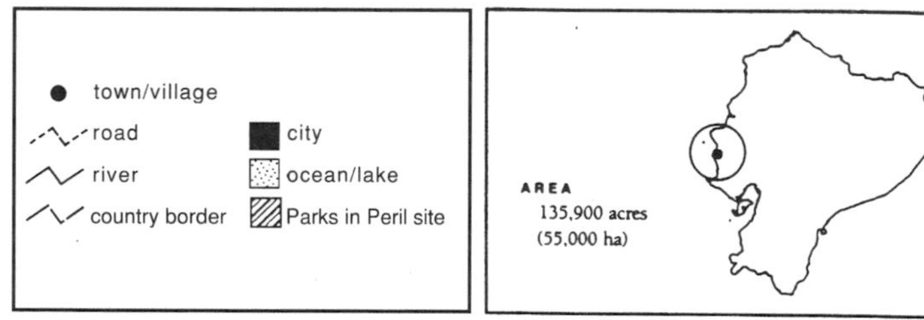

9. Ecuador: Machalilla National Park

Elba Alissie Fiallo and Lisa Naughton-Treves

The forests of western, coastal Ecuador have been almost entirely converted to agriculture and today represent one of the world's most highly threatened ecosystems (Dodson and Gentry 1991). Much of this conversion occurred between 1960 and 1980, leaving less than 6 percent of original lowland forest standing (Parker and Carr 1992).

Machalilla National Park (MNP) was established in 1979 in the coastal province of Manabí to protect remnants of dry and moist forest as well as coastal and marine ecosystems lying within an expanse of fifty-five thousand hectares. While the natural history of the park has scarcely been researched, high levels of endemism have been recorded in birds and plants, and an extraordinary diversity of plant communities has been identified (Parker and Carr 1992). MNP also holds archaeological relics of coastal cultures, dating to 5,000 B.P. and extending through the colonial period (Norton, cited in Arnal 1993). Humans are likely to have exerted long-term influence on the composition and function of coastal ecosystems in this area, both terrestrial and aquatic (Josse and Balshev 1993). Human impact on the regional environment today is characterized by overexploitation of forest and marine resources (Fundación Natura 1992; Coello 1993; Josse and Balshev 1993). Driving the intensification of resource use are forces including infrastructure development, unequal land distribution, changes in technology, and integration with external, industrial markets.

Most national parks in Ecuador face similar conflicts between local resource needs and long-term conservation goals. MNP is exceptional in two regards:

- The park is relatively small and extensively inhabited (roughly sixteen hundred people live within the thirty-nine-thousand-hectare terrestrial zone).

- The entire park ecosystem has been altered by humans to such an extent that restoration measures are required in addition to urgent habitat protection (Parker and Carr 1992).

MNP is also unique due to its high level of tourism; it is second only to the Galápagos Islands in generating tourism revenue for the country. On the one hand, tourism offers possible alternative sources of income for local populations, while on the other, uncontrolled tourism potentially threatens the ecological integrity of the area.

Amid the context of heavy resource use and disappearing natural habitat, park managers face a considerable challenge in defining the appropriate resource rights of communities living within and adjacent to MNP. The issue of community access to park resources has generated intense debate. Social scientists emphasize ethical arguments concerning traditional rights of residents to maintain their homes and call for programs that will alleviate local poverty while promoting a harmonious relationship between the Park and the people (Silva and McEwan, n.d.; Fiallo 1994). Biologists warn that under current regimes of livestock grazing, agriculture, and selective logging, local species extirpation and local extinction of rare plant and animal communities are imminent within the park. Also at risk is basic watershed function (Parker and Carr 1992; C. Josse, pers. comm.; H. Arnal, pers. comm.).

Attempts to achieve a balance between local demands and long-term conservation goals have been constrained by political and economic factors. During the 1980s until as recently as 1992, relations between MNP authorities and local residents were primarily antagonistic and at times resulted in violent confrontation (Cuellar et al. 1992; Fiallo 1994; C. Zambrano, pers. comm.). Despite efforts to enforce park protection, habitat degradation continued during this period. Very recently, a dramatic shift in conservation strategies has occurred in both public agencies and NGOs. Park authorities now avoid public confrontation with residents and instead attempt to control resource use via dialogue and education. Where before park authorities destroyed charcoal ovens and confiscated timber, today they work with technicians to help communities create tree nurseries or improve agricultural production. The NGO Fundación Natura (FN) has also made considerable progress in establishing constructive dialogue between park managers, communities, and regional institutions.

Improving park-community relations opens the possibility of participatory management and is thus essential for the long-term survival of the park. However, the short term survival of species and plant communities remains at risk, particularly in the biologically richest areas of the park (Parker and Carr 1992). Goat eradication has been attempted on the uninhabited Isla de la Plata, with partial success (151 eliminated in 1993–95) (Zambrano 1995). It remains unclear, however, whether habitat disturbance from livestock grazing, logging, or fuelwood collection has significantly diminished on the mainland portion of MNP. Empirical data on rates of resource use are lacking. It is difficult to compare the results of the few existing studies, as their methods and sample size are often vague (see section titled "Resource Use," p. 262). Furthermore, park residents may be reluctant to report personal information on farm size, number of livestock, etc., for fear of taxation or legal penalties (Fiallo 1994). Generally, informal, verbal reports indicate declining rates of hunting and logging in the park, although the rate of decline reported varies with the individual's perspective.

In the following case study, we discuss the progress and obstacles encountered by state and private organizations striving to protect MNP. In documenting the case of Machalilla, we reveal profound dilemmas that arise whenever sustainable development is paired with biodiversity conservation in a fragile environment. For example, the resident population within MNP has traditionally suffered from greater poverty and political marginalization than average as compared to other regions or country wide (Fiallo 1994). These conditions, combined with severe drought, led to a 50 percent decline in population between 1986 and 1992 due to emigration (Cuellar et al. 1992).

It should be noted that current estimates (C. Zambrano, pers. comm., 1995) of the resident population within MNP exceed those of Cuellar et al. by a factor of 2.4 (1,600 compared with 664). In 1992, Sylva and León estimated a population of 5,000. Therefore, it is difficult to make conclusions about the demographics of MNP based on studies that do not clearly define census methods. Moreover, the population level appears to be dynamic, perhaps seasonally and annually. Meanwhile, Ecuadorian Institute of Foresty, Natural Areas, and Wildlife (INEFAN) and FN staff agree that the park population is steadily declining. This attrition reflects an unstated goal of park management authorities (Paucar 1987). Current projects designed to improve the quality of life among resident communities could reverse emigration patterns, potentially imposing renewed pressure on the park's ecosystem. An equally difficult question is how to assist park residents to become economically autonomous while at the same time controlling their traditional extractive activities. Moreover, deciding whom to work with among park residents or neighbors is difficult; budget and time constraints preclude working with all individuals whose behavior affects the park. A diverse approach is warranted because of the sociopolitical and ecological heterogeneity of MNP.

Public authorities and NGO representatives working at MNP are rising to these challenges and difficult decisions. Their experiences provide valuable insight for conservation managers elsewhere. Comparing results of participatory development activities between sites at MNP should permit a rigorous evaluation of the relative costs and benefits of each effort.

Park Establishment and Management

In 1974, a small team of Ecuadorian and international biologists nominated Machalilla to a list of ninety areas within Ecuador deserving conservation attention. Machalilla emerged from this list as a priority site for protection due to the presence of a substantial tract of coastal dry forest (already a threatened habitat in the 1970s), pristine and scenic coastal resources, and a major pre-Columbian archaeological site. After subsequent surveys and legal procedures, Machalilla was declared a national park in 1979. MNP has a total area of 55,095 ha including terrestrial habitat, a coastal zone, and two islands (see map p. 248).

Park authorities at the time deemed it best to avoid including the major towns of Puerto López and Machalilla in the park. Hence MNP was established in three sections, known as the north block (10,570 ha), the south block (29,580 ha), and Sector Los Piqueros (900 ha). (Note: These figures do not include the two-mile marine corridor of the park, which follows the coastline and encircles the two islands of La Plata y Salango.) The resulting, fragmented form of the park runs counter to the guidelines for ideal reserve design prescribed by conservation biologists, as it maximizes edge effects. Yet theoretical discussion of reserve design assumes that there exists an abrupt change in resource use at the edge of a protected area—i.e., intensive use outside the park, minimal disturbance within the park. It is difficult to identify such an abrupt edge at MNP, as grazing, fuelwood extraction, and other resource uses proceed across park boundaries. Likewise, biologically rich fog forest lies just outside the Park, along Río Ayampe. The only physical edges readily identified at MNP are edges between vegetation types created by natural physical conditions, and edges around moist forest remnants at upper altitudes created by recent logging activities. The former edge type is likely to provide important habitat for certain species and enhance biodiversity (Parker and Carr 1992). Edge effects around forest remnants are cause for concern, as they could amplify deleterious effects of habitat loss for species surviving in the remnants.

The boundaries of the park were adjusted in June of 1994. In the northern zone (Salaite), the boundaries were officially increased by 1,125 ha, which improved the integrity of the park since it incorporated the watershed of Río Jipijapa. However, both to the north and to the south of the town of Machalilla and in the southern part of MNP near Salango, the boundaries were reduced. The exact amount of the overall change in land area, given the increase in one area and reduction in the other, is unknown, but park administrators indicate that the total size of the park was increased with these changes.

The park director describes a future plan to connect the northern and southern block of the park via the purchase of a 2,000-ha corridor. Creating a corridor is an important step to reduce habitat fragmentation and allow wildlife greater movement. However, such a plan requires effective control of livestock grazing, fuelwood collection, etc., within the proposed corridor in order to significantly improve biodiversity protection.

Key to understanding current relations between communities and the park is a brief description of how local communities viewed the park establishment process. Park establishment was characterized by a lack of public participation, and local residents had little understanding of the concept of a national park (Fiallo 1994). According to INEFAN staff, community meetings were organized at the time of park establishment, and public relations were by and large peaceful (J. Vizcarra, pers. comm., 1995; A. Ponce, pers. comm., 1995). Fiallo (1994) discovered a very different account of park public relations in her interviews with community members. Park residents viewed the park as an alien, imposing force on their lives and livelihoods. During the period of park establishment many families temporarily migrated to other provinces because of drought. Upon their return, they found their land within park boundaries and park staff attempting to restrict their use of resources and traditional activities. During the next decade, efforts by park staff to prohibit or restrict the use of timber, fuelwood, wildlife, and arable land heightened local resentment of the park. Unfulfilled promises by park staff to bring water, infrastructure, and schools to resident communities have left many residents disillusioned with the park (see "Conflict Management and Resolution," p. 274).

Confusion exists among both official and unofficial reports regarding the government's position on expropriating or purchasing land within the park. A former director of Ecuador's parks states that land acquisition was a consistent goal that was never operationalized due to lack of funds and political will. Others state that it was never seriously considered. Likewise, there is debate over local residents' attitude toward selling their land to the park. Some sources say that many individuals initially wanted to sell their land. Others recall steadfast, and at times, violent local opposition to land purchases or expropriations. Many people want to remain on their lands, but they do not want the land to be part of the park. In all probability, the residents of MNP do not unanimously agree on the desirability of remaining within the park. Indeed, Fiallo found diverse attitudes toward the park among residents according to their age, land tenure, education level, and other attributes.

In the end, despite the illegal status of humans residing in a national park, institutions working at MNP have adopted the policy of allowing people to remain there as an integral element of the ecosystem (see "Land and Resource Tenure" below). Some have suggested that this decision requires a change in the protected-area category of Machalilla from a national park to a multiple-use area. A more likely step will be the establishment of zones within MNP where resource extraction of varying intensities is permitted. Such zoning should consider both biodiversity protection and socioeconomic interests. Some habitat within the park appears to be resilient in the face of human use (e.g. tropical desert scrub). Other habitats, such as the premontane forest at upper altitudes, will require protection and even restoration to persist into the next century.

Land and Resource Tenure

Ecuadorian law states that land within national parks is national patrimony and the sole property of the state. Private property is illegal within parks, and the Forestry Law (article 73) states that at the time of park establishment, land should be expropriated or reverted to state property. Contrary to legal proclamations, nearly all of Ecuador's parks are inhabited by people claiming informal and formal rights to land and resources.

"When communal land is legally recognized, it is recognized by INEFAN, and activities are permitted subject to existing zoning and management categories. In cases where communal land has not been recognized (through official documents), INEFAN does not recognize rights to land which communities claim as theirs" (F. Bucheli, pers. comm., 1995). Within protected areas, there is an unofficial policy in INEFAN to cooperate with *comunas* (community landholding systems) before individual landholders and to recognize the rights of comunas to both the use and the management of the natural resources. The Law of the Comunas (1936) formalized traditional land tenure arrangements in Ecuador by giving comunas power to own and control land as a community and to allocate it to members for their use. Members cannot sell land to outsiders, nor can they rent land without comuna permission (Becker and Gibson 1996).

Where land ownership is contested, it is difficult to enforce regulations prohibiting or restricting resource use. However, throughout the history of Ecuador's national parks, no large-scale projects to expropriate land or evict squatters have been attempted, due to lack of funds and/or unclear policies (A. Ponce, pers. comm.). Fundación Natura questions the merit of land purchases, because such actions:

- create land rushes and speculation
- produce new processes of colonization
- disrupt social structures
- violate traditional rights
- are expensive
- are bureaucratically time-consuming

Within MNP, five basic types of land tenure exist:

- individuals holding formal titles (e.g., in Pueblo Nuevo)
- individuals holding informal documents indicating land ownership prior to park establishment (e.g., in Salaite)
- individuals occupying land without documentation (e.g., also in Salaite)
- titled communal land ownership (e.g., in Agua Blanca)
- untitled communal ownership (e.g., in Casas Viejas).

(Note: Examples are named where the tenure type is found but do not indicate a single tenure form for each site.) Land and resource tenure is further complicated by the fact that several landholdings (both private and communal) span the park boundary.

Detailed and complete data on tenure within MNP are lacking due to absentee owners and a general reluctance on the part of residents to explain or prove their legal status within the park (Cuellar et al. 1992). Estimates of private landholdings within the park range from 20 percent to 75 percent of the terrestrial area (IUCN 1982; Arriaga 1987). The variable estimates have stymied the formation of clear policies on land acquisition and resource use. To remedy this confusion, FN and INEFAN have each sponsored recent studies of tenure in the park. The following figures (unless noted otherwise) come from a 1992 study sponsored by FN (Cuellar et al. 1992). Unfortunately, the study does not always have comparable information for all landholding categories; the information included here is all that is currently known.

In a survey of 163 residents of MNP and its "zone of influence" (which is unfortunately not clearly defined, but is assumed to be adjacent to park boundaries), 115 families held private land (70 percent) and 48 communal (30 percent). (Note: Fiallo [1994] obtained similar figures in 1993.) As shown in figure 9-1, of the private owners surveyed, 12 were absentee owners (10 percent). Each of the absentee owners held legal

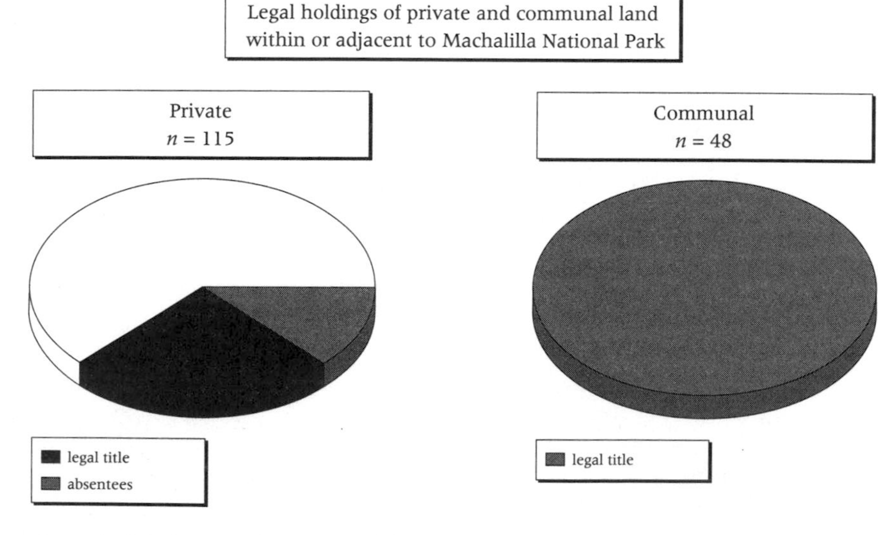

Figure 9-1. Private and communal lands within or adjacent to Machalilla National Park, and legal status ($n = 163$). *Source:* Cuellar et al. 1992.

title, whereas only 44 percent of the residents did ($n = 45$). All communal landholders surveyed held legal titles. Agua Blanca has not had adjudicated all of the communal land that it claims (S. Martínez, pers. comm., 1995).

Regarding farm size, among private residents ($n = 115$), 80 percent held fewer than 20 ha, 11 percent held 20–50 ha, and 9 percent held more than 50 ha., as shown in figure 9-2. The majority of absentee holders held fewer than 20 ha each, but one individual had more than 50 ha. The smaller the farm, the more intensely it was used (Cuellar et al. 1992; Forster 1992). Pueblo Nuevo has 205 ha adjudicated by the Ecuadorian Institute of Agrarian Reform and Colonization (IERAC), 13.5 of which is in the populated zone. One comuna consisting of forty-one families (Agua Blanca) claims ownership of 10,000 ha of park land (approximately 250 ha per family) but has legal title for only 12 ha. Casas Viejas claims ownership of 2,000 ha and El Pital 1,200 ha. Within comunas, families hold private farms in addition to sharing communal lands (S. González, pers. comm., 1995). One problem with the comunas in MNP is that while they are legally recognized, their lands have not been defined. When we asked representatives of three different comunas about land availability for agriculture within the park, all three answered that land was abundant but a shortage of labor or rainfall limited production. These results refute assumptions that all park inhabitants are small landholders (Cuellar et al. 1992). San Francisco is a large farm estate with 905 ha, of which 430 ha. are within the MNP. IERAC recognized and granted them only the portions located outside the park.

Methods of land acquisition varied among park residents. Cuellar et al. (1992) reported that the majority of park residents have inhabited the area between eleven and 41 years (56 percent), 38 percent have been present for over 41 years, and 6 percent less than 10 years, as shown in

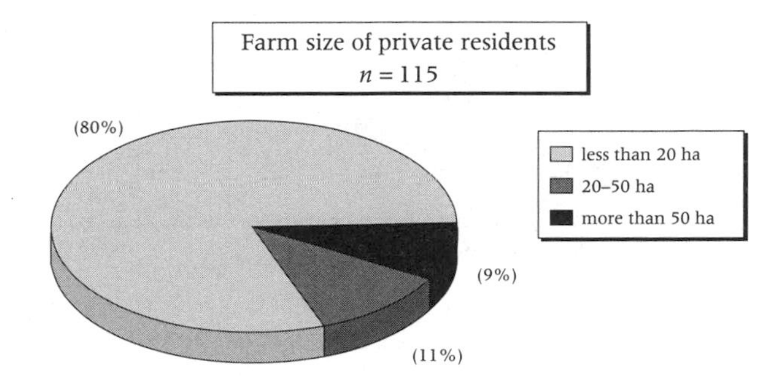

Figure 9-2. Size of private (noncommunal) farms claimed by park residents ($n = 115$). *Source:* Cuellar et al. 1992.

figure 9–3. Similarly, Fiallo (1994) found that park residents had lived in the area on average 37.4 years (a range of 2 to 71 years).

Of the private resident landholders surveyed, 43 percent inherited their land, 22 percent occupied public land prior to park establishment, 13 percent purchased their land, 12 percent received land as a donation, 2 percent occupied public land after Park establishment, and the means for 8 percent is unknown (see figure 9-4). Most of the absentee landholders purchased their land (58 percent). The comuna Agua Blanca has historical recognition dating back to 1541 (Huber, cited in Cuellar et al. 1992) but was not recognized by the Ecuadorian state until 1965 (Fiallo 1994).

Cuellar et al. (1992) estimated the total population within the park to be 684 individuals (p. 41) which translates to roughly 126 families (p. 37), 82 of which were individual landholders (p. 107). According to the study's

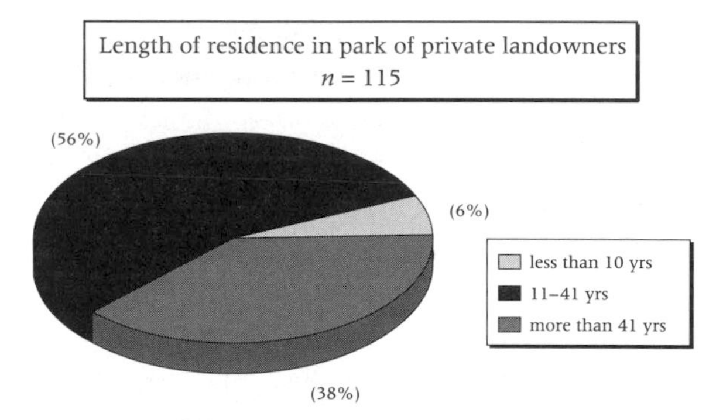

Figure 9-3. Length of residence in the park (n = 115). *Source:* Cuellar et al. 1992.

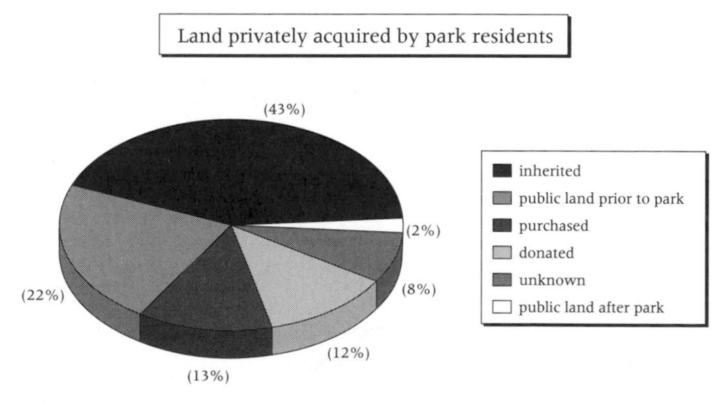

Figure 9-4. Methods of private land acquisition by park residents (n = 115). *Source:* Cuellar et al. 1992.

data on landholding size, this translates to approximately 1,312 ha under private holding, or 3.4 percent of the terrestrial portion of the park (vastly below the IUCN or Arriaga estimate). If 49 percent of those private lands are formally titled, that figure is reduced to a mere 643 ha. (Note: Communal lands are not included in that figure.) The figure is subject to considerable modification, as recent estimates of internal population are around 1,600. Informal reports describe a single 500-ha private landholding in San Francisco (C. Zambrano, pers. comm., 1995). This could more than double the proportion of private land in the park based on the estimates of Cuellar et al. Comparing the extent of private and communal landholdings (30 percent of the land-based part of MNP) and the extent of habitat disturbance (100 percent of the park's forests, according to Parker and Carr 1992), it is clear that individuals are using park resources beyond the boundaries of their landholdings.

The disparity between survey results regarding demographics and land ownership within MNP limits clear analysis; however, the following conclusions emerge:

- Comparing MNP to adjacent land, one finds a lower human population density, a greater predominance of communal land ownership, and a lower level of formalized land titles.
- The presence of comunas within MNP is an important factor affecting resource use.
- Comunas are unlikely to disappear over time due to attrition (Cuellar et al. 1992), nor can their titles be purchased.
- Promoting communal projects may be more efficient or politically appealing than working with individuals, although such an approach does not always reflect the true nature of the social organization of a community (Dove 1982).
- Many analysts view comunas as a promising organization for managing resources (J. Black, pers. comm., 1995; A. Martínez, pers. comm., 1995; Cuellar et al. 1992; Fiallo 1994), and, indeed, comunas of MNP have recently begun to defend their resources from outsiders.

More difficult perhaps will be their ability to regulate the resource use of comuna members in accordance with park mandates (see "Resource Use," p. 262).

Expropriation of private land was considered at the time MNP was established (J. Gallardo, pers. comm., 1995). In fact, throughout the 1980s, park directors annually drafted plans for expropriating land. Various factors prevented the execution of these plans, including insufficient funds, inadequate information on landholdings and legal status, and lack of political will. Local communities vehemently protested eviction during the 1980s, and at times threatened park authorities with violence.

It is unclear now, as it was then, whether local residents are unanimously opposed to expropriation. Some sources report that 30 to 50 percent of private landholders would be willing to sell their land "given a good price" (S. Martínez, C. Zambrano, J. Gallardo, pers. comm., 1995). Fiallo (1994) found that 52.7 percent of park residents would prefer to live outside the park, but she also found strong attachment to place among many individuals, particularly among members of Agua Blanca comuna. In many cases, when an individual states that he or she wants to "leave the park," what is wanted is the exclusion of his or her land from the park. At the first meeting of the Inter-institutional Committee for the Support of MNP, the announcement by the director of MNP that people would be allowed to remain in the park was met with a burst of applause from park residents.

Park authorities cautiously state that expropriation of private lands is "still an option." However, the park director predicts that the persistent lack of funds, information, and political inclination will prevent action, as will the position of FN in reference to land purchases or evictions (C. Zambrano, pers. comm., 1995). Assuming that individuals are using park resources beyond the scale of their individual or communal landholdings, expropriating land alone would not protect the park. Rather, land expropriation would have to be combined with concerted efforts to control grazing and other extractive activities by remaining residents.

FN is currently addressing land tenure conflicts in protected areas by sponsoring legal reform measures within Congress to clarify that

- activities within protected areas must comply with the proposed Law of Protected Areas and Wildlife;
- where titled private property precedes protected-area establishment, land could be expropriated according to appropriate norms;
- property holders should be reimbursed for investments in property preceding protected-area establishment;

- expropriated property owners have rights to land in buffer zones or other areas; and
- traditional settlements preceding park establishment should receive preferential treatment in access to resources and benefits of protected areas (FN-ESTADE 1993).

The proposed Law of Protected Areas and Wildlife, which was never passed by the Congress, is likely to be superseded by a proposed new Forestry, Protected Areas, and Wildlife Law that is currently being drafted by a multisectoral working group that includes government and NGO representation (see "National Policy Framework," p. 277).

Compared to land tenure, resource tenure is less easily categorized, as it depends on the nature of the resource and the organization level and political power of its users. The partitioning of resources by individuals and groups within MNP will be described in the following section on resource use; however, a brief introduction here is appropriate. Fish and shellfish of the park's marine corridor are primarily an open-access resource. The recent arrival of industrial shrimp larva collectors has provoked some of the local, artisanal fishers to defend their access to portions of the coastal park.

Wildlife has been traditionally an open-access resource at MNP, although in recent years, some comunas have begun to defend hunting rights from outsiders (S. Martínez, pers. comm., 1995). Grazing rights are poorly defined among park residents. Within comunas, rules originating at the national level prohibit an individual from grazing livestock on another individual's land. At the local level, those rules are subject to diverse interpretation, however.

Comuna leaders report that typically comuna members graze their goats on the common and their cattle on "individual farms" within the comuna (S. González, pers. comm., 1995). Fuelwood (for charcoal, brick making, or cooking) has become coveted as it has become scarce. While farmers in Salaite told us that each farmer freely collects a little fuelwood from the park, we were not able to clearly ascertain property rights to firewood. Park authorities have attempted to curtail firewood collection, but no specific zones or harvest rates have yet been established. Timber extraction is illegal and subject to enforcement, thus much of the activity is illicit. One comuna has attempted to limit timber extraction by its members by prohibiting the use of chainsaws.

Resource Use

Local resource use is perhaps the major management concern for MNP. INEFAN's decision to allow residents to remain within the park has shifted management objectives to center on "sustainable resource use" (or "rational resource use" or "ordered resource use") within MNP. This shift reflects a national (and international) trend to integrate socioeconomic development with biodiversity conservation. Lacking in the discussion is a definition of sustainable use, or at least the identification of clear criteria for evaluating sustainability. A resident of a local comuna within MNP was overheard talking to his neighbor after the first meeting of the Inter-institutional Committee in Support of MNP. He asked, "Just what does the man from Natura mean by 'reasonable use [*uso razonable*]'?"

In the following discussion of resource use within MNP, we attempt to briefly describe the major resource uses in the park in terms of: historical traditions, insider versus outsider access, changes in use, changes in control, and ecological impact.

Timber Extraction

Timber extraction for lumber and charcoal production is a traditional activity for many park residents. A gradual change in the species exploited has occurred during recent decades, from selective logging of a few, large hardwood species to the logging of young trees of numerous species. During the 1960s and 1970s, charcoal production reached a level of 500,000 sacks per year (Arnal 1993). Traditionally, timber extraction has been largely a local activity in the park's higher altitudes due to the inaccessibility of the forests (Cuellar et al. 1992). With the construction of roads through the park, regional merchants (apparently one in particular) have gained access to these forests. A 1992 study estimated that 82 percent of the wood products extracted in the park are sold, typically through an intermediary (Cuellar et al. 1992). The near exhaustion of forest resources appears to have inspired some communities to defend their resources from extraction by outsiders. Park guards, as well, tend to focus their patrolling efforts on defending timber from outsiders. This was not always the case, as park officials during the 1980s confiscated timber and charcoal from local residents and even went as far as destroying charcoal ovens. Eventually, quotas were established for charcoal production. For example, the eleven charcoal makers in the comuna Agua Blanca now produce approximately one bag a week each, down from six bags per week prior to the quota (C. Zambrano, pers. comm., 1995).

Local authorities describe a reduction in the rate of logging since the park was created. Fiallo observed declines of greater than 50 percent in the number of residents participating in timber extraction or charcoal making. The comuna of Agua Blanca and some residents of the comuna of El Pital have decided to stop using chainsaws to extract lumber. This is likely to be due in part to official restrictions on logging, to greater environmental awareness, and/or to the creation of alternative sources of income. A more discouraging explanation is suggested by some who state that all valuable trees nearby have already been cut (C. Josse, pers. comm., 1995; M. Sigle, pers. comm.). The impact of timber extraction (combined with grazing) has been so severe that "the Machalilla dry forest hardly exists as a 'forest' anymore" (Parker and Carr 1992; Arnal 1993; C. Josse, pers. comm., 1995). In 1992, Dodson and Gentry (1991) observed *Cedrela* spp. to be extinct or near extinct in the park. He warned that forest fragments within MNP would soon be exhausted unless strict protection measures were implemented. Without baseline data or field monitoring, it is difficult to judge whether the fragments are now securely protected. Josse and Balshev (1993) concluded that dry forests are able to recover more quickly after disturbance than moist forests, based on the study of a single hectare of forest near Agua Blanca. The potential recovery of dry forest habitat ought to encourage managers to protect and monitor forest patches.

Livestock

The majority of park residents raise some type of livestock (see figure 9-5). Most animals are allowed to graze freely, although pasture access is sometimes controlled in comunas (see "Land and Resource Tenure," p. 254).

Park authorities assert that cattle raising within the park has declined. Based on interviews with park residents, Fiallo (1994) reports a 66 percent decline in the number of families raising cattle before and after MNP was established. A 1993 estimate of a 90 percent decline in five years is likely to be inaccurate. The inaccuracy derives from an inappropriate comparison between a park-wide count of livestock (1,000 cows and 1,500 goats: Paucar et al. 1987) and a restricted study of seventeen families devoted to livestock raising (87 cows and 609 goats: Cuellar et al. 1992). Since the Cuellar et al. data refer only to seventeen families (p. 136) and not to a park-wide census, direct comparison of the two figures is inappropriate. Without consistent survey methods, it is risky to assume livestock numbers have drastically declined, particularly in light of

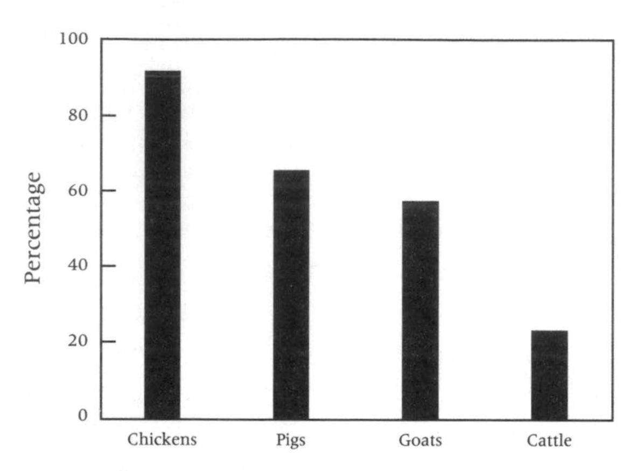

Figure 9-5. Percent of park residents surveyed who raise livestock, by type of livestock. *Source:* Fiallo 1994.

a recent report that three families in Agua Blanca own a total of 800 goats (S. Martínez, cited by M. Sigle, pers. comm., 1995).

Three of the four park staff interviewed in July 1995 identified grazing as the most difficult resource use to control in MNP because of the abundance of free-ranging animals and the formidable job of identifying and locating their owners. Parker and Carr (1992) state that grazing by goats and cattle represents a greater threat to forests than even timber extraction. They describe severe damage by goats and cattle throughout Cerro San Sébastián, the most biologically diverse part of the park. There is an urgent need to establish livestock exclosures around the most vulnerable forest fragments to allow rare and endangered species to regenerate. Exclosures are by no means a long-term solution but rather a means of ensuring survival for threatened plant–animal communities. Ideally, they would be designed so that empirical data on regeneration could be collected and later applied to other sites in the park. Experience from elsewhere should be considered in undertaking such an initiative (e.g., Galápagos with goat exclosures, and Guanacaste, Costa Rica, dry forest regeneration in pastures).

Agriculture

The overall estimate of the number of families working in agriculture in the park varies from 19 percent (Cuellar et al. 1992) to greater than 50 percent (Arnal 1993) to 83 percent (Fiallo 1994). This disparity may be due to seasonal fluctuations or different sample groups. Cuellar et al. report that agriculture is more common in semi-humid zones of the park than elsewhere, where it is limited by recurrent drought. Principal crops include maize, cassava, plantains, pumpkins, watermelons, and coffee. Farms of fewer than ten hectares are cultivated more intensely than larger farms. Sixty percent of respondents in Fiallo's (1994) study reported that agricultural production was for home consumption; 38 percent sold surplus for cash income.

Park staff attempt to control agriculture in the park by prohibiting the expansion of individual farms. Nevertheless, Josse reported observing some farmers in Agua Blanca expand their fenced property during the time of her fieldwork. The construction of roads in and around MNP has improved market opportunities for agriculturists. In informal interviews, respondents stated that production was limited by drought or labor shortage and that agricultural land was abundant in the park. The cultivation of plantains and coffee, in what Parker and Carr (1992) call fog forest, together with livestock grazing, threatens the long-term survival of montane birds and several large species of raptors.

Hunting

The fauna of MNP suffered heavy hunting pressure until the 1990s (S. Martínez, pers. comm., 1995). Commercial hunting was also common, and some local residents periodically earned wages as hunting guides. Park authorities report that hunting has greatly diminished due to patrol operations. They also report that commercial hunting has been completely eliminated or has disappeared due to scarcity of game. Park rangers unanimously agreed that wildlife populations have increased in recent years. Meanwhile, Fiallo (1994) found that most park residents perceive wildlife to be less abundant now than ten years ago. Two farmers from El Pital recently reported a scarcity of wildlife in the park except in upper-altitude forests. They also discussed the recent success of a hunter from their community who, using four dogs, shot five peccaries in San Sebastián. Without field monitoring, it is difficult to ascertain the status of MNP's wildlife. In 1992, Emmons and Albuja observed that native mammals had been reduced to small, fragmentary populations by hunting and severe forest degradation (apparently agouti populations were high, however). They also indicated that these populations could potentially recover if hunting were controlled and vegetation protected.

Fishing

Fishing has been a primary subsistence activity for coastal communities in the area of MNP throughout history (Coello 1993). Commercial fishing appeared in the latter half of this century. During the past two decades, fishing in the area has intensified with the use of larger boats. The heavy demand for nontraditional species of high value (e.g., post-larval shrimp and sea cucumbers) has also fundamentally changed traditional fishing activity.

Currently between 950 and 1,250 fishers reside in the zone of influence of MNP. The population appears to be stable, although each family has five to seven children (Coello 1993). Their activities and technologies are diverse, ranging from those who catch octopus by hook to industrial pelagic fishers. Legally, an eight mile zone along the coast exists which is exclusively reserved for artisanal fishermen. Between 1,165 and 1,514 people work in the fishing industry in the area, including fishers, merchants, processing workers, and longshoremen. Slightly less than half of the approximately one hundred merchants are outsiders who sell local products in other zones.

The park has never attempted to protect marine resources, and fishing resources are declining (Coello 1993). The most important types of fishing resources in the zone are: gravid shrimp, *corvina de roca* (*Brotula clarkae*), and demersal fish, and large and small pelagic fish. Artisanal fishers blame the decline on overfishing by *larveros,* who harvest shrimp larvae, and industrial fishers, who are also violating the eight-mile artisanal zone with increasing frequency. Conflicts between artisanal fishers (mostly local residents) and larveros (primarily immigrants or outsiders) are increasing. Likewise, conflicts arise between tourism operators and fishers over cleaning fish and contaminating the beach. Fishing regulations are by and large ignored. Most fishers know that MNP exists; however, they have little understanding of the objectives of the park. Most fishers report a neutral attitude to the park or state that they have no relation to the park, despite the fact that they fish in its waters. The new park management plan that will be prepared for MNP with financing from the Global Environmental Facility will give equal weight to management of terrestrial and marine resources.

Other Resource Uses

Other resource uses occur in MNP, including brick making, fuelwood collection, and tagua nut collection. As in the cases described above (with the exception of fishing), extraction levels are generally considered to be declining, although, again, it is difficult to judge the sustainability of use without monitoring the resource base.

Organizational Roles

The management of protected areas in Ecuador is formally assigned to the Ecuadorian Institute of Forestry, Natural Areas, and Wildlife, a state agency that is a branch of the Ministry of Agriculture. Under new leadership, INEFAN has greatly improved its flexibility, efficiency, and commitment to protecting Ecuador's parks and reserves. The administrative obstacles it faces are formidable, however. Conservation action is often paralyzed by a centralized, hierarchical bureaucracy. The dominance of industrial forestry interests within the agency saps INEFAN's strength. The current official policy to reduce the size of the bureaucracy (see "National Policy Framework," p. 277) threatens the ability of INEFAN to work effectively in the field on protection, education, and monitoring activities.

An increasingly important source of support for managing Ecuador's protected areas comes from the rapidly growing, private, nonprofit sector. NGOs, such as Fundación Natura, have played increasingly pivotal roles in environmental education, environmental lobbying, conflict resolution, research, and the management of protected areas.

The recent transition to collaboration with the public sector in park management has not always been smooth. On one hand, some public employees resented the intrusion of the private sector into traditional realms of public responsibility. This was closely linked to resentment over the capture of foreign funds by NGOs and the brain drain (hiring of top professionals away from public service posts). On the other hand, NGOs often complained that urgent conservation action was stymied by bureaucratic hassles and corrupt or incompetent state agencies.

With time and experience, relations between public agencies and NGOs have improved. Formal agreements helped delineate roles and responsibilities. Some conservation leaders have worked for both the state and NGOs. Such individuals, with experience in both the state and NGO sectors, have helped create cooperation between those sectors. In that vein, FN appointed a full-time liaison officer to coordinate the daily working of FN-INEFAN activities.

Legally, the management of MNP is the sole responsibility of INEFAN, and during the first decade of the park, INEFAN (then DINAF) authorities acted in relative isolation from other public and private agencies. Over the past five years this has changed dramatically. INEFAN continues to retain legal authority over MNP; however, it is now attempting to cooperate with a complex constellation of institutions ranging from national and international conservation NGOs to development agencies, from the military forces to regional development agencies, and from local comunas to national tourism agencies. Some of these institutions offer INEFAN vital support (FN, The Nature Conservancy, the German Development Assistance Agency, or DeD), while others potentially conflict with Park protection (e.g., Petroecuador, by constructing an oil pipeline through the Park, and the Ministry of Public Works). An analysis of institutional roles requires a stratified approach by level of organization, keeping in mind that these institutions all have cross-scale linkages.

Local and Regional Institutions

Centralized, top-down management of protected areas is widely criticized as inefficient and impractical over the long term. Building local support for protected areas has emerged as a central goal of most conservation programs. In Ecuador, this goal is particularly important, given the dominant policy mandate to shrink the state and decentralize (see "National Policy Framework," p. 277). INEFAN is currently in an exploratory phase of building local support, through formal contracts, informal agreements, and dialogue. So pervasive is this goal that when asked what they do when they encounter timber or game poachers, three park guards independently answered, "We enter into dialogue."

Despite a history of antagonistic relations with local communities, park authorities are gradually building alliances with some of the comunas and communities. The long-term aims of participatory activities are to ameliorate local impact on the ecosystem and, ideally, to convert communities into protectors of the park. The nature of relations between INEFAN and local communities varies considerably by site, ranging from cooperative to antagonistic (Fiallo 1994). In general, comunas have a more advanced dialogue with MNP authorities than do individual residents. Moreover, communities causing the worst impact on the park ecosystem are subject to increased outreach efforts (e.g., Salaite and El Pital).

The comuna Agua Blanca is much celebrated as a successful example of community participation in park management. Due to its sophisticated leadership, employment with MNP, and status as the foremost archaeological site in the region, it has managed to benefit from tourism and the presence of international development and research projects. In terms of specific support for the park, Agua Blanca members have attempted to prevent the construction of an oil pipeline across their land, have protected archaeological relics from pillage, have hosted research biologists, and have spoken to other comunas about the importance of conservation. Agua Blanca members are strong leaders in a newly formed confederation, the Confederation of Comunas from the South of Manabí. This confederation unites several communities and comunas that previously were part of the Peasant Committee to Support MNP, an initiative promoted and financed by FN with the purpose of helping inhabitants to understand the importance and finality of MNP and involve them in activities that benefit the park. The confederation is an initiative of the local residents that brings together a greater number of communities and deals with broader themes of greater interest to local communities. This confederation is likely to play a key role in negotiating the relationship between INEFAN and comunas in the future. At the other extreme is the Matapalo community within the comuna El Pital, which persists in threatening MNP employees who attempt to control its use of park resources (e.g., by inserting nails in tires of MNP vehicles).

The population of local fishers is weakly organized (Coello 1993) and has not participated in dialogue with INEFAN to the same extent as local agriculturists. Nevertheless, local fishers exert a significant impact on the park and will be important to work with. FN is currently organizing further research into marine resource management alternatives in MNP.

Neighboring municipal populations play a potentially important role in MNP, particularly in the development of tourism. An Association of Tour Guides has formed in Puerto López. INEFAN is attempting to work with it to establish regulations controlling tourist activities, standardize fees, and improve the natural history knowledge of guides. An agreement has been signed between INEFAN, the Municipality of Puerto López, the German Development Assistance Agency, and FN to carry out various actions in support of the park, initially including environmental education.

Regional technical universities such as the University of Manabí have participated in studies of timber species in MNP, supported by PiP funds. They, along with other organizations and institutions that conduct research in the country, could play an important role in monitoring the MNP ecosystem.

Among the several regional conservation NGOs, the Guayaquil chapter of FN is the strongest. The current technical supervisor for FN projects in MNP is based in Guayaquil. When the Provincial Council of Manabí wanted to develop a project to encourage shipping to the Isla de la Plata, there was conflict with FN. Fortunately there was no follow-through on this proposed project, which would have negatively affected local biodiversity. Since then, there have been no major difficulties in relations between the administration of MNP and this council.

At the first meeting of the Inter-institutional Committee for Support of MNP, the director of MNP declared the park to be the property of Manabí, and emphasized that parks are intended for human use. The ensuing, animated discussion focused on the park as a source of revenue and social development. Through this meeting, the first steps toward regional cooperation were achieved. A regional INEFAN representative with twenty-five years of experience in the area warned that the demands of regional public agencies for revenue will outweigh their investment in park protection. While this may be true, ignoring regional institutions carries an even greater risk to the park. Regional agencies in charge of constructing infrastructure will obviously continue to affect the park. What's more, the regional office of INEFAN itself has at times undermined park protection by selling individuals timber licenses for forests adjacent to MNP (MNP staff, pers. comm., 1995). Ideally, environmental enlightenment of regional institutions will be a by-product of such cooperation, so that MNP is not viewed simply as a source of income. (See the case study of Podocarpus National Park, chapter 10, for an example of regional support for a park.)

National Institutions

A debt swap was arranged with help from WWF and The Nature Conservancy, and the funds were used to support the management of eight protected areas in Ecuador, including MNP. In administering debt-swap and PiP funds, FN is far and away the dominant NGO working at MNP. Regarding debt-swap funds, the operational plans were developed between FN and INEFAN, and biannual reports are sent on completed activities to all donors.

Regarding PiP funds, a formal agreement between FN and INEFAN and between the Conservancy and FN outlines the working relationship in which every year the Conservancy provides a quantity of funds allocated by the four PiP program components (on-site protection and management, community outreach and compatible development, long-term security of protected areas, and project management). The operational plans are developed between FN and INEFAN, and reports are sent to the Conservancy every three months. In consultation with INEFAN authorities (both national and at MNP), FN recommends specific activities within programs or suggests adjustments in allocation of funds between programs. These plans are ultimately presented to the National Directorate of Natural Areas and Wildlife (within INEFAN) for approval.

The national recognition of FN allows it to effectively organize policy discussion among institutions at various levels. However, MNP management could be enhanced if a local, field-oriented NGO would participate in day-to-day activities at MNP (see example of Arcoiris at Podocarpus, chapter 10). Unfortunately, no local conservation NGOs are currently working in the area of MNP. Even though FN is not a local NGO, it has been able to undertake numerous important field-based activities within MNP.

Another formal agreement involves the creation of a Forest Guard Force within the Armed Forces of Ecuador. The Forest Guard was first described in the 1981 Forest Law, but only in 1994–95 did it start to be implemented. The duty of the Forest Guard is to assist INEFAN in patrol and protection activities. INEFAN is obliged to provide field support for the Forest Guards, but it is constrained from doing so by lack of funds. Park guards report that being accompanied by soldiers makes people respect them more and is necessary for operations in areas where they "have enemies." Some analysts worry that the military presence alienates local communities and sets back efforts to build solidarity with the people.

Individual students from the Catholic University of Quito have carried out field research in MNP, but unfortunately the students have been few and their work has not been consistently applicable to management concerns. In general, it is difficult to attract field biologists (either national or international) to heavily disturbed ecosystems.

CETUR is the National Council of Tourism for Ecuador and as such should be playing a leadership role in organizing and regulating tourism in the region. To date its presence has been negligible as tourism has grown.

The state agency Petroecuador is infamous for undermining park protection in Ecuador. In the case of MNP, Petroecuador used its political might to override less powerful state institutions and build an oil pipeline through the length of the park (see "Conflict Management and Resolution," p. 274). Admittedly, Petroecuador did reforest the land adjacent to the pipeline and is now working with comuna El Pital on a reforestation project.

International Institutions

The international institutions working at MNP show no signs of the competition or backbiting often seen at other sites. Instead, a positive, collaborative spirit exists, which is a great advantage for MNP. Among international conservation NGOs supporting MNP, the Conservancy, working through FN, is the largest and most actively involved: Conservancy officers frequently visit the site. In sponsoring the rapid assessment program team of Ted Parker, Al Gentry, and others, Conservation International played a unique role in providing support for a rapid ecological appraisal exercise in the park. The current deficiency of field biology in the area makes the deaths of these scientists, which occurred in a light aircraft accident in Ecuador, all the more tragic.

USAID works via the Conservancy and FN. The German Development Assistance Agency is active in MNP and the zone of influence (up to eight kilometers from the boundary) by promoting projects in environmental education, agroforestry, and reforestation with native species. It has recently signed an agreement with FN, INEFAN, and the Municipality of Puerto López. Other institutions that have worked in rural development in MNP or are now implementing projects, include the Inter-American Foundation (with COMUNIDEC), the University of Denmark through the Catholic University of Quito, the University of Bristol, the Italian International Committee for Local Development, and the Fondo Ecuatoriano Populorum Progresso (FEPP).

Linkages between Park and Buffer Areas

Working with communities is obviously the central management challenge in MNP, considering that individuals are settled throughout the park. Previously, in the "Land and Resource Tenure" (p. 254) and "Resource Use" (p. 262) sections, we described the transition in the position of INEFAN toward communities, from one of confrontation and prohibition, to a cooperative, permissive stance. This change has allowed for improved public relations and reduced conflicts. In some cases, local resource use has also been affected (e.g., Agua Blanca working in tourism versus charcoal production). However, a cautious view of the long-term sustainability of community resources in MNP is appropriate until field monitoring provides appropriate data.

The basic operational objectives apparent in the outreach work of FN and other agencies in MNP such as the DeD are to:

- enlighten communities on the importance of conservation
- strengthen community organization
- provide ecologically sustainable alternative employment and economic activities.

Outreach work is promoted on a communal basis, and NGOs promote communication and cooperation between communities.

The ultimate goal of the outreach projects is not rural development, but rather the adequate management of MNP. Working with people is a means toward reaching that end. The conservation of MNP will be secured only when the people support it. Any project must be well thought out and in agreement with the conservation objectives of the park.

Projects are underway both within and outside the MNP and are similar in mission. Some people residing outside MNP perceive a preferential treatment for communities within the park (e.g., Salango). Examples of current or recent projects are: a tree nursery and reforestation project in comuna El Pital, support for the protection and maintenance of archaeological sites in Agua Blanca, and environmental education presentations in nearly all communities. The development of a fishing project for the community of Salaite is also proposed.

The obstacles and challenges encountered in these activities parallel the standard problems in rural development:

- providing assistance without developing dependence
- facing schisms within the community
- balancing short-term and long-term needs.

Most residents do not participate in communal systems of production (although they do work together on constructing schools, etc.). Thus it is especially difficult to promote a new economic activity that also involves a new form of social organization. There is a general tendency in development assistance to view peasant life in communal terms (Dove 1982). Forster (1992) describes the complaints of participants in some development projects elsewhere in Ecuador about imposed communal production when community leadership is weak and members' labor contributions are unequal. Beneficiaries may more readily accept credit and service co-ops than production co-ops. On top of these difficulties, conservationists are searching for activities that are both ecologically sustainable and of interest to the community. Finally, neither INEFAN nor FN is a development organization, thus they lack extensive experience and technical expertise in rural development. However, they have attempted to overcome these deficiencies through cooperative work with other organizations that specialize in these areas.

Conflict Management and Resolution

Conflicts over resource control and management in and around MNP occur at local, regional, and national levels. (For a discussion of the conflict between INEFAN and local communities, see "Resource Use," p. 262.)

Regional Level

In Salango (approximately five kilometers south of the park) there exists a large-scale fish processing and fish meal factory (Polar) that contaminates sea and air for several kilometers with its industrial waste (Coello 1993). Not only does the factory threaten marine habitats in the park, it has caused health problems in the surrounding population (e.g., skin disease and respiratory ailments). The contamination also threatens local tourism. Only five people from Salango work at the factory; most of the workers are from Platanales, a small town located near, but outside of, the park. The residents of Platanales avoid contact with park staff or residents of neighboring communities, especially those within the park.

The residents of Salango have fought for clean air and water for years and received support from conservation NGOs at the national level, but Polar has not reduced its environmental impact. The history of conflict with Polar is instructive as a case in which grassroots protest efforts combined with concerted lobbying by NGOs at a national level have failed to change the polluting activities of a private industry.

Polar Company owns eight boats, most of them Class 3 (71–105 tons), although two are Class 4 (more than 105 tons) (Coello 1993). During a typical month, these boats provide the factory with 3,454 tons of small pelagic and white fish. Independent fishers also sell their catches to the factory in an unknown quantity. Coello describes Polar as a source of local employment; however, Salango residents state that fewer than five individuals work for the factory (A. Pincay, pers. comm., 1995). The process of converting fish to fish meal yields gasses and wastewater, which are released directly into the atmosphere and the sea.

During the early 1990s, FN worked with the Peasant Committee to Support MNP in organizing a campaign to protest the pollution produced by Polar. Representatives traveled to Quito and registered their protest with the Ministries of Health and Welfare. Public authorities visited the site on several occasions and documented the pollution levels; however, no formal action was taken against Polar. Local protests resulted in Salango residents being threatened with incarceration if protests continued (A. Pincay, pers. comm., 1995).

FN called public attention to the problem by formally listing Polar among Ecuador's worst polluters. To date, Polar has not received a sanction, fine, or other penalty for its environmental damage. Ironically, the conflict may end if Polar closes, an outcome predicted by some due to the decline in pelagic fish stock, particularly sardines and mackerel (Coello 1993).

When asked why Polar escaped penalty, informants unanimously agreed that the factory had high connections within the Ecuadorian government. Unfortunately, this is not an unusual case, as the division between the state and private industry is not easily identified (T. Bustamante, pers. comm., 1995).

National Level

In 1983, Petroecuador initiated the construction of a coastal oil pipeline, planned to run through MNP and connect the Port of Manta with refineries to the south. Alarmed by the threat to the environment of construction and potential oil spills and the risk of destroying archaeological sites, MNP authorities organized an opposition effort at the regional level. First, they formed a Park Defense Commission composed of representatives of DINAF, Petroecuador-Guayaquil, archaeologists, and FN. The commission argued that the pipeline should be built around MNP and sent its petition to the Petroecuador office in Quito. FN specifically objected to the manner in which the environmental impact study was completed (FN 1992). Petroecuador rejected the commission's petition because the company considered the cost of an alternative course for the pipeline to be prohibitively expensive. (Unfortunately, cost estimates are not available for the two alternatives.)

The conflict was hence redefined to a debate over where the pipeline would be constructed in the park and how environmental and archaeological impact could be minimized. The subsecretary of the environment within the National Directorate of Mining (DINAMI) did not attempt to protest the pipeline crossing the park but instead attempted to keep its path from crossing the center of the park. It is unclear where exactly the original plan placed the pipeline, but the final plan had the pipeline approximate the course of the coastal highway for 32 km through MNP, buried at 1.5 m.

Local protests continued during the initial construction phase of the pipeline. MNP authorities paralyzed the construction for twenty days and demanded that Petroecuador reduce the width of the trench from 28 m to 10 m and invest in on-site environmental rehabilitation. Petroecuador agreed to limit the trench to 16 m and to later reforest the course of the pipeline with native species. The Quito office of DINAF then sent orders to field staff to allow construction to proceed. Construction resumed but was again stopped when it reached the boundary of the Agua Blanca comuna. There, comuna members physically paralyzed construction activities for eight more days while they presented various demands, including that Petroecuador pave their access road. Petroecuador agreed, signed a document, and construction resumed (S. Martínez, pers. comm., 1995).

Upon completion of the pipeline, Petroecuador reforested the ground above the pipeline with native species and watered the seedlings for a year. Since then, the survival of saplings has been variable, depending primarily on microhabitat. In some areas, vegetation recovery makes it difficult to identify the site of the pipeline. Elsewhere, seedlings have died, due to drought conditions (M. Sigle, pers. comm., 1995). Agua Blanca comuna members complain that Petroecuador never fulfilled the conditions it agreed to.

The pipeline in MNP is an interesting case, in which opposition to a project was solid across public and private lines at the regional level but was dismissed by national interests. MNP authorities deserve commendation for organizing the opposition and rallying local and regional support. In fact, multiple mechanisms of conflict resolution were in operation, although no sign of international support is indicated. The impact of the pipeline was lessened by those mechanisms. Ultimately, however, Petroecuador's considerable political power overruled opposition.

Large-Scale Threats

The foremost immediate threats to the terrestrial ecosystem of MNP are overgrazing and timber poaching. Aquatic ecosystems may be similarly threatened by overfishing; however, data are lacking on impacts of fishing on the park's marine system. (See "Resource Use" particularly in regard to livestock grazing.)

Uncontrolled tourism also represents a serious potential threat to MNP. Between 1988 and 1993, visitation rates for the park increased from 2,370 to 15,139 (Ortiz et al. 1995). Powerful companies within the tourism industry hope to increase that number to over 70,000, but their plans for expanding tourism and improving infrastructure ignore environmental concerns (Arnal 1993). Tourism is heavily concentrated on only two or three sites of scenic beauty along the park's coast (Los Frailes and Isla de la Plata). The tourism-carrying capacity for these sites has yet to be established. Recommendations from biologists for tourism management have generally not been implemented (D. Platt, pers. comm., 1995).

Institutional leadership is lacking in the development of tourism within the park. INEFAN staff complain that regulating tourism is very difficult due to the volume of visitors and the large number of formal and informal tour operators (L. Medina, pers. comm., 1995). Ideally, the newly formed Inter-institutional Committee for Support of MNP ought to fortify INEFAN's role in controlling tourism. INEFAN carried out a training course for local tour guides in 1993 but has not continued it. In 1995, FN prepared a detailed tourism management plan (Ortiz et al. 1995), but financial constraints limit its immediate implementation.

National Policy Framework

Environmental legislation in Ecuador is scattered through more than one hundred codes, laws, and rules and in numerous provincial and municipal regulations (FN 1991). This makes the application of law difficult.

Regarding the management of renewable resources, the most important law is the Forest and Natural Areas and Wildlife Conservation Law, passed in August 1981. Article 197 of the law establishes five basic objectives, to

- tend to the conservation of renewable natural resources according to the social, economic, and cultural interests of the country
- preserve the outstanding flora and fauna, landscapes, and historical and archaeological relicts, based on ecological principles
- perpetuate in a natural state representative samples of biotic communities, physiographic regions and biogeographic units, aquatic systems, genetic resources, and threatened species
- promote the integration of human beings with nature
- assure the conservation of wildlife for its rational utilization in benefit of the people.

In 1976, various studies were carried out to identify extraordinary areas in Ecuador deserving alternative management. Of the numerous areas considered, thirteen were selected to form part of the Minimum System of Natural Protected Areas of Ecuador. MNP just barely made it onto the list but was selected largely because of the great scenic value the area holds.

In 1989, FN and INEFAN carried out the second phase of the Strategy for a National Protected Areas System of Ecuador, from which various suggestions emerged to improve the management of the country's parks and reserves, and to include unrepresented ecosystems. In spite of the fact that INEFAN officially accepted the document, mid-level personnel of the agency did not consider the second strategy to have official standing. Some of the suggestions made in the document have been implemented but not according to the recommended priorities.

The planning, management, development, administration, protection, and control of the natural area patrimony of Ecuador is under the authority of INEFAN. Although INEFAN is charged with managing both forestry resources and natural areas, it invests much more in the forestry sector than in natural areas, reflecting patterns of political power.

Ecuadorian environmental institutions are perceived to be politically weak and subordinate to other sectors in terms of government priority. In cases in which there is an overlap of laws, hydrocarbon or mining will be considered the national priority. This has led to oil exploitation and pipeline and road construction in the center of national parks.

The lack of inter-institutional coordination is another factor that seriously affects the management of protected areas. During the 1970s and 1980s, there were few cases in which IERAC adjudicated land within protected areas. Sumaco-Napo Galeras National Park was created in 1994, and only a few months later, oil exploration was allowed within the territory of the new park.

Although the Forestry Law recognizes seven categories of protected areas, all protected areas in Ecuador are similarly managed in practice. Initially, the national park principles were applied to all protected areas, but the presence of human settlements, the use of resources, and the exploitation of nonrenewable resources in the interior of the areas have made it impossible to complete these conservation objectives. Budget limits, the lack of clear policies, limited technical capacity, and the limited number of personnel prevent protected areas from being managed according to their designated category.

The financing and field personnel provided by the government only enable the most basic protection activities to be carried out in parks and reserves. Until August 1994, FN, through the debt-swap program, financed the employment of an increased number of park guards and conservation officers for protected areas (although the number was still inadequate). INEFAN made a commitment to assume responsibility for these newly contracted personnel but was eventually unable to do so due to modernization policies of the current government, which aim among other things to reduce the number of public employees. No special considerations have been made for field personnel in protected areas, whose number has been significantly reduced. Currently, MNP has twelve park guards, two conservation officers, and a park director for an area greater than fifty-five thousand hectares.

In 1994, FN carried out a study regarding investments in biodiversity, with funding from the World Conservation Monitoring Centre. Results indicated that foreign investment and the number of implemented projects had increased between 1991 and 1994. However, data also show that funding from the Ecuadorian government has diminished and is currently very low.

In May of 1993, FN and ESTADE presented a proposal for a Protected Areas Law to Congress, which suggested various points to be included in the reform of the Forestry Law:

- to promote greater autonomy for the institution charged with managing protected areas in relation to the forestry sector
- to consecrate the supreme role of the state in caring for the national patrimony and regulating its administration and use, but at the same time, reduce operational activities, promoting a participatory system of managing protected areas; the legal proposal as outlined would permit civil society (popular organizations, NGOs, and other institutions) or sectional governments to manage protected areas under the supervision of the state
- to oblige the state to include the participation of various social actors in decision making regarding management policies for protected areas
- to establish buffer zones adjacent to protected areas, which should be subject to special management treatment (sustainable use)
- to recognize the right of local communities to make use of natural resources in protected areas under certain rules
- to apply the concept of decentralization of power, to achieve a regionalized system of management of natural areas, based in local and regional organizations
- to establish strict environmental regulations to prevent environmental impacts and restrict the use of nonrenewable resources.

In early 1994, the Environmental Advisory Committee to the Presidency of Ecuador, known as CAAM, formed a National Work Group on Biodiversity, which elaborated a Preliminary Strategy for the Conservation and Sustainable Use of Biological Diversity of Ecuador. This document includes a chapter exclusively devoted to natural protected areas. In addition, a sub-working group is elaborating a proposal for a biodiversity law, framed by the commitments of the signatory countries of the Biological Diversity Convention.

In 1995 the Core Multisectoral Group was formed with staff from INEFAN, CAAM, CEDENMA (the Ecuadorian Committee for the Defense of Wildlife and the Environment), the Environmental Commission of the Congress, and AIMA (the Association of Industrial Loggers). This group has drafted a new law called the Law of Forestry, Natural Protected Areas, and Wild Biodiversity. The group is coordinated by the Program for Forest Policy (PPF).

Ecuador was the first Latin American country to sign the Convention on Biological Diversity; it has also signed twenty-five other multilateral agreements regarding the environment, including Agenda 21, CITES, Ramsar, Montreal Protocol, and the Basilea Convention, among others. (Note: The number of bilateral agreements signed by the Ecuadorian government is not available.) Nonetheless, the monitoring of most agreements is very limited. The Ecuadorian government may recently be paying more attention to the Convention on Biological Diversity, especially because of the pressure NGOs exert on it to consider the implications regarding property rights and genetic resources.

Conclusion

INEFAN and FN face a number of vexing questions in defending MNP:

- Will MNP's fragile and biologically rich habitats (e.g., the fog forest) survive current grazing and timber harvesting? The dry forest, heralded as the most important habitat to protect in the park, hardly exists as a dry forest anymore.
- Will residents be willing (or able) to curtail their traditional activities (e.g., timber extraction and grazing) so as to live in harmony with the ecosystem? Who is going to define harmony and by what criteria? How will resource use be regulated?
- Who will control and regulate tourism activities?

Considerable technical knowledge and political savvy are required to reconcile economic development with biodiversity conservation in a small park holding both resilient and fragile habitat. Sustainable resource use often is elusive even under stable conditions; MNP's ecosystem is subject to unpredictable climatic fluctuations, the occasional emigration of the human population, and the vacillation of local resource use regimes with external market conditions (e.g., shrimp larveros).

FN and INEFAN have accomplished a great deal in a short time. Most notable perhaps is the revolution in relations between the park and the community. After years of antagonism and conflict, there is dialogue, cooperation, and, in some cases, local advocacy for MNP (e.g., the Confederation of Comunas from the South of Manabí). Establishing positive relations with local communities is essential for MNP's long-term survival. To foster this positive relationship, managers abandoned the effort to tightly control resource use. Rather they now emphasize preventing outsider exploitation of MNP, while allowing subsistence utilization by residents. In the absence of field data, it is inappropriate to make definitive conclusions regarding the ecological impact of current resource use. The park's fragile habitats appear to still be at risk.

Postscript

In July of 1996, a team of INEFAN staff and independent technical consultants launched the MNP management plan, with financing from the INEFAN/GEF project for the protection of biodiversity. The study is expected to be completed by October 1998. This new management plan will study the possibility of adding a marine zone to the park, which would extend from the shoreline of the park to the Isla de la Plata, thirty-seven kilometers from Puerto López. The goal would be to protect crustaceans, sea turtles, and benthic species. Both traditional and industrial fishing occur adjacent to the park, and neither form would be affected by the creation of this zone, since they are already regulated by government fishing laws. The park's administration finds it impractical to prohibit fishing in this zone. Not only is fishing a traditional activity, it also is an economic base for many of the local communities, which were established long before the park was created. Regulation and control of fishing in this area fall outside the jurisdiction and capacity of INEFAN. Nevertheless, there has been a certain level of coordination between the parties with interest in management of the park's marine and coastal resources. They have initiated a collaborative and coordinated effort through the Zoning Committee, which brings together all the institutions that work within the park and the buffer zone.

The management plan will recognize that there are populations living within the park. Although they do not have any legal rights, they are important to the area's participatory management style. This includes a community work program, which includes projects designed to generate additional income through productive activities that do not strain the park's natural resources. The success of the PiP program has been an important influence on these activities.

In 1996 and 1997, several projects were developed through the PiP program that involved local communities:

- chicken raising with the Women's Committee in the Río Blanco community
- honey production with the Casas Viejas Apiculture Committee
- family organic gardens in the Río Blanco and Casas Viejas communities
- artisanal fishing in the Salaite community
- solid-waste management in the Agua Blanca community, a cooperative project to improve local sanitation.

Participation in these projects has not only improved relations between the communities and the park, but has also demonstrated on a small scale how opportunities exist for communities to develop productive microenterprises that can improve their standard of living, reduce their dependence on activities that degrade the park's natural resources, and contribute to its conservation.

It is too early to evaluate the lasting impact of these projects on the use of the park's natural resources, but Fundación Natura has developed a monitoring methodology in order to evaluate these impacts. This methodology will be applied to the park's principal communities between 1997 and 1999. The conclusions drawn from these results will provide valuable information to the park administration and environmental organizations on best ways to collaborate and interact with the communities.

If the increasing rate of visitors to Machalilla National Park holds steady, it should become the most visited national park on mainland Ecuador within a couple of years. The park's new management plan will take into account the tourism management plan completed by Fundación Natura in 1995, a PiP-funded project. The plan proposes an organized approach to tourism and recommends limiting access to certain zones that are especially fragile while at the same time permitting the use of other areas that had not been available for use before.

One of the park's goals is to decentralize park management from the central INEFAN office in Quito. This would hopefully result in a more efficient handling of increasing resources. Since it would require the approval of a new protected-areas law, which has not yet been passed by the national Congress, decentralization has not yet been finalized.

Fundación Natura, which has participated in the PiP program since 1991, has a very important working commitment to the MNP. This partnership has worked to formulate new proposals that will support the management of the MNP through new projects during the coming years. Some of the ideas that FN is working on include developing participatory management projects with the communities living in and around the park, such as a possible carbon sequestration project. Such new efforts will be possible due to positive experiences working with those communities to this point.

Glossary

AIMA—Association of Industrial Loggers (Asociación de Industriales Madereros)

CAAM—Environmental Advisory Committee to the Presidency of Ecuador

CEDENMA—Ecuadorian Committee for the Defense of Wildlife and the Environment

CETUR—National Council of Tourism for Ecuador

comunas—community landholding system

Confederation of Comunas from the South of Manabí—Confederación de Comunas del Sur de Manabí

DeD—German Development Assistance Agency

DINAF—Dirección Nacional Forestal (National Directorate of Forestry)

DINAMI—Dirección Nacional de Minería (National Directorate of Mining)

ESTADE—NGO that works on environmental legislation

FEPP—Fondo Ecuatoriano Populorum Progresso (Ecuadorian Popular Progress Fund)

FN—Fundación Natura

IERAC—Instituto Ecuatoriano de Reforma Agraria y Colonización (Ecuadorian Institute of Agrarian Reform and Colonization)

INEFAN—Ecuadorian Institute of Forestry, Natural Areas, and Wildlife

larveros—people who harvest shrimp larvae
MNP—Machalilla National Park
Peasant Committee to Support MNP—Comité Campesino de Apoyo al MNP
PPF—Program for Forest Policy
RAMSAR—an international convention for the protection of wetlands

Podocarpus National Park

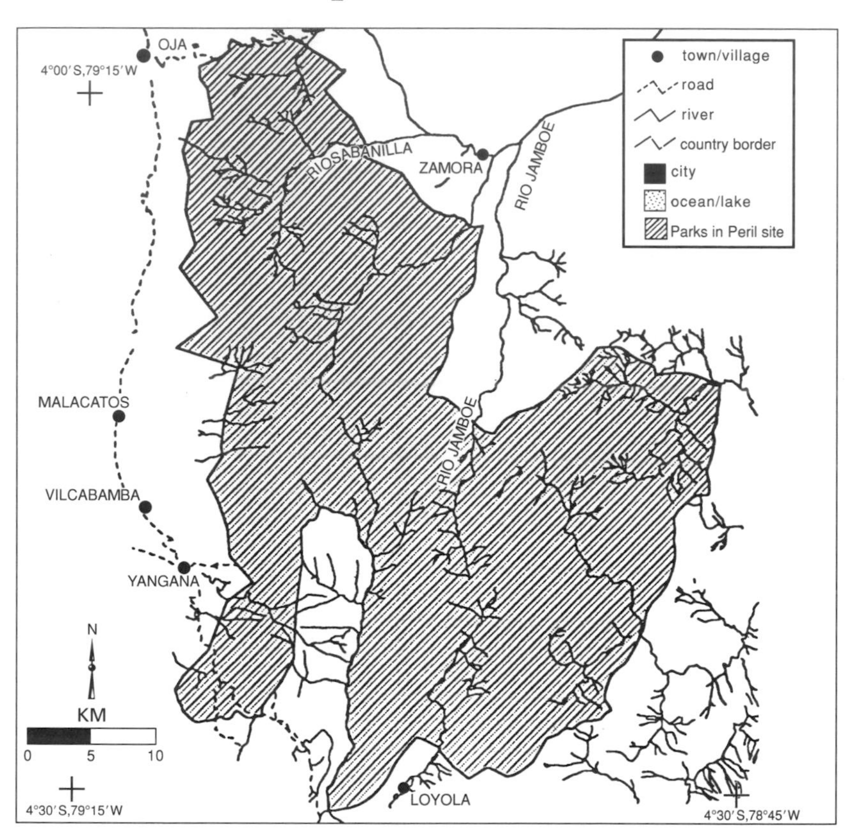

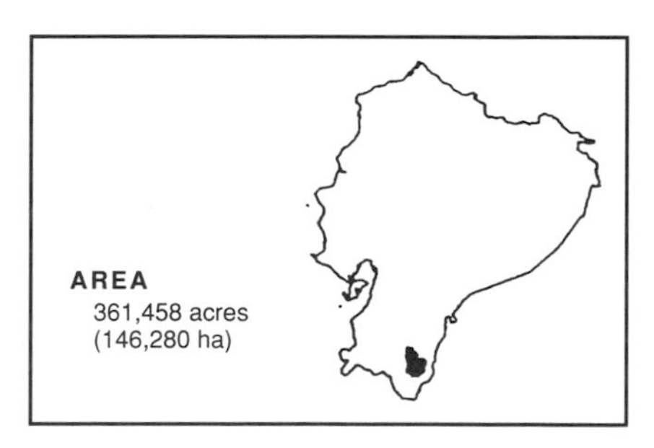

10. Ecuador: Podocarpus National Park

Bolívar Tello, Elba Alissie Fiallo, and
Lisa Naughton-Treves

Viewing Podocarpus National Park (PNP), one is immediately struck by its rugged contours. This steeply dissected topography has restricted agricultural expansion, a process that has converted much of the area surrounding the park from forest to arid farms. PNP is the only protected area in southern Ecuador and holds the region's last substantial tracts of pristine forest at mid to upper altitudes (900–3,400 meters above sea level). The park encompasses 146,200 ha and straddles the Andes in an area of exceptional endemism. Botanists estimate that PNP contains three to four thousand plant species (Madsen, cited in Fundación Arcoiris 1992). Within these forests, the fever-bark tree (*Chinchona succirubra*) was discovered and later exploited for quinine. The avian species richness of PNP is of international significance; perhaps 6 percent of the world's bird species are found in the park (Block et al., cited in Vallée 1992).

At national and international levels, conservationists celebrate the extraordinarily high levels of biotic diversity and endemism at PNP. At a local level, residents of the provinces of Loja and Zamora-Chinchipe, if aware, would be more likely to value PNP for its watershed protection. Several urban areas are entirely dependent on PNP as a water source, including Zamora (population 30,000) and the city of Loja (population 130,000).

Beyond biodiversity or plentiful water, it is the glimmer of gold in the park's mountains that has attracted people throughout history. Traces of Incan trails skirt the richest deposits in PNP (Vallée 1992). Local legends describe buried gold treasure, originally destined for Peru (Cajamarca) to pay ransom for the Incan king, Atahualpa (Apolo 1984). In the extreme south of the park, remnants of trails and walls spark speculation about a lost Spanish city. Today, the gold of PNP is coveted by a diverse collection of people and corporations, ranging from artisanal miners to regional merchants and international mining companies. The battle to protect PNP from mining interests has likewise involved a diverse collection of individuals and organizations, who have used lobbying, environmental education, and research to build a constituency to defend the park.

Mining in protected areas is not an unusual problem. In a 1985 survey, nearly 40 percent of national parks in lesser developed countries recorded problems with illegal mining activities (Machlis and Tichnell 1985). In this case study, we hope to illustrate the distinct nature of gold mining as a threat to protected areas, and to describe the results of various defensive strategies. In the process we will reveal the unique cooperative arrangements between nongovernment organizations (NGOs), state agencies, and local organizations concerning PNP. The presence of a highly motivated, local NGO, Fundación Arcoiris (FAI), has been critical to the defense of PNP, as has the support of the national Fundación Natura, and international groups such as The Nature Conservancy. Meanwhile, the state agency legally responsible for park management, the Ecuadorian Institute of Forestry, Natural Areas, and Wildlife (INEFAN) is severely constrained by budget and personnel shortages. The creation of a regionally based Inter-institutional Committee in Support of PNP has broadened support for the park, but its influence is stronger in Loja than in Zamora-Chinchipe, the province holding most of PNP's area and the richest gold deposits. The conflict over gold mining in PNP reveals a hierarchy of claims to the site, ranging from the distinct interests of the two provinces, to the interests of national institutions competing to manage the land for biodiversity (INEFAN) or gold (since 1992, DINAMI, the National Directorate of Mining; prior to that, INEMIN, the Ecuadorian Institute of Energy and Mines).

Park Establishment

Some of Loja's citizens recognized the natural value of the province's forests and lakes before any government agency in Quito formally identified the area's conservation importance (PREDESUR 1978). In 1976, a regional employee of the Ministry of Agriculture (W. Apolo) was assigned the task of surveying the Loja and Zamora-Chinchipe provinces to evaluate the possibility of establishing a national park. Apolo identified a forested area as a strong candidate site for protection. Initially, the proposed "Podocarpus National Park" was bypassed in favor of sites elsewhere in Ecuador. Regional institutions (e.g., PREDESUR, the Ecuadorian Subcommission Development Program of the Southern Region) subsequently campaigned for the creation of a park in southern Ecuador, and in 1982, an area of 146,280 ha was officially established as Podocarpus National Park.

The official objectives of PNP were defined as follows (Apolo 1984):

- to preserve a pristine sample of montane and premontane ecosystems
- to maintain *Podocarpus* forests in a natural state
- to conserve geomorphologic features, vegetation, and soil cover of the upper watersheds of the rivers Jamboe, Sabanilla, Bombuscaro, Numbala, Loyola, Nangaritza, Quebrada de Campana, and Vilcabamba, among others
- to conserve scenic páramo (humid, tropical, alpine grasslands) and lake sites for tourism visitation
- to provide opportunities for open-air recreation and environmental education for the region's growing urban populations, particularly of Loja and Zamora
- to support an integrated land-use system compatible within the region, providing opportunities for community development and the preservation and sustained, economical use of the area's resources.

More recently, the extraordinarily high levels of endemism and species richness of PNP have become its celebrated features, as international and Ecuadorian scientists began documenting the site's biodiversity and discovering new species. Hence, biodiversity protection is now identified as a key goal of conservation at PNP within management plans and public documents.

The rugged terrain of PNP and the paucity of cartographic information for the area hindered the exact delineation of park boundaries. The decision whether to include occupied lands in the park was made on a case-by-case basis (Apolo, cited in Kapila 1993). The western boundary of the park was drawn to include several farms on the assumption that pasture expansion was improbable in this area due to severe climatic conditions and steep topography. On the other hand, recently colonized lands near Zamora and Loyola were excluded from the park, as they formed part of an actively expanding frontier. Plans to protect the forests lying south of the current boundary were abandoned because of their proximity to the

Peruvian border and problems with colonization in the area. Finally, extensive, flat, pristine forest to the east was excluded because park planners assumed it would be quickly colonized.

The first attempt to physically demarcate PNP was ineffectual. Working without maps or compass, the park staff simply hiked up slopes until they reached dense forest and erected signposts (S. Hoffman, pers. comm., 1995). Ill-defined boundaries led to conflict between park managers and a community clearing forest near the southern boundary. While the park staff threatened them with eviction, the colonists insisted that they were outside the park. The conflict stemmed from uncertainty regarding the position of the park boundary in relation to Río Numpatacaime. In 1995, FAI used a geographic position system (GPS) to ascertain that the community was indeed outside the park.

With the support of PiP, FAI is currently demarcating PNP boundaries using GPS and current maps. The work often entails entering remote, inaccessible regions. Attempting to draw precise lines around a protected area according to altitudinal lines is at times paradoxical. In some places, the field technician finds himself defining the boundary through vast forest, leaving pristine habitat outside the park (R. Tapia, pers. comm., 1995). At other times, he finds himself defining the boundary in the middle of a pasture. The terrain is so steeply dissected that the result is a very fine-grained, serpentine boundary. Greater technical training would likely improve the pace and accuracy of park delineation (H. Arnal, pers. comm., 1995). However, Tapia reports that the most challenging part of his work is not the grueling hikes nor the garbled maps, but rather dealing with local farmers who often object to the location of park boundaries (see "Linkages between Park and Buffer Areas," p. 308).

Land and Resource Tenure

Within PNP and its adjacent zone, land is held in a variety of ways, both formal and informal, according to length of residence, access to government technical and legal services, ethnicity, and economic level. To understand variation in land tenure, it is useful to divide the population neighboring the park into three socioeconomic categories: long-term residents, recent colonists, and indigenous residents (as per Hernández et al. 1995). Our analysis does not include land tenure within urban areas;

rather, we focus on rural settlements adjacent to or within PNP. Concessionary rights to gold will be discussed in the next section. Unless otherwise noted, the data presented below come from the socioeconomic study carried out by FN in preparation of the management plan for PNP (Hernández et al. 1995).

Long-Term Residents

Along the western boundary of PNP, from the city of Loja extending south to Valladolid, is an agricultural population whose residence in the area dates from colonial times. The farms along the western boundary of PNP fall into two categories: large properties belonging to absentee owners, and small farms belonging to occupants. Informal reports indicate that the majority of farms along the western face of PNP are formally titled properties (R. Tapia, pers. comm.). A sample of the size of landholdings censured and titled by the Ecuadorian Institute of Agrarian Reform and Colonization (IERAC, now INDA, the National Institute of Agrarian Development) during the agrarian reform (in the 1970s) averaged from 173 ha (in Malacatos) to 1,047 (in Vilcabamba). These farms were obtained at minimal cost, and many were later sold for profit or subdivided into parcels.

Currently, farms in this area are being purchased for vacation homes at a steady rate. There is a slow decrease in population (approximately 0.5–1 percent per year) due to emigration to urban areas or to other regions of Ecuador. Absentee landholders are common in this area. The farms lying immediately adjacent to PNP are devoted primarily to livestock production. During recent years, pastures have expanded up slopes into PNP.

A unique example of land tenure is found at Vilcabamba, where a North American family owns 2,500 ha of land, 75 percent of which lies within PNP. The family obtained Bosque Protector status for their forest before PNP was established by petitioning the Dirección Nacional Forestal (DINAF, now INEFAN) and submitting a formal management plan. In such a case of overlapping categories of forest management, the more restrictive category carries greater legal weight; i.e., the land is managed as a national park. In fact, the 2,500 ha in question are better protected than many other sites in PNP. To date, the owners have not been granted tax exemption on their property, despite the fact that the state offers such incentives for those who retain forest on their land (*Forestry Law,* chapter 8, article 53). Achieving tax-free status requires a field survey of the property by a government technician from Quito. Given that property tax collection in the area is sporadic at best, there is little motivation for owners of forested property to tackle relevant bureaucratic procedures.

An unusual turn of events is occurring in the southeastern section of PNP, where the owner of 700 ha within the park complains that National Park status has not protected her forested property from colonists. Currently, she is seeking Bosque Protector status for her land in the hope that it will somehow deter colonists. However, National Park is currently the strictest conservation category within Ecuador.

South of Loja, along the park's western boundary, population densities decrease (Loja canton, about 32 sq. km.; Malacatos, about 27 sq. km.; Vilcabamba, about 18 sq. km.; Yangana, about 4 sq. km.; and Vallodolid, about 2.3 sq. km.). Along with this trend, the population becomes younger in age, and cash crop plantations are replaced by subsistence crops or cattle production. Although population densities are lower around the southern reaches of the park, INEFAN and conservation NGOs consider the active colonization process in this area to be a potential threat to park protection. This is due to the nature of the colonists' land use (steady conversion of forest to pasture) and to the fact that their populations are rapidly growing (e.g., El Porvenir del Carmen grew 5.2 percent between 1974 and 1990) (Hernández et al. 1995). The stable or declining population conditions of Loja province are reversed in Zamora-Chinchipe, a province known for its colonists and gold miners.

Recent Colonists

On a national level, agrarian reform in Ecuador during the 1960s and 1970s accelerated spontaneous colonization due to national policies that favored industrial export production and granted ownership to individuals who cleared forested public land (*tierras baldías*). The impact of the agrarian reform reached the region of PNP in the 1970s. Driven by shortages of arable land in the Sierra and drawn by the promise of owning large farms on good soil, Andean farmers began clearing forested lands lying to the southwest, south, and northeast of the park (in Numbala Alto, Numbala Bajo, El Porvenir del Carmen, Loyola, and Romerillos). All of these sites are located in Zamora-Chinchipe, a province targeted by the Ecuadorian government for colonization.

During recent decades, Zamora-Chinchipe has undergone rapid, extensive colonization. The province's population grew 4 percent between 1974 and 1990, a much higher rate than that of the neighboring province of Loja (0.7 percent). Nonetheless, the population density of Zamora-Chinchipe remains low at 3.4 people per square kilometer (PREDESUR, cited in Hernández et. al. 1995). Like agrarian frontiers elsewhere in the Amazon basin, it is the extensive, temporary quality of the colonists' land use that leads to deforestation, not the density of their population.

The pattern and rate of colonization in Zamora-Chinchipe is complex. Colonization is not a steady process, rather it is intermittent as immigrants arrive in pulses and are drawn to certain areas. For example, during the last decade, population growth in the province reached 10 percent due to the influx of gold miners who were responding to the elevated price of gold on the world market (see "Gold," p. 300). Government intervention has also facilitated colonization. Between 1985 and 1986, IERAC adjudicated land titles for 34,508 ha in the province, divided among 758 families (average farm size 46 ha/family). By 1992, 367,300 additional ha were allocated to 7,529 beneficiaries by IERAC. Another 637 beneficiaries received title for 46,910 ha under special national agrarian reform initiatives (average farm size 73 ha/family).

Classic problems of rapid colonization of the forest frontier are manifested in Zamora-Chinchipe. Much forest has been lost to pastures, which support a limited number of cattle (less than 0.7 head/ha) for a period of only a few years. The indigenous Shuar population residing in the province has lost traditional lands, and their cultural customs and land-use practices are eroding. Absentee landholdings are common at the forest frontier, but as yet there has not been extensive fragmentation of landholdings. Instead, at some sites land concentration is occurring (e.g., at Romerillos). Most of the recent settlers along the park's boundary in the province of Zamora-Chinchipe originate from the neighboring province of Loja. More specifically, the colonists of Loyola and Numbala come from Yangana, Vilcabamba, Malacatos, and the canton of Loja, while those of Bombuscaro, Romerillos, and Curintza come from the canton of Zamora and the parishes of Loja.

Over seventy farms in these new settlements are partially or completely within the borders of PNP. Whether these farms predate the park's establishment is subject to debate. The first colonists to arrive in the area secured landholdings simply by occupying and working the land (i.e., clearing the forest). More recent colonists have resorted to buying land parcels from first-generation colonists. In a 1995 survey of 182 families at these sites (both within and outside the park), only 46 (20 percent) held legal title. Of the 73 partly or wholly within park boundaries, 15 percent had title. The majority of colonists own only one piece of land. Out of a sample of 228 individuals residing along the park's boundary in the province of Zamora-Chinchipe, 18 percent owned less than 20 ha, 53 percent between 21 and 50 ha, 18 percent between 51 and 100 ha, and 11 percent more than 100 ha. Those individuals holding farms of more than 100 ha (11 percent) together control 36 percent of the land sampled. Among the same 228 farmers surveyed, 99 (43 percent) farmers resided on their land, while 129 were absentees (57 percent). The proportion of absentee owners varies among communities. The communities having the highest number of residents are Loyola and Numbala Alto. The farms of Romerillos, Curintza, Numbala Bajo, and Bombuscaro belong primarily to absentee owners. The level of absenteeism affects the type of land use by a community (see "Resource Use," p. 295).

Hernández et al. (1995) identify several factors influencing the rate of expansion of the agricultural frontier at the park's edge, including level of infrastructure, government restrictions on land use, and land markets. For example, Romerillos is located at a formerly active front of colonization in the northeastern sector of the park, but the population is currently declining, and expansion has also ceased at Loyola. It is unclear whether INEFAN is influencing the dynamics of frontier expansion by restricting forest clearing and other practices, or whether, instead, regional or national factors are in operation. The fact that frontier settlements in the area are dispersed and small might deter colonists from defying INEFAN to invade PNP in the forceful manner of colonists invading protected areas elsewhere in Ecuador (e.g., Cuyabeno Reserve). With the completion of the road flanking the eastern boundary of PNP (Proyecto Carretera Marginal de la Selva), colonization is likely to greatly accelerate around the park.

Most colonists neighboring PNP work independently, but forty-five colonists clearing land at Numpatacaime have formed an association (Asociación de Colonos de Nangaritza) and obtained a collective legal title for 14,700 ha. Within their property, the association intends to designate communal land as well as private farms of about 150 ha each. The association's property contains extensive primary forest, which protects wildlife and an important watershed. The members intend not to clear forest along the slopes of the watershed or within five hundred meters of the rivers. Currently, the association is discussing other sustainable land management practices with FAI (see "Resource Use" below).

Shuar Population

Historically, the Shuar people inhabited the southern Ecuadorian Amazon in widely dispersed, small communities (about 1.4 to 4.3 sq. km. per family) (PREDESUR, cited in Hernández et al. 1995). Nine Shuar settlements, or *centros*, are situated outside PNP along the Río Nangaritza (M. Chiriapo, pers. comm., 1995). Centros range in size from ten to eighty families (approximately 36 to 380 people). The nine centros proximate to PNP occupy an area of 20,400 ha. Each holds a legal, collective land title. Individuals do not own private land titles, although each family uses a delineated area (Naichap, cited in Hernández et al. 1995).

Describing the impact of colonization on Shuar populations in southern Ecuador, PREDESUR (1978) identifies three stages of cultural disintegration : (1) traditional lands are lost to colonists in a formal or informal process sanctioned by the state, (2) Shuar communities are relegated to reserves, and (3) there is an obligatory imposition of cooperatives as formal productive organizations on Shuar communities. This latter stage derives from a strategy used by the Federación Shuar to legally secure land title from IERAC (now INDA).

Resource Use

There are a variety of resource uses in and around PNP. The area to the west of PNP has been extensively deforested. Yet the park is presently connected to relatively intact forest extending several kilometers to the south and southeast. Unfortunately, this large expanse of forest is threatened by growing colonization pressure. A discussion of resource use around PNP is best organized by comparing the two provinces, Loja and Zamora-Chinchipe. Again, unless otherwise noted, the data presented in this section come from the socioeconomic study carried out by Fundación Natura in preparation of the management plan for PNP (Hernández et al. 1995).

Loja Province

Most natural habitat in Loja province has already been converted to agriculture. The population of the province earns its income primarily through agriculture (50 percent in 1990) or employment in public service or industry. The production of cash crops such as sugar cane, coffee, and tobacco dominates agricultural activity in the vicinity of the canton of Loja. Only 0.7 percent of Loja's population works in mining, but employment in tourism is growing steadily in Vilcabamba. To the south of Loja, beyond Vilcabamba, subsistence agriculture and cattle raising are of increasing importance.

Along the western boundary of PNP, edge habitat is affected by fire and grazing. Fire started by one individual can spread to several neighboring farms and into park lands during times of drought (S. Hoffman, pers. comm., 1995). During the 1970s and 1980s, the law recommended penalties for those burning open lands without a burning permit. Attempts to levy burning penalties were ineffectual (S. Hoffman, pers. comm.), and now there is no official intervention. One of the major impacts of fire is that it changes the composition of the vegetation, including making it easier for exotic grasses to invade. Some long-term residents, and park staff as well, report that the rate of burning has recently declined (S. Hoffman, pers. comm., 1995; S. Calderón, pers. comm., 1995).

Grazing is a persistent management problem along the western boundary of PNP. Individuals frequently transport their cattle by truck to the park's edge at Cajanuma (P. Cabrera, pers. comm.). Although the stocking rate is low, cattle suppress habitat regeneration along much of the western fringes of PNP and cause serious soil erosion. Assuming individuals maintain current herds of cattle, pasture degradation will result in increased grazing within fragile park lands.

Zamora-Chinchipe Province

In 1990, approximately 50 percent of Zamora's population were described as agriculturalists and 13 percent as miners. Only sixteen years earlier, 76 percent of the population farmed and only 0.1 percent mined (INEC, cited in Hernández et al. 1995). These figures reflect the impact of the gold rush on the province during the eighties (see "Gold," pp. 300, 311). When considering land use in Zamora-Chinchipe province in terms of area, cattle raising is by far the most extensive activity. This is not surprising given the local conditions of scarce labor, abundant land, and remote markets. Crop production covers only a limited area and is primarily for subsistence (maize, yucca, and plantains predominate).

Within Zamora-Chinchipe province, approximately 550 people reside in or own land adjacent to PNP. Most of these residents established their farms in the past two decades. The population is young (60 percent are less than 19 years old) and predominantly male (women move to urban areas in search of employment or to raise children). Absentee farms are common in this area.

The main economic activities of residents neighboring the park are livestock (77 percent), agriculture (11 percent), and timber (11 percent). A typical farm in this area covers fifty-four hectares; most of which is still under forest (60 percent), while the remaining area is allocated to pasture (39 percent), and cultivated fields (1 percent). These proportions vary, however, according to whether the farm is in or outside of PNP and whether the owner is an absentee or a resident. Colonists are not a homogeneous group; rather, they have variable access to resources and markets. Individuals owning farms within the park boundaries on average leave 77 percent of their land in forest, compared to 52 percent for farms outside the park. Farmers residing on their land are likely to use it more intensively than absentees, who use their land almost exclusively for cattle.

Most of the cattle in the area are criollo (native stock mixes) and are raised for meat. At the time information was collected for the 1995 management plan, there were a total of 591 cattle (approximately eight per family) within the park and 3,310 (twenty-three per family) head of cattle on farms adjacent to the park. While the average number of cattle per family is modest, certain individuals own large herds (e.g., more than 300 on a farm adjacent to PNP). The average stocking rate is slightly lower in the park (0.7 animal/ha) than outside, reportedly due to INEFAN's restrictions. However, greater variation exists between stocking rates at each settlement than between farms inside and outside the park. Most ranchers use fencing and practice rotational grazing. *Cetaria* spp. is the most popular exotic grass. Technical or veterinary support is scarce, but colonists do treat their animals for parasites. Most families also have chickens, and slightly over half have pigs. Colonists sell their cattle in "local" markets (those from El Porvenir del Carmen, Loyola, Numbala Alto, and Numbala Bajo have to walk six to ten hours to reach a market).

The Shuar are known as subsistence hunters and fishers, although this is changing as colonists and miners enter their territory. They utilize modern technology in hunting (shotguns) and in clearing the forest (chain saws). Only 5.6 percent of Shuar land is under agricultural production. They typically plant subsistence crops (maize, cassava, yam, and plantains) in plots of less than one hectare. This low-intensity resource use will inevitably change as Shuar populations grow (they are presently increasing at 4.3 percent and are expected to double by 2012), and as they become integrated into commercial production systems.

Colonists have begun to invade Shuar lands (forested land in particular), bringing with them commercial markets, new human diseases, and infrastructure. Their arrival is transforming Shuar land-use systems and their relationship with their environment. No longer do the Shuar have unlimited land available for hunting or shifting agriculture. Likewise, the proportion of their land converted to agriculture is steadily increasing. Perhaps the most dramatic change is in the Shuars' increasing investment in cattle raising. Currently, their stocking rate is equivalent to that of recent colonists (0.7 cattle/ha). Hernández et al. (1995) predict cattle raising will soon become the primary economic activity among Shuar settlements.

According to the representative of the Shuar Federation in Zamora, it has been more difficult to defend Shuar land from gold miners than from colonists (M. Chiriapo, pers. comm.). He also stated that Shuar youth are entering into mining activities.

Resource Use within PNP

Traditionally, resource use within the area of PNP was limited by the area's inaccessibility and the ready availability of wildlife, timber, farmland, etc., in other parts of the region. New roads are now facilitating access to PNP, and natural resources (e.g., timber and wildlife) are becoming scarce at the regional level. Some resources are affected more than others by national or international markets. For example, the increase in the international price of gold over recent decades has intensified the demand for access to PNP's gold.

All resource extraction is illegal within PNP; however, legal prohibition has not prevented various extractive activities in the park, ranging from sport hunting by individuals to commercial timber extraction, to artisanal and industrial mining. The efforts of INEFAN to control such activities to date have been seriously limited by: (1) a severe shortage of personnel (currently, seven park guards are assigned to patrol 146,280 ha); (2) a lack of support from provincial law enforcement (e.g., the police in Zamora); (3) managerial limitations of INEFAN at the regional level; and (4) ambiguous government policies and/or lack of coordination between public agencies. Until 1994, FN contracted park guards for the eight areas where they worked, including PNP, using funds from the debt swap. The government has since been unable or unwilling to support an adequate number of guards.

NGOs such as Fundación Arcoiris, Fundación Natura, Ecociencia, etc., have used lobbying, policy dialogue, field research, and environmental education to support INEFAN in protecting PNP from illegal resource use. Fundación Arcoiris even joins park rangers in patrol operations on a periodic basis. Some resource use (e.g., hunting or timber poaching) is best controlled at a local level. Others, gold mining in particular, require regional, national, and even international intervention due to the political and economic stakes involved.

Hunting, Fishing, and Wildlife Traffic

Most local informants report an increasing scarcity of wildlife within and around PNP, although there is no baseline data on wildlife populations or monitoring (Hernández et al. 1995; Suárez and Mena 1992). INEFAN's ability to control hunting is limited by its shortage of park rangers. Rifles and dogs are confiscated from poachers, as is any quarry they may have hunted. Park guards are often intimidated, however, as they are unarmed on their patrols. Approximately three to four times a year PNP staff find individuals attempting to sell birds or monkeys as pets (S. Calderón, pers. comm., 1995). No facilities exist for caring for or rehabilitating confiscated animals. Confiscation records of wildlife (dead or alive) should not be interpreted as estimates of the volume of wildlife trade or hunting because INEFAN conducts patrols only sporadically.

Little research has been carried out in the eastern flanks of the park, where Shuar may occasionally hunt. Illegal fishing occurs in PNP. The use of dynamite has largely been controlled; however, some individuals continue to use barbasco, a plant-based poison, for fishing.

Timber Extraction

Problems in the control of timber extraction parallel those related to hunting. *Podocarpus* spp. and other fine timber species are illegally extracted from PNP and sold to a handful of merchants in Loja. Timber poaching is a highly localized activity. Paths in Romerillos and Curintza allow colonists to extract trees. In Numbala Alto, Numbala Bajo, El Porvenir del Carmen, and Loyola, the absence of paths and the distance to markets preclude selling timber.

When park staff encounter timber poachers, the cut wood is confiscated and a court case is prepared. INEFAN currently takes this action approximately two to four times a year (S. Calderón, pers. comm., 1995). In some cases, a fine of six to nine days' wages has been paid; in other cases, INEFAN lacked the support of local law enforcement (the police and army) to enforce the penalty. Likewise, most court cases are not tried. Timber extraction rates in PNP are reported to be declining due to INEFAN's restrictions and the growing scarcity of large commercially valuable species (Hernández et al. 1995). Without data from field monitoring, it is inappropriate to assume that timber extraction is not a persistent threat to PNP's forests.

Orchids

Both international and local citizens have been found illegally collecting orchids in PNP. In one such case, the park director attempted to confiscate an international collector's illegal harvest by confronting him at the airport. In an ironic turn of events, the director wound up incarcerated by local authorities. The vulnerability of PNP's orchids (and other genetic resources) to overharvesting by foreign collectors reflects a national problem. Seldom do collectors follow the legal procedures required to obtain collecting permits or licenses. Even when permits are issued, there is minimal oversight or control by the Ecuadorian government over the collection of genetic resources. FN is currently addressing national policy on genetic resources and property rights through participation in discussions for the Régimen Común de Acceso a Recursos Genéticos as part of the Cartagena Accord, participation in the National Biodiversity Group of the CAAM (Comisión Asesora Ambiental de la Presidencia de la República), and generation of institutional policies on this theme.

Gold

The Forestry Law of Ecuador does not allow mining within any natural protected areas that are part of the National System of Protected Areas, including parks. Article 87 of the Mining Law restates these obligations with an additional condition:

> The State will not foster mining activity within the limits of a national forest or protected area. Only for reasons of national interest will such activities be permitted, provided they comply with all the regulations stated herein.

To be able to prospect, explore, or exploit gold in a national park, a mining company must obtain a permit from INEFAN. Mining legislation recommends that an exploitation system of tunnels should be used, with the fewest possible incursions on the park. Ordinary citizens are strictly forbidden to undertake mining activities within the limits of a national park. (Note: Unless otherwise noted, the information in this section comes from Vallée et al. 1992, and FAI 1992.)

In the case of PNP, mining companies and artisanal miners alike have explored or exploited gold. (See also "Large-Scale Threats," p. 311) Public institutions have played contradictory roles, both facilitating mining activity in the park and restricting it. A brief chronological account of mining in PNP during the past decade reveals a complex constellation of economic and political interests currently surrounding the park.

The increase in gold prices during the 1970s led to a gold rush in southern Ecuador. By the mid 1980s, fifteen to twenty thousand miners were working at Nambija, a site to the northeast of PNP. In 1985, INEMIN granted mining concessions to various national and international companies for over 95 percent of PNP's area. In 1986 a Norwegian-Ecuadorian mining company called Cumbinamasa (also known as EcuaNor) obtained an exploration concession for an area of 16,875 ha at San Luis, in the core of PNP. The area of interest, surrounded by primary forest and mountainous topography, was very difficult to access. In 1987 a thirty-two-kilometer trail for mules was constructed to San Luis on the trace of an existing Inca trail. This trail was paved with the park's tree trunks, including *Podocarpus* spp., *Clusia* spp., and *Weinmannia* spp. Cumbinamasa eventually signed a contract with the British company Rio Tinto Zinc (RTZ), one of the largest mining companies in the world, which supported the construction of three camps and a heliport. During this period, Cumbinamasa erected signs prohibiting public entrance to the concession. Even park guards had to ask permission to enter San Luis.

The newly constructed trail allowed artisanal miners improved access to the core of the park. Since 1984, a group of approximately 120 farmers from Zamora have mined in the park as a subsidiary source of income. Eventually, they formed a cooperative (Cooperative San Luis), and in 1988 they petitioned for a five-thousand-hectare concession at the ridge of San Luis. DINAMI refused their application on the grounds that the concession was inside a national park. The artisanal miners' resentment was palpable, given that Cumbinamasa had been granted exploratory concessions that same year. Despite the rejection of their petition, these artisanal miners continued to work in PNP.

Artisanal miners and employees of Cumbinamasa clashed in the heart of the park, at times even firing shots at one another over access to mining sites. In 1990, a three-party agreement was reached between Cumbinamasa, RTZ, and the Cooperative San Luis; the two companies prospected the Romerillos concession, and the cooperative mined in an area between that concession and the Augusta concession. To avoid the entrance of additional artisanal miners, Cumbinamasa paid for two guards to help INEFAN restrict access to San Luis.

During the late 1980s and the early 1990s, there was growing concern among public and private institutions involved in environmental conservation regarding ecological damage to PNP by mining activity. Local and national NGOs (e.g., FAI, DANTA, FN, Ecociencia) mounted environmental education campaigns, supported field research and monitoring, and worked with public agencies to improve PNP infrastructure. The field staff of the park had to face the mining problem most directly. Unarmed and few in number, park guards suffered the hostile response of miners to their efforts to prevent mining or confiscate mercury or mining equipment. Two park guards were physically beaten by miners and suffered death threats. They received minimal backup from authorities in Zamora. In fact, the police from Zamora in one case arrested and incarcerated a park guard for attempting to confiscate a miner's property just outside PNP (L. Tambo, pers. comm., 1995). Threats from miners eventually resulted in park staff entering PNP on patrols only when accompanied by members of the armed forces.

It was reported that DINAF staff in Quito were overwhelmed by the problem and felt unable to confront the powerful political and economic interests behind the mining industry. During a 1990 interview regarding mining concessions in PNP, a high-ranking INEMIN authority in Quito asked, "What park?" and claimed there was no record of a protected area in the concessionary maps in INEMIN's files.

By 1990, FAI had organized an arduous campaign against mining in PNP at the local, national, and international levels. RTZ withdrew from the site in late 1990. Fundación Arcoiris filed a suit against INEMIN. In March of 1993, the Tribunal de Garantías Constitucionales (TGC, or the Court of Constitutional Guarantees) responded to FAI's suit, calling attention to the fact that the Ministry of Agriculture and the national director of mines had permitted mining in the park, and ordering INEMIN not to authorize any mining in PNP in the future. Furthermore, the TGC resolution prohibited the construction of a road between Cerro Toledo and La Esmeralda (Numbala) until environmental impact assessments were undertaken. Cumbinamasa finally left the park in 1993. With the departure of international companies, artisanal miners entered PNP in greater numbers. By the end of 1993, approximately 800 or more miners were working in three sites in the area of San Luis, with attendant social problems such as drug use and prostitution.

Artisanal mining is a term that encompasses a broad variety of mining operations, from individual panners to contracted teams of miners using pumps, pressure hoses, dredges, rock crushers (*chanchos*), and stamp mills (*chancadores*). Despite the picturesque name, artisanal mining (both placer and hard-rock) has caused substantial damage to several sites within PNP. Miners removed vegetation and topsoil, then used pressure hoses to wash away rocks and sand. In a steep environment such as San Luis, the erosion was severe. Over 150 tunnels were excavated to depths of up to thirty meters and lengths of two hundred meters. Gold was recovered using the process of sedimentation amalgamation and burning, in which mercury is released into the environment in liquid and vapor form. Estimates range from one to five grams of mercury released for each gram of gold extracted. Alarmingly high levels of mercury have been recorded in streams at mining sites within the park (4.51 ug/g), equivalent to those recorded at mines in Nambija. It is difficult to ascertain how great a mercury exposure was suffered by communities downstream. Beyond mercury contamination, the area's ecosystem has suffered deforestation along trails and at mining sites (e.g., twenty-three hectares at one mining site, Curishiro), erosion, and loss of wildlife due to hunting (Suárez and Mena 1992).

The Inter-institutional Committee for the Defense of PNP (ICD-PNP) began to take on the defense of the park against mining interests (see "Organizational Roles," p. 305). Meanwhile, the miners in the park also organized and created the Association of Artisanal Miners of San Luis to promote mining activities within the park. In an effort to resolve the mining conflict, NGOs together with INEFAN initiated a long process of dialogue between the mining leaders, ICD-PNP, and the principal civil service agencies. After lengthy, heated negotiations, the leaders of the Association of Artisanal Miners of San Luis agreed that the miners would leave the park on two conditions: (1) that they be allowed two months more in San Luis to transport their heavy machinery out of the park, and (2) that they be given a mining concession outside the park boundaries. In March of 1994, the miners peacefully exited from the park. They were granted a 420-ha concession at Chumbiriatza, which later proved to contain minimal gold reserves. The support of the armed forces in preventing the subsequent entrance of other miners to San Luis was essential. Rotations of four soldiers were stationed within the park at the former mining sites.

In January of 1995, the military conflict between Ecuador and Peru caused the soldiers guarding the mining sites to be recalled. Immediately following this, miners entered PNP, among them members of the Association San Luis as well as many new miners from other regions of Ecuador. Again, the number of miners working in the park reached four hundred individuals, and the investment in equipment and transport was indicative of economic interests beyond the scale of single individuals (INEFAN 1995). The miners were notified in April of 1995 that they must exit the park, but the order was ignored. In June of 1995, the 62nd Battalion of the Armed Forces (from Zamora) carried out a census of miners and found that the miners were aware of their illegal presence and that while they appeared to be poor, they were working with equipment valued at "dozens of millions of sucres" (INEFAN 1995). After further negotiation, in July of 1995, a meeting occurred at the 62nd Battalion between the leaders of the newly formed Association of Sentinel Miners of Ecuador located in PNP, the subsecretary of mines, the director of natural areas and wildlife, and the ICD-PNP. During hours of debate, various perspectives became clear, roughly characterized as follows:

INEFAN, civil authorities of Loja province, and conservation NGOs: Mining should not occur within PNP, as it is illegal and ecologically destructive and poses a health risk (that of mercury contamination) to several communities downstream.

Subsecretary of mines: Artisanal mining should not occur within PNP because it is environmentally destructive, illegal, dangerous, and an inefficient way to extract gold.

Miners' association: Miners deserve the right to make a living in the best way that they can; they receive no support from government agencies like DINAMI. If they are to be forced out of PNP, no one else, not even international companies, should be allowed to mine in the park.

Civil authorities of Zamora province: Zamorans ought not to suffer the imposition of the political will of Loja authorities (nor the will of national authorities). Mining is essential to the economy of the province.

Armed forces: The law of Ecuador must be enforced as it is interpreted by national institutions.

A final agreement was reached that the miners would leave the park after a two-month grace period; they would not be legally punished if they left by September 30, 1995. No new miners could enter the park.

Organizational Roles

The public agency responsible for protecting PNP (INEFAN) is limited by a shortage of funding and personnel, low managerial and technical capacity, and a lack of political power. Meanwhile, an unusually numerous and active collection of local and national NGOs are working to defend PNP. PNP is thus an interesting case for exploring the questions of what responsibilities are appropriate for NGOs (both local and national) in the context of weak public environmental agencies; how and if NGOs can strengthen public environmental agencies; and how local and national NGOs can work together. A subsample of NGOs are described here to demonstrate the variety of institutions, both national and local, working at PNP.

Fundación Natura. FN is a national NGO that supports management actions in eight protected areas throughout Ecuador, among them PNP. FN had been working in PNP for several years when the debt-for-nature swap was carried out in 1989. As it was already dealing with a high volume of projects, FN signed agreements and hired consultants to do some of the in situ work for it (Arnal 1993). Strengthening and working with local NGOs has been a Fundación Natura policy applied in several protected areas throughout Ecuador, including strengthening local groups by hiring them as consultants. In PNP, Fundación Natura started working with Fundación Arcoiris in 1992. In the early 1990s, local NGOs became active. At the moment, FN mainly does national-level work while local NGOs do most of the work on site.

Fundación Arcoiris. FAI was founded in 1989 and is based in the city of Loja. Its main focus is the protection of PNP and other natural areas of southern Ecuador. The members of FAI work primarily on a voluntary basis for the benefit of the park. FAI's activities are concentrated in four main areas: environmental education, scientific research, community development, and public awareness. FAI has advanced rapidly (as evidenced by its recent national and international awards and honors) and now is a direct partner of The Nature Conservancy and INEFAN. FAI (along with FN) has played a key role in creating and maintaining the ICD-PNP.

Fundación Maquipucuna. FM is working in the area of Jamboe and Loyola to improve farm management, reforest slopes, inform communities regarding the importance of conservation, and support local organizations. Its focus is on residents who live within approximately one kilometer of the park boundary. Its policy is not to work directly with people living illegally within the park. Thirty-five out of 110 families at Jamboe are actively participating in FM's agricultural and forestry projects (i.e., 35 families have demonstrably changed their land practices in some way). This is an unusually high participation level. FM once intended to work in park delimitation, but work in that area appears to be suspended.

Fundación Colinas Verdes. FCV is formed of twenty-five farmers and intends to work to reduce erosion and improve irrigation by demonstrating alternative farming and pasture management techniques on a one-hundred-hectare trial plot and working with individual farmers on production problems. The foundation tries to provide options for farmers with poor-quality pastures, and it has an elaborate extension program of incentives, credit and technical assistance. It is also developing basic research facilities in the twenty-five-hundred-hectare Bosque Protector and in the future hopes to work on slopes outside PNP in other areas of the buffer zone.

Ecociencia. Ecociencia focuses on field research related to conservation and management of protected areas. At times it was the only field research presence at PNP. It initiated a training and monitoring project financed by the Wildlife Conservation Society (WCS), which drew international attention to the site and resulted in the training of other investigators and spinoff projects on hunting and forest disturbance. The monitoring component of the project at PNP was not designed to meet any management need of INEFAN (Arnal 1993). FAI has continued the monitoring project and has improved its design with the help of international expertise (P. Feinsinger, pers. comm., 1995).

By and large INEFAN has positive working relations with local and national NGOs. It is in many ways dependent on NGO support. Some INEFAN staff expressed resentment of the high level of funding NGOs can attract. Others complained that more emphasis should be placed on implementation versus planning ("What use is a sophisticated management plan if INEFAN has only four park guards?"). Referring to one particular case, INEFAN felt that the NGO approached INEFAN only because it was necessary for submitting a proposal for funding, but that once funding was in hand, INEFAN was ignored. Although INEFAN may feel ignored, FN and FAI elaborate their annual plan of activities with the INEFAN park administration and formally seek the approval of the Directorate of Natural Areas and Wildlife.

The citizens of towns are more likely than residents in rural communities to be aware of the differences between the staff of INEFAN or NGOs in particular and the distinct roles of public agencies and conservation NGOs more generally (S. Calderón, pers. comm., 1995). Calderón explained that these confusions can cause conflict. For example, recently a team of consultants working with FN to develop the management plan for PNP conducted participatory rural appraisals (PRAs) and open discussions with several communities on the park's edge, including those in conflict with the park over timber and hunting. Calderón complained that subsequently some community members interpreted statements made during those meetings as sanctions to freely use PNP's resources. Other sources countered that park staff, including the chief of the park, were in attendance at those meetings and were free to clarify any debate. Invitations to the workshops were signed by INEFAN; however, it does not actively participate in such discussions.

A key achievement in protecting PNP has been the creation of the ICD-PNP. FN and FAI worked together to create this group, which comprises representatives of public and private institutions in the region. Not only has creating this committee helped expand advocacy for park protection, the committee itself has acted as a facilitator in resolving conflicts related to the park (e.g., mining). Integrating representatives of both Loja and Zamora provinces within the committee has been difficult; at one time

Zamora wanted to created its own committee. Membership in the committee is made up of PREDESUR and representatives of Loja and Zamora, including mayoral and provincial representation, and other government agencies. The original ideal behind creation of the committee was that it could provide political support to INEFAN to give them greater strength in their decisions and actions to protect PNP. In reality, it should be an advisory rather than an executing agency. More recently, the original committee has indicated that it wishes to receive funds for managing conservation projects. Originally, it was not envisioned to be an implementor of projects. Maintaining the committee's enthusiasm while dissuading it from actively entering into project administration might prove difficult.

Linkages between Park and Buffer Areas

During recent years, NGOs have played an increasingly central role in various activities at PNP, including environmental education, public awareness campaigns, patrols against timber and wildlife poaching, infrastructure construction, research, and boundary delimitation. The work of greatest importance is incorporating communities into the management of PNP, while at the same time finding solutions to the problems affecting them. Each NGO operates according to an independent mission, yet certain basic objectives are shared by the NGOs working in community outreach around PNP:

- to inform communities about goals of the park via environmental education, both formal and informal
- to enter into dialogue with communities about their development goals and resource needs
- to provide for certain demands of communities like water pumps, latrines, and health care classes (these are not directly related to conservation but are a means to establish positive relations with a community)
- to provide alternative economic activities to substitute for current destructive land-use practices.

To date, NGOs in the area have made the greatest achievements in the first and second objectives. With regard to the third objective, NGOs report that they have built positive relations with communities due in part to answering the needs identified by community members. The greatest challenges experienced thus far are related to substantially influencing economic activities of a community (the fourth objective). In evaluating the strategies and progress of NGOs working with communities around PNP, it is important to remember that most projects have been underway for only a year or two. Therefore, it is premature to categorically describe any of these activities as successes or failures. An additional critical point is that the objectives behind many of these activities are so exceptionally broad (e.g., to harmonize the relationship between a community and PNP) that evaluation is difficult.

There is great variation between the communities neighboring PNP. They each hold a unique history, land use, and socioeconomic context. Some communities were established way before the creation of PNP. Park establishment has affected each community differently, and within each community families are affected differently. In some cases, communities have felt rejected for being near or within PNP (e.g., Loyola). The relationship between park administration and the communities hasn't improved greatly since the creation of the park, due in part to the lack of clear policies by INEFAN for resolving land tenure or resource-use conflicts.

The city of Loja is the largest urban settlement in the region and is the site of greatest park advocacy. Government and nongovernment organizations have cooperated in the defense of the natural resources of PNP (see "Organizational Roles," p. 305). The western zone of influence of PNP is made up of numerous towns (e.g.,Vilcabamba, San Pedro de Vilcabamba, Sacapo, Chalaca, Rumishitana) that generally have very positive attitudes to PNP. Tour operators from Vilcabamba realize that PNP represents an important tourist attraction. To date, however, there is no control of visitors to PNP or of their impact on the park's ecosystem. Nevertheless, tour operators have requested training from park authorities in ecotourism management and are willing to actively participate in the regulation of tourism activities.

An outstanding case of community outreach is in Sacapo, a longstanding rural settlement. FAI has worked in environmental education there for an extended period (more than five years) and, at the request of the community, has now initiated a series of lectures focused on environmental protection. This has served as a first step in forming a link between the community, the park, and FAI. Current projects include the development of a tree nursery, small animal husbandry, beekeeping, and organic agriculture. Outreach work at Sacapo has been overwhelmingly successful due partly to the positive attitude of the community members and the fact that PNP has not been an obstacle to the community's development activities (Sacapo lies five kilometers from the park boundary). On the contrary, the presence of PNP has opened the opportunity for receiving extension and education services from NGOs such as FAI and public agencies.

A very different situation is found in Loyola, a recent colonist settlement to the south of PNP. A challenge to working with neighboring communities in remote areas is that most near the park have only recently initiated the process of internal organization. The associations established to date are first-level organizations—e.g., school-focused Parents' Committees, Committees of Promejoras (improvements), or sports and culture clubs. Loyola has a history of antagonistic relations with INEFAN. In an effort to restrict or prohibit the community's access to park resources, park guards have used aggressive tactics, confiscating timber and reportedly even burning a house built within what rangers believe are park boundaries. Some Loyola citizens have refused to participate in any kind of dialogue with park staff or conservation NGOs until boundaries are officially defined so that their farms lie outside of the park. Boundary signposts have been defaced or removed. The director of PNP (S. Calderón) insists that park boundaries, which exist only on an outdated map, not be altered in response to local demand because he claims it would subsequently lead to ceding park land to communities all around the perimeter.

The city of Zamora is largely devoted to mining activities. This is reflected in the positions of local political authorities in relation to the park (i.e., that mining should be allowed to proceed in the park). Many farmers in the region also depend on mining as an additional source of income. This creates a delicate situation for conservation NGOs working in community outreach. For example, the staff of Fundación Maquipucuna were challenged to maintain positive relations with farmers in the zone of Jomboe who earned income from renting mules to miners traveling to San Luis. The community perceived the park as a threat to that income; therefore, some were reluctant to participate in agroforestry or reforestation activities sponsored by an environmental NGO. A solid working relationship with key members of the community allowed FM to advance with its work despite the troubled political atmosphere.

No significant level of outreach work has occurred thus far with the Shuar populations residing in the remote, eastern region of PNP. Initial discussions with Shuar leaders have been positive, however, and the opportunity to work in this area is promising. Shuar leaders have requested support to resist the invasion of their traditional lands by mining groups.

Large-Scale Threats

Gold mining and road construction, with the potential for subsequent colonization, constitute the two large scale-threats to PNP.

Gold

There is no doubt that there is a rich deposit of gold within PNP (about thirty grams per ton of rock, according to S. Cordovez, the subsecretary of mines). The controversy is over whether the gold ought to be mined, and if so, who should be allowed concessionary rights in the park—industrial mining companies or artisanal miners. These questions are debated in every protected area where highly valued minerals are present.

Many conservationists adamantly oppose any mining in national parks on the grounds that it is environmentally destructive, typically illegal, and leads to unacceptable compromise for other protected areas within a national system (FAI 1992, Naughton 1993). Others argue that in a developing country, a rich gold deposit within a park will inevitably be mined, due to the urgent need to generate foreign revenue. This latter argument bears weight in Ecuador, where mining has been identified as the next primary source of foreign revenue when oil reserves are exhausted in the near future.

If mining is to be allowed in a park, some recommend that allowing international industrial companies exclusive concessionary rights is the best option because: (1) their technical capacity is so advanced that environmental damage can be ameliorated (e.g., release of mercury can be tightly controlled); and (2) their presence is formal in nature, allowing taxes to be levied and destructive activities (e.g., hunting by miners) controlled (J. Black, pers. comm., 1995; S. Cordovez, pers. comm., 1995). Others vehemently disagree and argue that

- The heavy machinery used by international industries will cause greater damage than the operations of artisanal miners;
- Even "green" mining companies can accidentally cause massive chemical spills (e.g., Canadian Mining Company spilling cyanide in Guyana); and
- The rural poor deserve preferential access rights to local resources.

Gold may indeed represent a supplemental source of income for subsistence farmers. However, often miners invading a park originate from other provinces. Arguments concerning the local benefit of informal mining must also address the fact that local economies dependent on mining suffer from boom-and-bust cycles and high prices of basic commodities. There is also an array of social problems that often accompany gold mining, such as violence, alchoholism, and prostitution. Gold is often removed from the region and/or country without taxes or other compensation being paid. Therefore, there is typically very little investment in community infrastructure (schools, health clinics, etc.).

The recent experience of conservationists struggling to protect Ecuador's national parks and indigenous peoples from oil drilling is highly relevant to the mining controversy at PNP. Despite an intense, five-year campaign waged by conservationists and human rights activists at national and international levels, oil drilling is now occurring in three protected areas in Ecuador, and oil exploration is occurring in a fourth. Negotiated, formal agreements between oil companies and conservation NGOs that originally aimed to ameliorate and monitor the ecological impact of drilling activities in protected areas have been ignored by oil companies (D. Silva, pers. comm., 1995). Legal regulations to control pollution and other environmental damage associated with drilling activities have by and large been ignored, due to a lack of political will by oil companies (foreign and national) and the limited technical and political capacity of INEFAN to enforce environmental legislation. These experiences cast doubt on expectations that foreign, industrial companies will voluntarily comply with strict environmental regulations, or that public environmental agencies will be able to successfully police the companies' activities.

In the case of PNP, both local and national conservation NGOs have assumed the position that no mining should be allowed within PNP. Their arguments emphasize the exceptional value of PNP's biodiversity, its fragile nature, the threat of mercury contamination for communities downstream, and the illegal nature of mining in a national park. They also point to the environmental and health disaster of Nambija, a gold mining site just northeast of PNP. To date, these conservationists have seriously curtailed mining activities within PNP (see "Resource Use," p. 298). Through an arduous campaign of lobbying, environmental education, and research, they managed to pressure international companies to leave PNP. A very different sort of battle has been necessary to evict artisanal miners from the park. Artisanal miners themselves may be a marginalized group without political power, but they are connected to wealthy merchants and provincial politicians. NGO collaboration with public institutions, including the armed forces, has been essential to protecting PNP from mining.

Preventing gold mining in PNP in the future will require persistent vigilance, lobbying, and activism by NGOs. The director of natural areas and wildlife within INEFAN at the time of the study demonstrated his commitment to defending PNP by spending ten days in the southern region seeking resolution to the mining conflict. The political will of the Ecuadorian government to protect its national parks will continue to be tested. Protecting PNP from mining not only entails an opportunity cost of forgoing mining, it will also require investment in protection. Following the eviction of gold miners from Corcovado National Park, the Costa Rican Park Service spent $290,000 in a year on patrols and park infrastructure. Stabilizing the situation at San Luis in PNP may also require substantial investment.

Road Construction

The forests lying to the south and east of PNP have been maintained in a relatively pristine state, due in large part to their inaccessibility. With the completion of the road flanking the eastern boundary of PNP (Proyecto Carretera Marginal de la Selva), colonization will greatly accelerate around the park. FAI has successfully stopped road construction through PNP on two previous occasions, by public awareness campaigns and lobbying of regional and national authorities. However, given that the current construction is beyond the park boundaries and that the idea of a living frontier (*fronteras vivas*) is a prominent national policy, it is unlikely that conservationists could deter the completion of the road. Instead, they will be challenged to ameliorate the environmental consequences (e.g., colonization) of improving access to the region.

National Policy Framework

(Refer to the Machalilla National Park case study in chapter 9 for detailed discussion and description of environmental and protected-area management policies and legal reforms in Ecuador.)

People from Loja and Zamora-Chinchipe resent the centralization of economic and political power in Quito and Guayaquil. Due to their location in the southernmost part of Ecuador, they have been marginalized within the national framework (Hernández et al. 1995). Government services have been neglected: e.g., a road connecting the two provincial capitals, Loja and Zamora, has been under construction for more than twenty years, and very few major development programs have been developed in the south.

Regional antagonism also exists between these two provinces due to the different economic activities in which people are engaged (agriculture in Loja and mining in Zamora-Chinchipe) and the roles each plays in Ecuadorian politics. The Zamora governor complains about the Loja governor's authoritarian attitudes within the ICD-PNP, explaining, "It is always the same. They make decisions for us, and we are in the same condition." This rivalry has seriously affected park management since politicians in Zamora feel they have more rights over the park because most of the park is located within their province. Several decisions regarding park management and other issues are made in Loja, a more developed province than Zamora-Chinchipe.

Lack of coordination between government institutions has brought problems to park management throughout Ecuador. The Ecuadorian Institute for Agrarian Reform and Colonization gave land out within some protected areas after they were already established. PNP was not an exception; activities, legally authorized, such as mining and road construction were carried out within park boundaries.

The National Directorate of Natural Areas and Wildlife within INEFAN, as with all government institutions related to environmental protection, has less political power than those engaged in productive activities. Authorizations to exploit nonrenewable natural resources have been granted in several protected areas in Ecuador: gold mining concessions to transnational companies were granted within PNP. Gold mining has always been a serious threat to PNP, and in the future the situation will worsen since the Ecuadorian government is looking for new sources of revenue to replace those generated by oil exploitation. Gold mining is a very strong option, as evidenced by the $14 million loan from the World Bank to the Ministry of Energy and Mines to support mining activities.

Although gold mining is legally prohibited in protected areas, if the Ecuadorian government considers it a national priority, mining activities will be allowed throughout the country without considering whether an area is protected or not. The government will encourage the participation of transnational companies because artisanal mining techniques recover low amounts of gold and the government does not receive revenue from this informal exploitation. On the contrary, transnational companies will have to pay for the concessions.

From an environmental perspective, experiences with transnational companies exploiting nonrenewable resources have been overwhelmingly negative in Ecuador. Oil drilling by Oryx, Occidental, City Investing, and Maxus has significantly altered natural environments in Yasuní National Park, Limoncocha Biological Reserve, and Cuyabeno Fauna Production Reserve. Neither national nor international companies have followed environmental regulations for oil exploitation, and INEFAN has insufficient employees and training to monitor such activity within protected areas. Unless gold mining is strictly controlled and monitored, mining within PNP is likely to be environmentally disastrous.

INEFAN's limited capacity to manage protected areas in Ecuador has led NGOs to carry out several of the necessary activities. Ecuadorian law designates INEFAN as legally responsible for the planning and administration of protected areas in the country; therefore, NGOs must work in tandem with INEFAN. The recent shift in policy of the present government to "shrink the state" has diminished markedly the number of park rangers in all protected areas, further limiting INEFAN's management capacity. Seven park rangers, two conservation officers, and the chief of the park are currently working in PNP. In PNP, NGO collaboration has been vitally important in park administration.

In 1994, FN carried out a study regarding investments in biodiversity with funding from the World Conservation Monitoring Centre. Results indicated that foreign investment and the number of implemented projects had increased between 1991 and 1994. However, more recent data show that funding from the Ecuadorian government has diminished and is currently very low. This low investment in biodiversity conservation contradicts the Ecuadorian predisposition to sign international agreements related to environmental protection. Ecuador was the first Latin American country to sign the Convention on Biological Diversity, and it has signed twenty-five other multilateral agreements regarding the environment, including Agenda 21, CITES, Ramsar, the Montreal Protocol, and the Brasilea Convention, among others. (The number of bilateral agreements signed by the Ecuadorian government is not available.) Nonetheless, the monitoring of most agreements is very limited. Possibly, the Ecuadorian government is paying more attention to the Convention on Biological Diversity, especially because of the pressure NGOs exert on it to consider the implications regarding property rights and genetic resources, a hot topic throughout the world.

Transboundary Issues

Transboundary issues do not directly influence PNP, other than the fact that proximity to Peru is at times a destabilizing factor (e.g., during the armed conflict). During times of international conflict, management activities at PNP are constrained, as they are in other protected areas of Ecuador.

A potentially more serious problem to be considered is the political strategy to build live boundaries (*"construyamos las fronteras vivas"*), or in other words, promote settlement and development along the Peruvian boundary so as to lower the risk of losing "empty," forested territory.

Conclusion

PNP suffers from having a poorly defined, serpentine boundary. PiP is currently supporting the formal delineation of the boundary. Some ecologists describe park boundaries as ecologically irrelevant and, instead of defending parks, would urge conservationists to manage biological diversity at the ecosystem level using flexible, integrated strategies (Light, Gunderson, and Holling 1995). Yet indefinite boundaries often create conflict between park managers and local communities, which is evident in PNP. Where boundaries are contested or absent, open-access resource systems are created that are highly vulnerable to overexploitation. This is especially true in unstable situations, e.g., at a rapidly expanding agricultural frontier. In such cases it is necessary to delineate areas for protection from short-term threats. Fostering sustainable resource use at a regional level is, however, a highly appropriate long-term goal. NGOs have initiated a variety of outreach projects at PNP, each of which reflects a particular view of the appropriate role of conservation NGOs in community development. The obstacles and successes encountered in these projects are instructive for those working in ICDPs (integrated conservation and development projects) elsewhere (Wells and Brandon 1992).

In working with settlements within and communities around PNP, conservation NGOs have entered the labyrinth of development assistance. Finding projects that improve biodiversity protection while meeting the economic aspirations of communities is difficult. Some projects are designed primarily to garner a community's good will so as to improve local attitudes toward PNP, and often it is difficult to discern a direct link to biodiversity protection in such projects (e.g., providing potable water, health, or veterinary services). Conservationists working in these projects would argue that they are responding to the priorities identified by participating communities. Others would argue that such projects are beyond the mandate of a conservation organization. Missing from the debate is clear evidence whether a community with "positive attitudes" toward the park or its managers will support active park managment, which may include restrictions on resource use. Projects directly linked to biodiversity protection (e.g., community-based reforestation) may have minimal impact, however, because of poor acceptance by a community. It is especially difficult to foster active community participation in projects involving untested markets or collective production systems. In most cases, although present land uses may be environmentally destructive, they reflect a rational response by individuals to political and economic conditions.

National policies often underlie the political and economic conditions that drive local land uses and practices. In the case of PNP, such policies have driven rapid and extensive colonization of surrounding regions. Of more immediate direct threat, and closely linked with national policy, is the presence of gold within the park. Gold is not the only threat to the ecological integrity of the park; illegal timber harvest, hunting, and agricultural encroachment all affect PNP. Park staff and NGOs are working to control these problems over the short term via patrols and over the long term via environmental education and sustainable development projects. What is remarkable about this case is the importance of dialogue and negotiation and the range of defensive strategies adopted to minimize the threat of mining. Creation and maintenance of a locally based park advocacy group, the Inter-institutional Committee for Defense of PNP, have been important accomplishments of FAI and FN. Not only has creating this committee helped expand advocacy for park protection, the committee has itself acted as a facilitator in resolving conflicts related to the park. Another key element is the relationship between FAI and FN, one that can deal with the local and regional issues facing the park, and one that can address national policy concerns that affect PNP and other protected areas in Ecuador.

Postscript

Several significant events have happened concerning Podocarpus National Park between 1995 and 1998. First, small-scale mining has continued. Second, studies have revealed high levels of mercury contamination. Third, there has been a reorganization of environmental management in Ecuador; and fourth, there have been changes in NGOs involved in Podocarpus.

The case study describes a July 1995 meeting between the leaders of the Association of Sentinel Miners of Ecuador in PNP, the subsecretary of mines, the director of natural areas and wildlife, and the ICD-PNP. At that meeting, a final agreement was reached that the miners would leave the park after two months, that miners leaving by September 30, 1995, would not be legally punished, and that no new miners could enter the Park. Miners did follow this accord and left the park.

However, in October of 1995, some mining leaders promised that they could legalize mining in the San Luis area, although this was a false promise since laws explicitly prohibit it. Negotiations began anew, and in July 1996, miners agreed to abandon the area—this time in six months. However, the fall of President Bucaram's government and the consequent alert status of the armed forces made it impossible for the armed forces to provide support at the time the miners were supposed to be evicted. While there was conflict among their leaders, miners refused to leave as they had agreed. Although the government of Zamora-Chinchipe had often sided with conservationists in past disputes, in a sudden about-face it supported a petition the miners sent to DINAMI to grant them permission to mine in the San Luis area. Miners also tried to raise their issues in the national Congress, sought the support of the country's vice president, and made false allegations to the press about the situation. The eighty or so miners petitioning for legal approval of mining in San Luis consist of those who had invaded the park initially and additional miners from Nambija.

Meanwhile, a study (1996) in the San Luis mining area in Podocarpus found that there were extremely high levels of mercury contamination in stream sediments—ranging from a low of 2.3 mg/kg to one exceptional sample that was 32 mg/kg dry weight (Haworth et al. 1996). As a reference, the World Health Organization has established the following maximum acceptable limits: potable water 0.1 mg/m^3; air 0.1 mg/m^3; fruits and vegetables 0.03 mg/kg. These results, along with other materials, were forwarded to DINAMI by FAI, urging DINAMI to turn down the miners' petition. In August of 1997, DINAMI declined the petition. Furthermore, DINAMI ordered that mining in the area be stopped and that miners immediately leave the park. In its response, DINAMI referred to FAI's information and viewpoint.

While DINAMI's verdict was cause for some celebration, the political will to remove the miners has been lacking, and mining continues. In response, FAI began (in August 1997) a letter-writing campaign asking the minister of the environment to halt mining within the park. FAI printed postcards that describe the writer's opposition to mining on one side and carry a picture showing the destruction caused by mining on the other. It hoped an international letter-writing campaign would force the government to take action to enforce its own policies.

There have been allegations that foreign mining companies have provided artisanal miners with machinery. The miners themselves have stated that an Australian mining company had helped provide them with mining equipment. The road to reach San Luis is in poor condition and can be traversed only on foot, which makes cargo costs very high. People generally charge $1.50–$2.00 per pound to transport things on this path, amounts that individual miners lack.

A recent socioeconomic study undertaken by Arcoiris in the Shuar community demonstrates that the Shuar contribute to the conservation of the area because they receive benefits from the park, such as medicinal plants, occasional hunting of mammals for subsistence, and fishing. Problems and conflicts have resulted from the presence of mining enterprises and colonist groups (Schulenberg and Awbrey 1997). As of May 1998, there had been little change in the mining situation.

In terms of policy changes, the Ministry of the Environment was established by decree in October of 1996. INEFAN, the CAAM, the Fund for Amazonian Ecodevelopment (ECORAE), and the Atomic Energy Commission were all incorporated into this new ministry.

In terms of the role of NGOs, Fundación Natura is no longer actively involved in PNP. As FAI has matured as an NGO, with the capacity to influence activities at local and national levels, it has received support from FN only as needed for specific issues.

One significant change is that the government of the Netherlands established what is known as the Podocarpus Program in November of 1997. This program will provide approximately five million dollars to support park management. A technical review panel has been established to review project proposals over a five-year period. A consortium of NGOs, including FAI, and governmental organizations, such as PREDESUR, will be submitting proposals for financing.

Glossary

barbasco—a plant used to make a poison for fishing
CAAM—Comisión Asesora Ambiental de la Presidencia de la República
centro—center; village or small town
chancadores—stamp mills
chanchos—rock crushers
Cumbinamasa—Norwegian-Ecuadorian mining company (also known as EcuaNor)
DINAF—Dirección Nacional Forestal
DINAMI—National Directorate of Mining
FAI—Fundación Arcoiris
FN—Fundación Natura
fronteras vivas—living boundaries
GPS—Geographic Positioning System
ICD-PNP—Inter-institutional Committee for the Defense of PNP
IERAC—Ecuadorian Institute of Agrarian Reform and Colonization
INDA—National Institute of Agrarian Development (Instituto Nacional de Desarrollo Agrario)

INEC—National Institute of Statistics and Census (Instituto Nacional de Estadística y Censo)
INEFAN—Ecuadorian Institute of Forestry, Natural Areas, and Wildlife
INEMIN—Ecuadorian Institute of Energy and Mines
páramo—humid, tropical, alpine grasslands
PNP—Podocarpus National Park
PRA—participator rural appraisal
PREDESUR—Subcomisión Ecuatoriana Programa de Desarrollo de la Rcgión Sur (Ecuadorian Subcommission Development Program of the Southern Region)
promejoras—improvements
RTZ—Rio Tinto Zinc

Amboró National Park

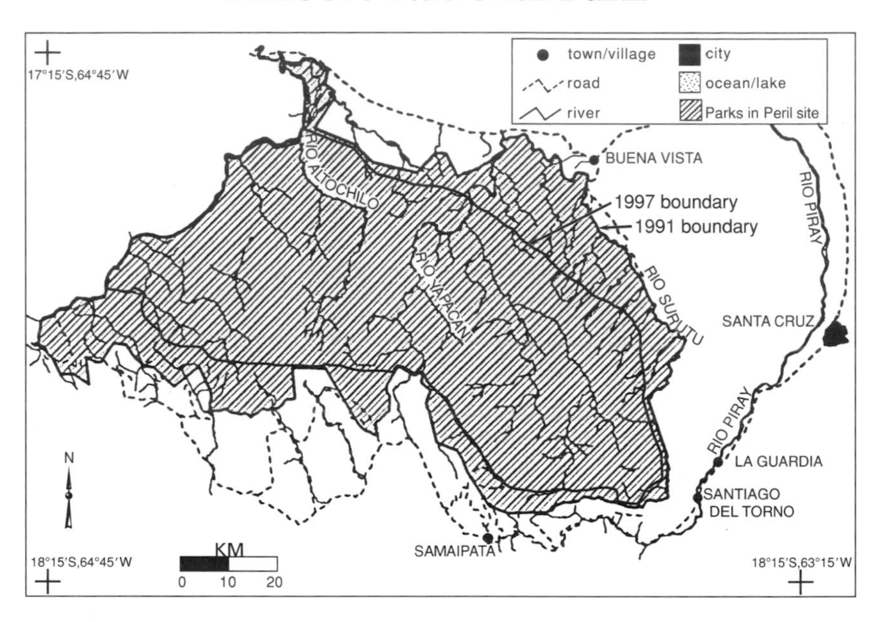

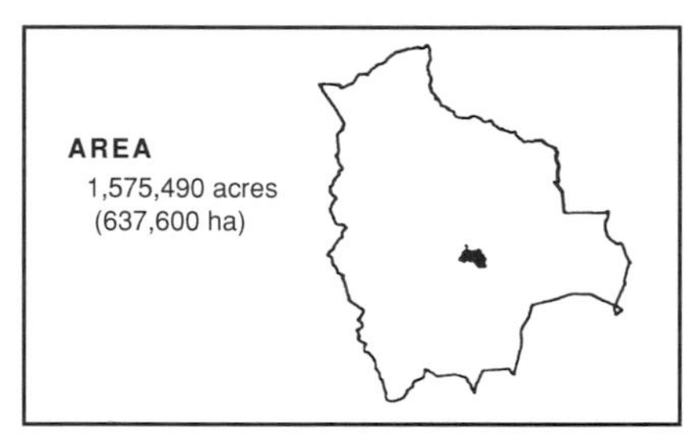

11. Bolivia: Amboró National Park

Adolfo Moreno, Richard Margoluis, and Katrina Brandon

This case study describes Amboró National Park, one of the most biologically diverse areas in the world. Most of the issues that threaten the park are rooted in government policies, which are largely the result of where the park boundaries were placed at the time of the park's establishment. The case describes the conflicts that have resulted from superimposing the park over existing communities. It also describes the role of Fundación Amigos de la Naturaleza, which has been instrumental in providing technical support and protection for the park.

Amboró National Park is located in central Bolivia, in the mountainous sector between the departments of Santa Cruz and Cochabamba. It is near the city of Santa Cruz in one of the regions of most rapid growth and of greatest economic prosperity. Amboró is a 637,000-hectare park that extends from the Andes in the west in the direction of the Chaco plains to the east. It is one of the parks with greatest topographic variation and biological richness in Bolivia. The park has both physical and biological importance, in that it straddles the divide between the temperate zone of the south and the northern tropical forests.

Amboró is located in an important zone of biogeographic transition known as the elbow of the Andes, an arm of land that extends east from the Andes mountain range to the eastern plains of Bolivia. The topography is varied, beginning with rugged peaks and abrupt cliffs reaching 2,700 m (9,000 feet) above sea level in the southern region and dropping perpendicularly to fertile and alluvial floodplains at 275 m (900 feet) above sea level in the northern zone. Amboró protects the upper basins of many important rivers, including the Yapacaní and the Ichilo, both tributaries of the Mamoré. Many of the rivers that provide water to the northeast region of Santa Cruz are supplied from sources within the park. As a result of efforts to protect the park, there is less risk of flooding to important agricultural and livestock areas. This in turn reduces the likelihood of erosion and sedimentation in the riverbeds.

Amboró Park is classified as an area of Yungas and important Amazonian lowlands because of the range of biotas it presents between the dry area of the Chaco to the east and the humid zone and cold of the Andes mountains to the west. Within the park's boundaries Holdridge (1967) describes eight life zones, the largest of which are the wet subtropical forest, the humid subtropical forest, and the humid subtropical forest of the mountainous lowlands. Annual precipitation in the Amboró region and surrounding areas fluctuates from 2,000 to 2,500 mm (80–100 inches). The average annual temperature is 24°C (75°F) with highs of 30°C (86°F) and lows of 17°C (63°F). The average humidity is 80 percent. The area's importance lies in the way it captures and retains atmospheric humidity. It is the source of many rivers and streams that have ecological and economic importance to the environment and the human settlements in the departments of Santa Cruz and Cochabamba.

From a biological standpoint, Amboró is one of the richest places on the planet. Preliminary research indicates the presence of 818 species of birds, 145 species of mammals, 105 species of reptiles, and 73 of amphibians. Mammals include collared and white-lipped peccaries; the *taitetú* (*Tayassu tajacu*) and the *tropero* (*Tayassus pecari*); various species of felines including the jaguar (*Panthera onca*), the ocelot (*Felis pardalis*), the gray cat (*Felis yagouaroundi*), and the *tigrillo* (*Felis wiedii*); eight species of monkeys; the spectacled bear (*Tremarctos ornatos*); the *pacarana* (*Dinomys branickii*); the *oso bandera* (*Myrmecophaga tridactyla*); the *perro de monte* (*Speothos vanaticus*); and rare sloths (*Bradypus variegatus*).

Species of the abundant avefauna that have been documented in the park include the Andean species of *gallito de la roca* (*Rupicola peruviana*), the endemic species of *paraba frente roja* (*Ara rubrogenys*), the crested *pava copete de piedra* (*Crax unicornis*), the harpy eagle (*Harpia harpyja*), species of the *trogón* genus, and the condor. The park also is home to important nesting areas for eight species from the family of guans and currasows and other species of parrots.

Park Establishment and Management

In December of 1973 the Germán Busch Natural Reserve was created as a preliminary step toward transforming the area into a national park

(Supreme Decree no. 11254). The reserve included about 180,000 hectares, covering the mountainous sector of Amboró, the Yapacaní and Surutú river basins to their confluences, and the 17,51′18″ of southern latitude in the Ichilo and Florida provinces of the department of Santa Cruz. The declaration of this area as a reserve was not accompanied by the initiation of any kind of management, nor did it modify local patterns of use, which included resource extraction and human settlement. What was created, in effect, was a paper park.

In August of 1984, the reserve was reclassified and its status elevated to create Amboró National Park. The park's established objectives were to conserve biological diversity; to protect the original forest cover, watersheds, ecosystems, and scenic beauty; to promote scientific research; and to create a suitable environment for recreation. The decree charged the park's organization, administration, and management to the Department of Wildlife of the Decentralized Technical Unit of the Center for Forestry Development of Santa Cruz (UTD-CDF-SC). It also categorically prohibited commercial and sport hunting and fishing and the extraction of wood; it annulled existing forestry concessions and ordered that all personal property and settlements found within the park would be subject to the limitations and rules dictated, prohibiting new settlements and logging. It ordered, on the other hand, that indigenous peasants with traditional settlements would enjoy special treatment and would be enlisted as active elements in the conservation of the area.

The UTD-CDF-SC, which was responsible for the park, contracted as its first director an English biologist who worked for the British Mission for Tropical Agriculture (MBAT). Also contracted were three peasants from neighboring communities to serve as park guards. UTD-CDF-SC then opened an office in the Buena Vista community on the north side of the reserve. No guidelines were established for work in the park or to plan its development. Nor were financial or technical means provided for planning, and only limited economic resources were designated for its management.

The first park administration began working in what is now the northeast part of Amboró National Park without official terms of reference for management or a clear idea of what was legally expected. Single-minded priority was given to protection, in spite of the prior presence of people in the reserve. What then developed were politics of exclusion to repel any kind of potential or effective threat. According to reports of people who observed the work of the first park authority firsthand, the methods for achieving park protection were not well received by the people living in the zone. They recall that the rights of the residents were not respected and that various people were treated badly. For that reason, the administration was flatly rejected by the peasant communities in the zone. Conservationists interviewed, however, recognized in the first administration the merit of having established some sort of management of the area as a national park despite the lack of official support and the constant pressure from established settlements as well as new settlements, which, despite efforts to thwart them, kept being established.

With the change of government in 1989, the National Secretariat of the Environment was created at the ministerial level. This new national ministry assumed responsibility for all protected areas in Bolivia. Within the region, however, Amboró National Park continued under the financing of the UTD-CDF-SC. At the same time, a new Environment Law was passed following a participatory process. This law came to replace, in great part, the existing Wildlife Law, which included protected areas. The new Environment Law also introduced modern environmental concepts and likewise adopted internationally accepted categories for protected areas and their management.

In October of 1991, the government of Bolivia issued another decree, which enlarged the size of the Amboró National Park to some 637,000 ha, making the park three times its original size. This expansion incorporated watersheds and ecosystems that were not previously represented but also meant that numerous established communities were incorporated into the park. Also, studies and recommendations conducted by the Santa Cruz Regional Development Corporation (CORDECRUZ) were disregarded in the expansion process. Social factors were not considered, nor was an analysis of existing land tenure completed.

In 1991, park management capacity was expanded through the Subproject for the Protection of Ethnic Resources and Renewable Natural Resources (SPERNR) of the National Environment Secretariat, financed by the Inter-American Development Bank. The SPERNR was the environmental component of a highway construction project linking Santa Cruz and Cochabamba that had just been completed to the north of the park. The SPERNR began financing part of the management and implementation of the larger park, placing on the north side, for the use of the park administration, a technical team, ten park guards, materials, and equipment.

At the end of the same year, through the Parks in Peril program, Fundación Amigos de la Naturaleza (Friends of Nature Foundation, or FAN) began supporting the management of the south side of the park by committing itself to the development of a formal plan for technical support and future financing for a minimum of three years. FAN opened offices in Samaipata, hired a responsible technician, and provided an agricultural outreach worker and seven park guards to serve the park, all appropriately equipped and with the capacity to begin protection and management.

In 1991, an Inter-institutional Coordinating Committee was established, with representatives from CDF, SPERNR, and FAN. This committee became responsible for making management decisions in the absence of a governing document. The CDF continued to be charged with official responsibility for the area and, in turn, gave responsibility for park directorship to different professionals.

The Red Line

Because the 1991 Park Enlargement decree was issued without the participation of the communities that lived around the park, serious problems arose. First, the new boundaries incorporated previously established communities into the park, implicitly contradicting the law that prohibits resource use in parks. Second, the communities that were incorporated into the park, as well as people in nearby communities, became hostile to the park for having usurped their lands.

Some attributed the current park management problems to the "anti-community" administration of the first park administrator. Others noted the exclusion of local people in decision making related to the management of the park and its surroundings. The incorporation of communities into the park is clearly a third key reason. Some key informants and written reports note that increased population pressure—an invasion of people around the park—was making matters worse. However, more than one of the local people interviewed during the study said, "We are not invading the park. It is the park that is invading us."

In order to reduce the tension over the new park boundaries instituted in 1991, a program was initiated to negotiate new limits for "internal protection" of the park that were acceptable to communities, government, and conservation organizations alike. Under a project financed by the Program for Alternative Regional Development (PDAR), the Red Line project was initiated. The name "Red Line" comes from census work done by the park administration in 1989, which resulted in a path marking off the properties of the original park residents. In this case, the demarcation was made by a technical group organized for this purpose and through committees formed of the peasants themselves in each sector.

The Red Line is a four-foot-wide path. In order to clear the path, committees were formed by the official project director, an agriculturist, two assistants, and representatives from the peasant unions. After agreeing on the process and procedures for establishing a new Red Line, community members participated in the clearing of the path. Many institutions were involved in this delimitation, including the National Environment Secretariat, the CDF (official head of the project), the SPERNR, FAN, and PDAR. During the Red Line project, those agencies worked with the residents of the communities affected by the park decrees. Although many community residents did not hold legal titles to the lands they claimed, the project negotiated with them to establish boundaries that were recognized by both parties. The Red Line strategy established a new internal line of protection within the park and raised the possibility that this new line would come to constitute the new park boundaries.

According to an informant, the southern line was established more quickly than the line in the north. Nevertheless, work has yet to be completed, and in both sectors there are still stretches where agreement has not been reached on the location of the line. This stalemate has occurred because the park administration was unwilling to give way to peasant pressures to exclude areas deep inside the park where the peasants did not have title and that were ecologically sensitive. On the south side, for example, the conflict persists with some well-established communities who claim land deep within the park.

Negotiations to finalize the Red Line around the entire park have stalled because the government and peasants were unable to reach a compromise. The government considers the stretches of land still disagreed upon and claimed by the peasants to be non-negotiable. A technical study is underway that will provide the basis for the government to issue another decree, which is projected to divide the park into two management areas. The first area, called an Integrated Management Natural Area, would comprise approximately 195,000 ha, and the other, which would remain a national park, would comprise 442,600 ha.

Park Administration and Management

In 1993 there was another change in government structure, which included restructuring executive powers. The Ministry of Sustainable Development and the Environment (MDSMA) and a National Directorate of Biodiversity Conservation (DNCB) were created. The new ministry became responsible for all protected areas in Bolivia. Although the transition from the former organization to the new organization was difficult, projects initiated in the prior administration were continued, including the Bolivian Biodiversity Conservation Project (PCBB), financed by the World Bank. This project was designed to develop new biodiversity priority areas, including Amboró National Park.

Following the mandate of the Environment Law, in May 1994 the DNCB issued a public request for management proposals for Amboró National Park. The government sought proposals that included concrete plans for the protection and sustainable use of the park, including activities such as administration, protection, local participation, environmental education, research, and public use. FAN was the only organization to present a proposal by the deadline. The FAN proposal was based on the gradual participation of peasants to a level at which the peasants themselves would hold park management positions, after learning about the concepts of conservation and sustainability.

Although the deadline had passed, the government later accepted another proposal, jointly submitted by the peasant organizations and two NGOs, PROBIOMA (Productivity, Biosphere, and Environment) and ALAS (the legal advisory branch of the workers' union in the department of Santa Cruz). It included the participation of various civilian groups and was oriented more toward a community development plan than a park management plan. Instead of having a strict protection component, one of the government requirements, it proposed opening the park to resource exploitation. It proposed that all the natural resources in the park should be available to the communities around the park. The proposal did not contain a concrete plan for administration of the park as a protected area. Instead, it was more oriented toward managing a multiple-use zone. Perhaps because it did not include a clear method for conserving the park's natural resources, this proposal was rejected by the government.

At the moment, the park management is the responsibility of the DNCB. The DNCB is hoping to find alternatives that would allow FAN to apply its resources and programs while the DNCB decides if the park is to be recategorized and in what form. It also needs to decide what role the national park administration and management will play and what will be the future role of the inhabited area, which could also eventually be recategorized.

Despite the conflicts over the boundaries and management of the park, a substantial finding from the site visit was that all the people interviewed, from members of the community to civil servants and NGO representatives, agreed on the importance of the park. They all said it was very important to protect the area that forms the park. The principal differences lie in the concepts of what is a national park, the form of its establishment and management, and the acceptable boundaries. Given the substantial disagreement over these factors, it is nonetheless significant that everyone agreed on the importance of the park.

Land and Resource Tenure

Before the creation of the reserve and its later declaration and enlargement as a national park, the zone that is now the Amboró National Park comprised government property (such as 100,000 hectares that belong to the National Army), large private estates, and small and medium-size peasant holdings. The government properties are located in the least accessible areas, as are many of the large estates and some of the peasants' plots. Few of the flat or moderately sloped private areas have been cultivated, with the exception of some located in the southern part of the park where communities have long been present.

From the 1950s to the 1970s colonization was very slow, beginning in what is now the eastern part of the park, in the Surutú River zone. This colonization was encouraged by the Agrarian Reform and Institute of Colonization. Since the 1970s colonization has intensified and bit by bit has extended north into what is now the park. Over time, both the flat areas and the steep slopes were either allocated to settlers by the Agrarian Reform or occupied by colonists without titles who invaded government property or portions of large estates.

A census of the eastern sector of the park counted 908 families residing in the original zone. When the park was enlarged in 1991, it incorporated all of what is now considered the southern sector, which extends along the old Santa Cruz–Cochabamba highway, is practically lacking in flat land, and has been occupied since the last century by peasants who earn their livelihood primarily by raising livestock. By the end of the boundary expansion, nearly three thousand families, in all zones, were residing within the park.

Currently, the park is, for all practical and operational purposes, divided into two portions, north and south, each with its own land tenure patterns. The communities on the south side are older and more established, and for this reason, the land tenure situation is clearer. Many people hold titles to land that has been in their families for generations. Others may not hold legal title but have lived on or worked the land for more than forty or fifty years. A minority arrive annually to work as laborers and, in doing so, try to locate small pieces of vacant land on which to stay and establish their own activities. For the most part, the southern zone is stable and the agricultural frontier is not advancing rapidly.

The south side of the park includes only moderately productive lands. Most cultivation is on hillsides and is unirrigated. More intensive agriculture is practiced in terraces on the riverbanks, with irrigation, and a much colder and drier climate. For the peasant, this has both advantages and disadvantages. The advantages include fewer problems with pests and loss of soil from erosion, while the disadvantages include less certainty of rain for sowing, dependence on irrigation, and limitations on what can be sown. There are an estimated eighteen hundred established families living in this sector, not all of which are much affected by the park. They are principally *vallunos cruceños,* that is to say, people who have lived here since the colony was founded, with specific customs and traditions and an established system tied to land that has belonged to them for generations.

Land tenure on the north side of the park is less stable than on the south side. As mentioned above, the major part of this zone was colonized by indigenous people from the high valleys of Cochabamba and Potosí during the last ten to fifteen years. Therefore, many of the communities are relatively new and have not adapted to lowland tropical agriculture. Rapid forest clearing as part of slash-and-burn agricultural systems and the in-migration of people to the zone have led to the advancement of the agricultural frontier on the north side. On the north side, the average personal holding is twenty-five hectares.

Since the establishment of the current government, the Agricultural Reform and Colonization Agency (RAC), the official agency charged with providing lands, has had various problems and has suffered a total reorganization after the enactment of a much awaited new landholding law. Land reform provisions and the titling process have been interrupted. The agrarian laws in Bolivia, however, still require that within two years of receiving a land title from the government, the owner must show that he is making use of the land and "improving" it. This is mostly done by clearing the land and sowing crops. Unfortunately, this runs counter to the promotion of conservation and sustainable development in the neighboring zones of the park, particularly in the north.

Informants indicated that sections of the settled land in the zone have been distributed by local trade unions. Such unions come to possess RAC lands or have occupied both government and private lands by taking over land using pressure tactics. Informants also described the lack of governmental coordination on land tenure. For example, after the decree of 1991, some colonists almost obtained titles to land within the Amboró National Park. The land nearly granted included about thirty thousand hectares in the zone called Las Playas, in the source of the Ichilo River on the south side. The titles were printed, and the government was about to deliver them to the colonists when the park administration found out and halted the process. It was confirmed that almost all of those who would have received titles already had other properties in the area.

Such bureaucratic mismanagement has been repeated numerous times in diverse sectors of the park. This has forced park administrators to remain vigilant in order to immediately initiate work to refute such claims and actions. There has been increasing resentment by landless peasants who view the park as a source of available land that they cannot use. The lack of agrarian policies to offer these peasants lands in other suitable, unpopulated areas, combined with the dishonest actions of peasant leaders who offer land anywhere in exchange for money, have exacerbated feelings against the park. This is the main obstacle to a definitive consolidation of the park and positive relationships with its neighbors.

Resource Use

Bolivian law does not allow resource use in national parks, yet the land tenure situation, the sheer vastness of the land, and the lack of sufficient management to control resources have led to a very different reality. Levels of resource use range from intense to moderate, depending on the resource, throughout thc cntire inhabited area. Nevertheless, it is believed that use levels have dropped significantly since a serious park administration and presence was established in the area. Based on empirical observations of natural resource use in the park, various informants state there is relatively little use today. They likewise declare that before the administrative presence, there was more hunting of animals, extraction of wood and tree ferns, and rudimentary mining activity.

Deforestation

There is no doubt that forests are the most seriously affected resource. Deforestation is due to three distinct trends:

- the expansion of slash-and-burn agriculture
- small-scale timber extraction
- large-scale timber extraction.

Although there is no statistical information, according to FAN, routine aerial flights over the park during the slash-and-burn season indicate that new areas in excess of one hundred hectares are deforested annually. This figure, which is merely an approximation, is likely to be an under- rather than an overestimate. Furthermore, the effects of deforestation are extremely significant since much of it occurs on very steep slopes. This affects the soil resource in three ways:

- loss of soil to erosion
- rapid soil exhaustion due to the annual sowing of highly extractive crops
- infestation of aggressive weeds.

Timber extraction from the park is illegal. At the subsistence level, use of some species of palm trees for palm heart (*palmito*), for roofing, or for posts is significant, as is use of *motacú* (*Scheelea princeps*), *jatata* (*Geonoma deversa*), and *pachiuva* (*Socratea exorrhiza*). Also used at a subsistence level are numerous species for food and medicines like *jipijapa* (*Carludovica palmata*). Finally, commercially desired species (described below) are also used. Although timber extraction is supposedly under control, during the research for this study it was observed adjacent to the southern sector of the park. The wood may be cut on peasants' plots with the declared intent of using it to improve their living quarters or property. Nevertheless, a great deal of wood is taken in small pieces to be sold in neighboring towns. The site visit also confirmed that in the main villages near the highway on the south side of the park (for instance, in Samaipata), many houses have flower pots made from the trunks of tree ferns. It was also observed that park guards had some confiscated fern trunks in the Samaipata offices.

Great strides have been made to halt large-scale logging by groups that had traditionally operated in the park, even after its creation as such. The most coveted commercial timber species are the *mara* (*Swietenia macrophylla*), the cedar (*Cedrella fissilis*), and the oak (*Amburana cearensis*). Other species are of less value but are commonly used in construction and in furniture of lower quality, like the *verdolago* (*Terminalia* spp.) and the *ochoó* (*Hura crepitans*). In a very publicized case in 1992, CDF, FAN, and SPERNR confronted a very powerful and influential logging company that had entered along the gorges of the Ichilo River to illegally cut timber sold to them in a corrupt deal. CDF had previously seized these illegally cut logs from the company. They decided to sell the confiscated timber to the highest bidder—which turned out to be the company that had illegally cut the timber in the first place. The bidding took place amid the confusion of a transitional period when park boundaries were being expanded. This corrupt deal was detected only because the company began illegally cutting trees when it went to collect the wood it had bought (and cut in the first place). It mistakenly believed that the park administration would not notice that additional trees had been cut. The park administration, however, discovered what the company was doing and managed to halt the scandal, which involved many people from CDF, who were, in the end, dismissed. The loggers withdrew from the zone, and they were fined by the UTD-CDF, a fine that, in the end, they did not pay. As a consequence of this whole affair, the executive director of CDF was obliged to resign. That same year, the park administration managed to halt another timber raid of great proportions in the north sector near the River Moile. In both cases, peasant leaders were collaborating with loggers.

Hunting

According to informants, commercial hunting continues to be a localized problem in some areas on the north side of the park, especially in the town of Yapacaní, which is on the new road from Santa Cruz to Cochabamba. Restaurants serve diverse species of fish (predominately the *surubí* [*Pseudoplatystoma*], the *pacú* [*Colesoma macropomum*]) and the meat of mountain mammals (such as peccaries, deer, pacas, tapirs, and armadillo), which reportedly come from the park. Subsistence hunting is practiced by the numerous families that live inside the park in both sectors. It is likely that subsistence hunting is as important, in terms of its volume and impact, as the commercial hunting that still exists.

One of the park guards and several informants commented that sometimes people kill animals that destroy crops or that attack domestic animals or human beings. One gentleman from a community near the park told a story about a jaguar that ate two people and was finally hunted and killed by several people in the community. Park guards interviewed commented that they don't take action against people who kill nuisance animals because they understand that sometimes it is necessary for residents to protect themselves or their crops. The park is home to the spectacled bear, and while there are few documented cases of hunting this species, there have been many claims of bears attacking domestic animals.

Other Resources

National planning and development plans respect the establishment of the park in this area and do not foresee any different uses for it. The Santa Cruz Land Use Plan, however, has classified the populated area of the park as capable of moderate-intensity use. The potential impact of this plan is unknown. If the Red Line process continues and/or if an Integrated Management Natural Area is established, the variety and abundance of natural resources, even in the inhabited areas, would allow for a variety of uses and management alternatives. Some informants reported that gold mining exists on the north side of the park, as well as outside park boundaries. Mining is prevented by the park guards when it is discovered.

Organizational Roles

There are various NGOs currently working in the park and surrounding areas with a wide range of activities, approaches, and attitudes toward the park. These include the following:

La Fundación Amigos de la Naturaleza. Established in 1988, FAN is a private nonprofit organization with offices in Santa Cruz and Samaipata. Its general objectives are to assure the stewardship of protected areas in Bolivia, to initiate the establishment of new protected areas, to train professionals in the area of conservation, and to promote environmental education programs. In the area around Amboró, its specific goals include protection activities, conducting research to resolve the colonization problem, and developing a fundamental base of scientific information about the park. FAN has involved itself in social and community planning programs in order to include and take into account the people who live around the park.

FAN is the main NGO in the region and the only one that has effectively intervened in the management of the protected area. Since its beginnings in the park in 1990, it has worked closely with the government and communities to define ways of protecting the park that are integrated with the rest of the region. With the support of PiP through The Nature Conservancy FAN has focused on the protection and stewardship of the park on the south side and has contributed to the conservation of the park as well as its development through the construction of administrative and protection infrastructure. It has employed a technician, an outreach specialist, and seven park guards since 1990. Since 1993, FAN has also included a team of participatory planning specialists, outreach specialists, and extension agents, and has obtained very positive results in coming together with the communities. In this way the most urgent and highest-priority community needs have been identified.

Along with the Conservancy, FAN has been instrumental in the consolidation of long-term financial initiatives for this park and others through the National Environment Fund (FONAMA).

Productivity, Biosphere, and Environment. PROBIOMA is an NGO that focuses on community development around the Amboró National Park. According to information provided by PROBIOMA's management, its fundamental objective is to support an increase in productivity through the sustainable management of natural resources in depressed rural areas. Its activities are underway in three provinces (Florida, Caballero, and Campero) and in the departments of Santa Cruz and Cochabamba. It uses a research-action methodology, in which research, technical assistance, start-up, and communication funds, are all key elements of institutional action. PROBIOMA works closely with community organizations and labor unions in the zone. It completed a socioeconomic study on the south side of the park at the request of the Catholic Relief Services (CRS).

Caritas. This is an organization sponsored by the Catholic church that undertakes rural development projects in the area influenced by the park. It has also conducted an extensive socioeconomic study on the north side of the park.

Ecological Association of the West. ASEO is a watchdog organization of ecologists that monitors the state of the environment in both urban and rural spheres. It has a certain presence in the park's zone of influence through a small environmental education program.

Dominican Order. Although a religious organization is not a typical NGO, the work of the Dominican Order, especially that of Fray Andrés María Langer, in the southern zone of the park cannot be overlooked. Inspired by an environmental calling and a marked mission to support the most poor, Father Langer (known better in the zone simply as Fray Andrés) has become the principal park guard of the southern sector. Through strong local contacts and awareness of the movements of violators, he immediately directs reports to park and community authorities. He is also one of the principal donors to the park, periodically provisioning the park guards of the zone with equipment he gets from Germany and coordinating plans to acquire more international support with FAN and other organizations.

Unions. The unions are an important and active force in the park's area of influence. Hierarchically, they are organized in local units at lower levels; higher up, they are in federations. They are aligned with two groups: the Peasant Federation and the Colonization Federation. Both federations have presumptuously claimed legal and total representation of the rural population.

CARE. CARE is a U.S.-based NGO that hopes to establish an effective presence in the zone. Over a two-year period, CARE has held a series of visits, meetings, and workshops with all the institutions most directly linked to the park. As a result, it wants to launch a huge project to be financed by the British government through the Overseas Development Administration (ODA) in the inhabited zone of the park, and it recently presented a $7 million proposal for projects in the "buffer zone" of Amboró National Park. According to CARE representatives, it does not plan to implement all the projects itself but rather hopes to find other NGOs, which it will finance to carry out sustainable development projects in the region. Some institutions in the zone fear that this large influx of money may radically alter the local and native development process in the region. They also fear that given the sheer size of CARE, they will be displaced. Finally, they are unclear whether CARE shares the same perspective of the other NGOs regarding the park, which, despite their differences, all agree on the importance of the park.

There is little information about other NGOs working in the proximity of the park, but they primarily attempt to provide technical assistance in the populated zone in areas such as health, farming, and animal husbandry.

Other institutions operating in the park include:

The Subproject for the Protection of Ethnic Resources and Renewable Natural Resources. While SPERNR has ended its direct participation in supporting the park, it can still be found working with communities on activities related to watershed protection in the park's interior.

The Center for International Tropical Agriculture. One of the international agricultural research centers, CIAT is providing technical assistance to farmers in the form of alternative farming technologies.

Development Program for the Provinces of Ichilo and Sara. PRODISA is also an important player from a microregional planning standpoint. It is part of CORDECRUZ, the Santa Cruz Regional Development Corporation, which with financial cooperation from the German government and as part of the Natural Resources Protection Project in the Department of Santa Cruz, has created a Rural Land Use Plan (PLUS) that constitutes the fundamental base for future development and land use in the region.

There exist differences of opinion and conception about the management of the Amboró National Park among the NGOs working in and around the park. This has created a certain amount of tension. According to information obtained during field visits, great differences exist between the ways FAN and PROBIOMA visualize the conservation of the park.

The two institutions arrive at conservation from two distinct perspectives: FAN is, at its core, a conservation organization, while PROBIOMA is fundamentally a community development organization. Although both institutions realize they cannot conserve the park without the involvement of local people, FAN has the philosophy that it should work as much in protecting the park as in sustainable development of neighboring communities, while PROBIOMA seems to dedicate itself principally to social sustainable development. The conflict between these institutions, although not unsurmountable, is recognized not only by other institutions but also in the communities on the south side of the park.

Despite the considerable differences in the roles and institutional models of the NGOs in the region, it is apparent that they have filled a space left empty by the government and have been better equipped to work with the communities than centralized government has. In different ways, the NGOs have decentralized many of the government's functions and applied their own and national resources with greater efficiency and transparency.

Linkages between Park and Buffer Areas

Historically, there has been little formal relationship between the park and what is known as the buffer zone—that is, the populated zone of the park. In fact, an official buffer zone does not exist; the existing areas between the actual borders and the inconclusive Red Line are not, in any way, a technical or a formal buffer zone.

Nevertheless, as mentioned earlier in this report ("Park Establishment and Management," p. 324), there are plans to newly delimit the park boundaries, which would include reducing the existing boundaries and declaring another management category between the new and old boundaries. According to Bolivian law and the IUCN's Multiple Use Area definitions, this new category of management, an Integrated Natural Management Area, would serve as an external buffer zone. As such, it would be officially decreed and formally related to the park.

Until recently, there have not been solid, concrete, or definitive efforts to involve people in the direct and immediate management of natural resources. Since the beginning, the majority of park activities have focused on the exclusion of people and the protection of natural resources by control and surveillance. Over the past three years, FAN has tried to give a community focus to park management. With the help of FAN, for example, the communities have arrived at a point where they have identified their strengths, as well as what they lack and need, and have designed their own development plans without waiting for the central or regional governments to write them. The presence of a conservation organization as facilitator of the planning process has allowed a more positive union of community and park with ideas of sustainable development. As a result, some of these communities have voluntarily begun their own activities to

- repair damaged forests
- mobilize the community to halt external attempts to alter or damage their water sources
- become voluntary park guards in order to collaborate with the park administration in sectors away from their own populations.

More recent work by PROBIOMA on social development in communities may signify the beginning of changes in the communities' attitudes toward the resource base—and an understanding of the sustainability of resources.

The relation between the park and the "buffer zone" has been more formal on the north side. The planning and management of the area outside the park have been complicated by the lack of an official buffer zone designation. This has made relationships between the park and institutions arriving in the zone with other projects more difficult. The SPERNR, by virtue of being an official entity, has been the exception and has become part of the Inter-institutional Coordinating Committee for park management. The SPERNR works with communities in the north and northeast zones of the park, but the communities to the west have not been reached by this project.

The existing institutional presence on the north side is largely oriented toward technical assistance. However, some of the positive experiences from the south of the park could be replicated in the north through the work of park guards and the formation of a technical team. Such work is proposed by CARE.

In accordance with the recently enacted Environment Law, all protected areas in Bolivia should include neighboring communities in what are called area management committees, formed by representatives from the neighboring communities and by private institutions that may be working in the area, as well as park authorities. These committees are charged with consulting and supervising activities in and around protected areas. The FAN proposal to the government to manage the park included the shaping of such a committee and included a period of familiarization with technical management schemes of this category of protected areas. These management committees can constitute an excellent link and forum for the communities to communicate to the park their points of view in the development of annual plans. They can also be a vehicle for providing and receiving information, coordination, and environmental education from and to local communities.

The new Popular Participation Law, though neither well understood nor extensively applied, should also be taken into account when forming the management committees, since protected areas constitute territories of national jurisdiction. On the other hand, it is possible that being designated as a biosphere reserve, or bioreserve, would be an appropriate alternative for Amboró. The Conservancy's experience with this category of protected areas may be instrumental as the Bolivian government determines whether to redesign the area in question.

Conflict Management and Resolution

The 1991 decree to enlarge the park has been the most significant source of conflict. The new boundaries absorbed several communities, and overnight the residents lost some of the rights they had previously enjoyed, in particular, access to natural resources and new land. Various institutions, including FAN, realized that the only way to resolve this problem was by negotiation, and from that realization, the Red Line project was developed.

There have also been other conflicts in and around Amboró National Park. Fortunately, the last administration of the park, when faced with government disinterest and conflicts it could not resolve, was able to rely on various institutions to help recapture the government's attention and resolve conflicts. FAN and park administrators have successfully confronted a wide range of problems, for example:

- conflicts with powerful logging companies
- other state agencies claiming existing resources in the park

- landowners claiming the right to unworked land in the park
- hunters flaunting political power.

Such coordination between the park administration and FAN has worked well to quickly paralyze and resolve recent massive efforts to colonize the park. In the end, the park administration—previously seen as insignificant and lacking in prestige—is no longer an anonymous body but, rather, enjoys a new recognition by the public. It has also gained the immediate attention of the government, which at the unrelenting insistence of FAN, has lent support for some of the conflicts that required a strong application of the law.

One example was the conflict between the park administration and settlers who wanted to colonize the Las Playas area, which is located to the north of the community of Santa Rosa. When some of the settlers entered the park to colonize it, the coordinated effort of the administration and FAN insured that only those people who had the signature of the president would gain title. Repeated attempts by the same people to colonize without such authorization were dealt with by the government, which had to remove them by force.

Large-Scale Threats

Threats change with time. While clandestine commercial hunting continues in the north, according to FAN, it is becoming rarer. An unknown amount of subsistence hunting, however, is practiced. While large-scale industrial logging used to be a problem, it has practically disappeared, although the park is still rich in profitable commercial forests. Small-scale loggers and settlers extract some timber, but it is generally from outside the park.

According to informants interviewed during this study, there are few direct threats; the majority are considered indirect. In the past, some have indicated that the greatest threat to the northern zone is the invasion of settlers who arrive spontaneously, primarily from the altiplano (highlands) and two Andean valleys (Cochabamba, Potosí, and Chuquisaca). However, the majority of the informants felt that this situation is more or less under control at the present time. Although there is continued pressure by humans on the north side, it seems that settlers are not entering new park land, and that the border between the park and human settlements is stabilizing.

Deforestation

A continuing threat to the park is deforestation, by peasants and settlers, of dozens of hectares of primary forest to clear land. Ironically, this land is inappropriate for agriculture. According to several informants, there are around one thousand families in the northeast zone of the park. However, this population is formed not by new settlers but by people who lived in the region before the most recent park decree. However, the continual advancement of the agricultural frontier is a consequence of flawed park design and was aggravated by the enlargement of the park. This threat can be removed by recategorizing the area and promoting sustainable management of natural resource use in the resultant buffer area.

Santa Cruz–Cochabamba Highway

One of the principal indirect threats was the construction of a new Santa Cruz–Cochabamba highway, financed by the IDB, which allowed access and settlement of the northern zone. A corridor has quickly developed between the two cities, and this has made it possible for new settlers to arrive and consequently clear a lot of forest. Although this situation was foreseen and the national government, with the support of the IDB, invested funds in order to limit the financing by SPERNR, several ex-employees have commented that the project has not turned out as it should have and appears to have had little success.

Logging

While logging was once a substantial problem, large-scale industrial exploitation has practically disappeared. What continues, without a doubt, is small-scale activity by settlers and illegal loggers. The majority of these work outside the park, adjacent to its boundaries. The park is still very rich in profitable commercial tree species, which may be targeted as they become more scarce elsewhere.

Land Claims and Union Pressure

Unions have heavily influenced some of the communities located in and around Amboró National Park, and in some areas within the park union–community relations have polarized groups. Union leaders affirm that they represent peasants, even though many of them do not live in the communities near the park. Some NGOs working in the zone, PROBIOMA, for example, believe it is important to take the unions into account and work through them since these NGOs view them as legitimate representatives of the people living around the park. Other NGOs, such as FAN, claim that the unions do not represent the peasants, and the union leaders even less so. Union leaders are seen as causing problems between those who are charged with managing the park and the resident neighbors.

Currently, and in relation to the park conflicts, the unions have more power in the south than in the north, since they have concrete claims for land tenure in the south. The communities in the south are more stable, and while the union structure at the community level is relatively weak, union leaders have strong ties to the department-level leadership.

According to some of the informants who were interviewed, the majority of the communities located on the north side are composed of indigenous peoples, primarily Quechuas from Cochabamba. Although some of these communities are well organized, the unions here are very local and do not have powerful institutional or political links at this time. The peasant movement is divided into two peasant federations, and both claim to represent all peasants.

Opinions about the unions were always strong in interviews in the field and in Santa Cruz; some informants declared that the majority of the unions are antigovernment, anti-order, and pro-vigilance anarchists. Others commented that the unions are democratic and authentic representatives of the people. From the government's perspective, unions are important political supporters and votes not worth driving away. Unions have played an important role not only in the history of Amboró but also countrywide due to their close relation to political parties and the traditional political structure of the country. The lack of action and resolution of the Red Line process is closely related to the government's uncertainty over the potential conflict, which might emerge in resolving the Red Line process. Yet failure to resolve such a conflict may seriously compromise the future existence of the park.

Coca Production

Another potential threat that should be taken into account is the cultivation of coca and drug trafficking. The north of the park is located very near the Chapare zone, where the largest number of coca plantations and cocaine factories in the country are located. Conflict and notable peasant movements have developed in the zone, and the region is a kind of quasi-republic, for the most part inaccessible to national legislation and control.

The coca business is highly profitable and seeks out remote places—far from the reaches of government—to cultivate and process coca. Some of the informants are afraid that the coca growers will want to enter Amboró National Park to cultivate coca. They also believe that some of the *cocaleros* are behind the peasants who demand more land and do not respect the park limits. What is certain is that old and abandoned processing and storage facilities in the park indicate that drug trafficking can easily occur in the region. This activity, as we know, brings with it all sorts of negative consequences for the environment and society as a whole.

National Policy and Legislation

In an exemplary participatory process, Bolivia designed and enacted its Environment Law in 1992. The law incorporates modern conservation concepts and offers general guidelines that establish the rules of the game; however, the general nature of the law and the lack of specific guidance on many issues often leaves it open to interpretation. Therefore, knowing how to implement it can be difficult. The bodies charged with applying the law are still weak, not yet organized, and vulnerable to the periodic changes in political support that are part of the national democratic process. Although the Environment Law includes protected areas, a new Biodiversity Law is perhaps more relevant. This law was also created in a participatory fashion and is yet to be revised by the national Congress, having been marginalized by other important laws and "stars" in the new government. Forest and agrarian legislation also are related to conservation and protected areas, and new versions of these are also awaiting treatment by the Congress.

The government continues to navigate through this legal uncertainty despite its executive restructuring two years ago. It is still adapting and has yet to identify its most suitable players and representatives. Proof of this political instability and of the uncertainty that still exists can be seen in the changes in the title of the Ministry of Sustainable Development and the Environment, with three different titles in two years: the Secretariat of Natural Resources, the National Directorate of Biodiversity Conservation (DNCB), and the National Environment Fund (FONAMA).

According to various informants, another principal yet indirect threat, is the unwillingness of the government to resolve urgent and fundamental problems. We were told that while the government has emitted decrees and laws to protect the park, it has done little to ensure its protection. While the government technically holds the solutions and has publicly indicated its intention to implement them, action has not been forthcoming. A simple example of this is the problem of boundaries and the park's recategorization. The government is convinced that reducing the size of the park and creating an Integrated Natural Management Area is the most just and viable solution. Yet it has not decreed such changes for fear that peasant leaders will demand that park limits be greatly reduced and no management category be assigned outside this small park.

While the government holds the technical solutions to the issues surrounding Amboró, solutions that are socially just and will stand up in the long term, it abstains from applying them because they are not what the peasant leaders are hoping for.

Despite all these threats, Amboró presents a relatively stable site, in the sense that there will not be any interference from other big national projects. There are no conflicts with mining, oil, sanitation, or transport interests, and the Land Use Plan—which governs the ordered development of the whole region—has correctly determined the capacity of land use of Amboró and surrounding areas.

Indigenous Peoples and Social Change

In the north zone of the park, entire communities of indigenous—Quechua—settlers are arriving from the departments of Cochabamba and Potosí. While these are traditional communities, they are not traditional to the zone. The Quechua people who come down from the high valleys of Cochabamba and the altiplano of Potosí are not well adapted to the climate or to the land and the conditions of the zone's lowlands. After settling in the lowlands they plant their traditional crops using traditional methods. The process of adapting to new conditions takes time, and these societies are undergoing a process of transformation. The community of El Carmen, in the northern sector, is a small traditional community that is much older. Its inhabitants are of Chiquitano origin, and residents practice palm weaving and subsistence agriculture.

On the south side, the reality is very different since traditional communities with a history of more than a century predominate. The majority of these people come from the adjacent valleys. These communities are completely established and present great stability in spite of their poverty. Little change is being seen here.

There have been references to native nomad populations in the interior of the reserve, but there is no concrete evidence of their existence.

Resettlement

According to informants, there has not been, and will not be, forced resettlement of the established communities around the park. The informants report that this is not even an option, since laws and customs in Bolivia prohibit the obligatory resettlement of communities and settled populations.

In Amboró, the sheer number of people and their diverse histories, but above all, the century-old settlements in the south; the infrastructure of roads, schools, and health care; the memories, symbols, and family feelings; and, finally, their inalienable property rights, make it impossible to even consider moving such established communities.

Conclusion

Amboró National Park presents a complicated conservation case study. Many of the informants in this study attribute the park's current management problems to the "anticommunity" administrative style of the first park administrator, the absorption of established communities inside the limits due to the decree of 1991, and the exclusion of local people in the decisions related to the management of the park and its surroundings. Some people perceive the incursion of people around the park to be a problem, but, as stated earlier, more than one of the local people interviewed during this study said, "We are not invading the park. It is the park that has invaded us."

The impact of the Red Line project, initiated in 1991, has been notable. It has been a legitimate participatory effort and, for that reason, has satisfied the majority of the people affected by the 1991 decree. Union leaders trying to aggravate the conflict for their own political gains seem to be the primary factor in some places where negotiations could not be finalized.

However, in the greater part of the park, what could have been a grave problem was resolved by taking into account the rights and claims of local people.

Situations such as this, where there is direct conflict between those who want to conserve an area and those who live in it, are very common worldwide. The strategies employed to solve these problems, however, have not normally been successful. The Red Line around Amboró serves as a positive—although imperfect—example of conflict resolution around a protected area.

It is hoped that the new delimitation of the park will contain two models of management: a core zone and an Integrated Management Natural Area, which will act as an external buffer zone to the park. From an administrative point of view, the two management categories would be separate, and would have to coordinate and collaborate jointly on macroregional policy.

The park is supported by the SPERNR, which in turn is dependent on the Ministry of Sustainable Development and the Environment. SPERNR's expertise in watershed, forest, and community management, as well as economic and technical training, make it a strong candidate to coordinate the management of the Integrated Management Natural Area, the buffer zone for Amboró National Park. All that is needed for this to happen is the political attention of the government. With this action, the problem of representation for NGOs and peasants would be resolved, and the SPERNR could function as coordinator of activities and work that would be done by the NGOs with a presence in the zone, under a system of management appropriate for the category and dictated by the government.

In order to take advantage of the park's resources, close technical coordination would need to be established between the two administrations. A redefined park would be dependent on the neighboring populations, from which the park guards originate, where the access roads pass, and where the park provides all kinds of services.

The park has great political importance but could also be a model of conservation and development. FONAMA has already offered economic resources for various projects, SPERNR is already paying for its second phase, FAN and the Conservancy have planned a long-term intervention in the park, and the arrival of CARE and its technical and financial resources is imminent. When combined, the efforts of NGOs working in the zone would be remarkably strong.

Dedication and willingness on the part of the government may allow for the application of solutions that the government has already researched and designed. As mentioned before, various informants expressed frustration with the Bolivian government. Nevertheless, it is noted that the government, in some ways, has shown more support for Amboró than for other parts of Bolivia. The government has also taken an active role in the new delimitation of the park. The greatest threat now is that it is precisely these small advances that can be lost if action is not taken immediately.

Postscript

The changes that have occurred at Amboró since the completion of the case study are exceptional; major sections of the case contain only historical information. Key changes, described below, are: (1) the redefinition of park boundaries; (2) the creation of a buffer zone for the park; (3) the creation of a new government entity responsible for park administration; and (4) the withdrawal of FAN from park-focused activities to work in the buffer zone.

In 1996, the government of Bolivia issued a Supreme Decree that reduced the size of Amboró National Park from 637,000 ha. to 440,000 ha. Virtually all organizations believed that the Red Line process, and some reduction of boundaries, was necessary—some NGOs suggested decreasing the park size by roughly 100,000 ha. The extent of the reduction in the park boundary is a substantial increase over what some regarded as necessary.

The same decree that reduced the boundaries created a Natural Area for Integrated Management (ANMIA), which is, for the most part, the area between the old and new boundaries. FAN publicly stated that the ANMIA should consider zoning some areas as strictly protected, even if they were privately owned. This was not done in the Supreme Decree. Communities are apparently collaborating well in the demarcation of new boundaries and these boundaries are being rationally defined.

The government's decision to manage Amboró and other protected areas directly through its National Directorate for Biodiversity Conservation, instead of passing the management to FAN after FAN won the public bidding in 1995, is a departure from previous policy. As a result, FAN no longer has any official role in the management, administration, and development of the park. The one exception is a joint FAN–Conservancy effort to produce a management plan for the park. FAN's work in the area is now limited to the ANMIA area in the park's buffer zone. Its pronouncements and actions on protection have decreased since it is no longer involved in the park. FAN turned over vehicles, equipment, and buildings that it had acquired for park management to responsible government authorities to enable them to manage the park more efficiently.

FAN and other NGOs continue to work around the park with renewed financial resources from the British government's Overseas Development Agency. The funds are being channeled through CARE and are strengthening the capacities of NGOs and about half of the communities in the ANMIA to implement sustainable development measures for local resource use. CARE's Integrated Project for Conservation and Development plans on working with communities through the NGOs that were formerly working in what was the park. The initial phase of the project is to be three years, after which time another phase will be added, with the length to be determined on circumstances at that time. The initial amount to be disbursed has not yet been made public; rumors put the initial disbursement of funds at US$7 million. All of these activities blend well with the newly designed management plan by FAN and the Conservancy, which incorporates many of the programs envisioned to reduce the human impact on the park: environmental education, community support and development, human impact and biodiversity monitoring, ecotourism, and others. FAN is also supporting participatory community planning activities through workshops and technical support to and financing of the priority activities resulting from the planning process.

As of 1997, CARE had established its presence in the region and FAN had opened an office in Comarapa. CORDECRUZ has disbanded and its functions have been taken over by the prefecture of Santa Cruz.

Note: The three authors were unable to conduct the site visit at the same time; R. Margoluis and K. Brandon traveled independently from A. Moreno.

Glossary

Agrarian Reform and Institute of Colonization—Reforma Agraria y el Instituto de Colonización

ALAS—legal advisory branch of the worker's union in the department of Santa Cruz

altiplano—highlands

ASEO—Asociación Ecológica del Oriente (Ecological Association of the West)

Caritas—A charity of the Catholic church

CIAT—Centro de Investigaciones Agrícolas Tropicales (Center for International Tropical Agriculture)

CORDECRUZ—Corporación Regional de Desarrollo de Santa Cruz (Santa Cruz Regional Development Corporation)

CRS—Catholic Relief Services

DNCB—Dirección Nacional de Conservación de Biodiversidad (National Directorate of Biodiversity Conservation)

FAN—Fundación Amigos de la Naturaleza (Friends of Nature Foundation)

FONAMA—Fondo Nacional para el Medio Ambiente (National Environment Fund)

MBAT—Misión Británica de Agricultura Tropical (British Mission for Tropical Agriculture)

MDSMA—Ministerio de Desarrollo Sostenible y Medio Ambiente (Ministry of Sustainable Development and the Environment)

ODA—Overseas Development Administration of the British government

PCBB—Proyecto de Conservación de Biodiversidad de Bolivia (Bolivian Biodiversity Conservation Project)

PDAR—Programa de Desarrollo Alternativo Regional (Program for Alternative Regional Development)

PLUS—El Plan de Uso del Suelo de Santa Cruz (Santa Cruz Rural Land-Use Plan)

PROBIOMA—Productividad, Biósfera y Medio Ambiente (Productivity, Biosphere, and Environment)

PRODISA—Programa de Desarrollo de las Provincias Ichilo y Sara (Development Program for the Provinces of Ichilo and Sara)

RAC—Reforma Agraria y Colonización (Agricultural Reform and Colonization)

SPERNR—Subproyecto de Protección de Etnias y Recursos Naturales Renovables (Subproject for the Protection of Ethnic Resources and Renewable Natural Resources)

UTD-CDF-SC—Unidad Técnica Desconcentrada del Centro de Desarrollo Forestal de Santa Cruz (Decentralized Technical Unit of the Santa Cruz Center for Forestry Development of Santa Cruz)

vallunos cruceños—long-term residents near Santa Cruz

Yanachaga-Chemillén National Park

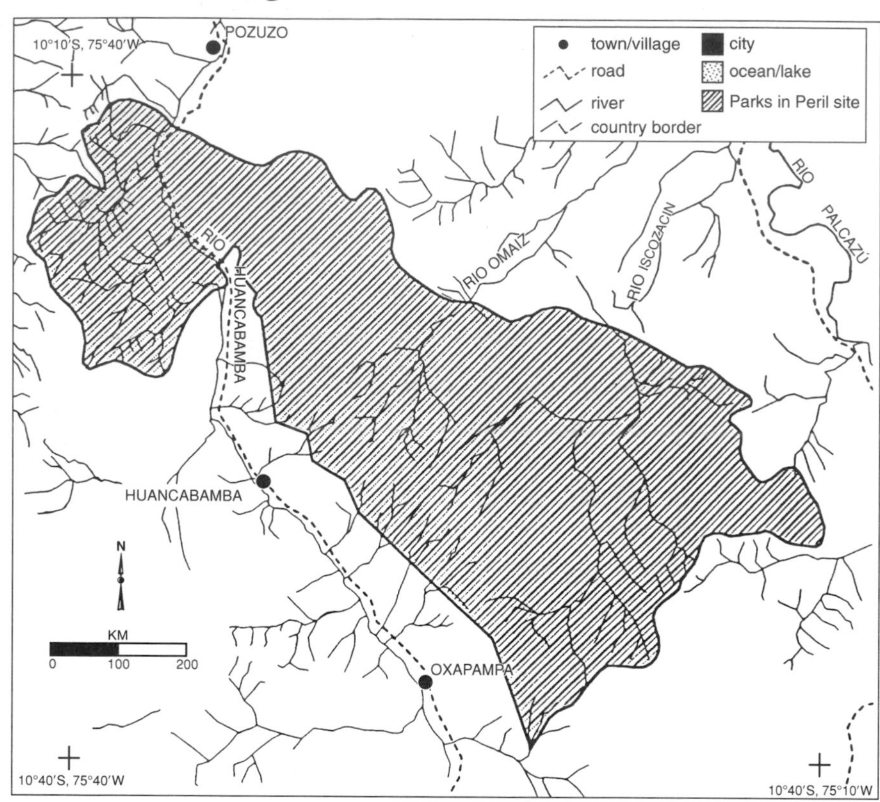

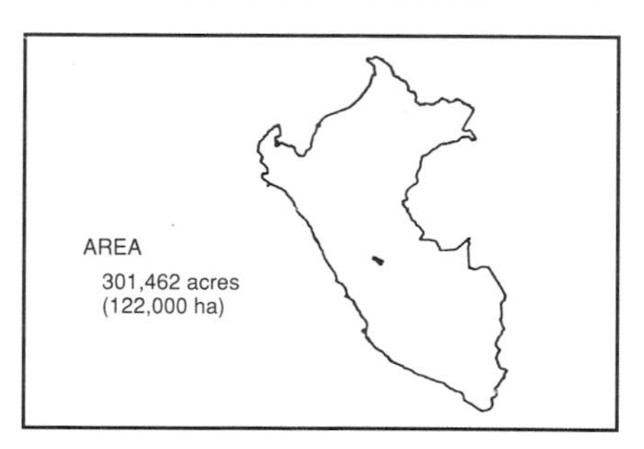

12. Peru: Yanachaga-Chemillén National Park

Luis Angel Yallico and Debra A. Rose

Yanachaga-Chemillén National Park, created in 1986, is located in the Palcazú River valley in the Amazon basin of central Peru. The park consists primarily of the Yanachaga Cordillera, which reaches a maximum altitude of 3,643 meters above sea level. On the east, the Yanachaga Mountains descend gradually into the valley of the Palcazú River. On the western flank of the park are the Santa Barbara Mountains, reaching 3,400 meters and separated from the Yanachaga Mountains by the deep Huancabamba Canyon. The park's diverse ecosystems include humid tropical forest, montane forest, and cloud forests. Although the park's fauna and flora have been little studied to date, species registered during preliminary surveys included 2,584 plants, 85 mammals, 427 birds, 16 reptiles, 2 amphibians, and 31 fish. Threatened species found within the park include the harpy eagle (*Harpia harpyia*), jaguar (*Felis onca*), and giant otter (*Pteronura brasiliensis*) (INRENA 1995c).

The name of the park reflects not only the ecological gradient of the park, from highland to lowland, but also the cultural history of the area that is reflected by this gradient. The Palcazú River valley, where the park is located, is also the traditional home of the Yanesha (also known as Amuesha) Indians, who have inhabited the Palcazú valley for an estimated thousand years (Stocks and Hartshorn 1989). *Yanachaga* is a word from the Quechua Indians, who inhabited the highland areas, while *Chemillén* is an Amuesha (lowland) Indian word. Others inhabiting the valley are mestizo colonists who are descended from a mix of European ancestry and Peruvian colonists who arrived during periods of oil exploration in the late 1950s and 1960s or more recently. Campa Indians reside in the valley as well. All of these groups reside outside of the park, in an area that has been undergoing exploration and change for centuries. Missionaries and colonists returned to the region in 1756, at the end of a rebellion that burned missions and haciendas. Austrian immigrants settled in the 1850s, and other immigrants arrived during a rubber boom in the late 1800s (Plan Maestro Parque Nacional Yanachaga-Chemillén 1987). Oil exploration in the 1950s sparked yet another period of migration to the region. Yet despite the diverse groups that have colonized the region, large expanses of forest are still intact, including the one that makes up the park.

Park Establishment and Management

Beginning in 1974, anthropologists and biologists submitted a number of proposals for the creation of a conservation unit in what is known as the Central Selva region, in order to protect Yanesha lands and culture, the watershed, and the region's diverse ecosystems. However, the creation of the Yanachaga-Chemillén National Park (YCNP) was finally the result of a colonization and development project initiated by the Peruvian government in the Central Selva region in 1980. The project was part of President Fernando Belaúnde's strategy to develop the Amazon as a solution to the nation's economic difficulties. It intended to settle 150,000 colonists in the Pichis, Palcazú, and Pachitea river valleys, construct a network of roads and highways, and encourage the development of extractive industries in the region. The U.S. Agency for International Development provided US$22 million in funding and technical assistance for the project from 1982 to 1987. However, following initial environmental and social studies and protests from Indians and anthropologists, the agency agreed to fund only specific components of the project in the Palcazú River valley. The USAID project, known within the agency as the Central Selva Resource Management Project, provided assistance for forest resource management, protected areas, and local participation, helping to redirect the focus of the Peruvian government's program in this region.

As a precondition for USAID involvement in the project, the agency required that the ten Yanesha communities living in the project area be given legal title to their lands. The Peruvian government conducted a land titling program that protected the land claims not only of the Yanesha, but also of European and national settlers in the region. The USAID project also included an environmental protection component, with the objectives of preserving an example of the Peruvian high jungle, preserving the hydrological and other ecological functions of the valley's upper watershed, contributing to the economic development of the Palcazú Valley through the production of renewable natural resources attractive to tourists, and preparing the studies preliminary to the YCNP and the San Matías–San Carlos Protection Forest (Aguilar 1990).

Implementation of the environmental component began in 1983 and resulted in the declaration of the YCNP in 1986, with an area of 122,000 hectares, and the San Matías–San Carlos Protection Forest in 1987, with an area of 145,818 hectares extending into the Pichis Valley and the traditional lands of the Ashaninca people. A five-year Master Plan for YCNP was developed in 1985 and approved by the Peruvian government in 1987. In 1987, field studies were begun to create the Yanesha Communal Reserve, which formed a buffer zone between YCNP and the Yanesha and colonist settlements east of the park. The communal reserve, the first of its kind in Peru, was declared in 1988 with an extension of 34,745 hectares.

Authority for management and conservation of all three protected areas is vested in the National Institute of Natural Resources (INRENA). The USAID project terminated in 1987, when the agency withdrew due to the increasingly unstable security situation associated with terrorist activities in the area. The Pichis-Palcazú Special Project (PEPP) continues to promote infrastructure and agricultural and livestock development in the region surrounding the park and adjacent protected areas. Its activities are now primarily development oriented and largely divorced from conservation activities in the three protected areas, although PEPP has recently provided limited support for infrastructure that is necessary to encourage visitors to the national park, and is assisting in planning and coordinating the development of a management plan for the Yanesha Communal Reserve. The Peruvian Foundation for Nature Conservation (FPCN), known as Pro Naturaleza, and The Nature Conservancy began to provide assistance to the park in 1986, under a USAID grant. Pro Naturaleza and the Conservancy have provided most of the financial and technical assistance available to the park since 1986 and have been the exclusive source of assistance since 1987. In 1991, Pro Naturaleza signed an agreement with INRENA for cooperation in park management and conservation and for the implementation of the PiP program funded by the Conservancy.

Land and Resource Tenure

A precondition for USAID involvement in the Pichis-Palcazú Special Project was the granting of legal land title to the ten Yanesha communities living in the project area. In response, the Peruvian government surveyed and titled lands occupied by both the Yanesha and the colonists. The land titling program, therefore, not only protected resident Yanesha against further settlement, but also allowed the boundaries of the park and the protection forest to be drawn around unoccupied lands, so that there are no legal human residents within any of the three protected areas. Tenure security in areas surrounding the park and other protected areas has also helped to discourage further colonization of the region. The land titling program conducted prior to the park's creation, therefore, prevented many of the land pressures and conflicts that have accompanied the creation of protected areas throughout the Americas.

Under the Forestry and Wildlife Law of 1975 and the Environmental Code of 1990, human activities permitted in national parks include only public visits for scientific, educational, recreational, and cultural use. Human settlement and natural resource use are prohibited in all Peruvian national parks, with the exception of ancestral indigenous populations. A new natural systems plan for the Peruvian protected areas currently being developed by INRENA with the support of the German Agency for Technical Cooperation (GTZ) would also in some cases permit the continuation of traditional resource uses and practices by indigenous peoples in national parks, to the extent that these are compatible with the objectives of the protected area. However, given the close proximity of both the San Matías–San Carlos Protection Forest and the Yanesha Communal Reserve, it is unlikely that traditional resource uses would be developed within the national park.

Resource tenure in areas adjacent to the park may, however, be more problematic. The traditional lands of the Yanesha incorporate the districts of Oxapampa, Villa Rica, Pozuzo, and parts of Chanchamayo. Land titling granted to the ten Yanesha communities in the Palcazú make the area the most important site of Yanesha settlement. Within the valley, the Yanesha make up 70 percent of the population but control only 40 percent of the land holdings. Furthermore, communal lands are located on the poorer soils of the valley's shallow tributaries and upper watershed, while colonists of European descent control large holdings for cattle ranching on the rich alluvial soils of the middle and lower valley as a result of the colonization process. The establishment of the communal reserve in 1988 nearly doubled the land area available to indigenous communities, and the region is rich in forestry resources. However, the communal reserve as well as the national park were created without significant input from the Yanesha and, indeed, engendered strident conflict with FECONAYA, the Federation of Native Yanesha Communities (Stocks 1990). This conflict has diminished, and the Yanesha themselves have created the Foundation for the Conservation of the Yanesha Communal Reserve and are working together with Pro Naturaleza toward this aim.

In the protection forest, human activities that affect the integrity of the forest cover are prohibited, precluding by definition human settlement, agricultural and livestock production, and commercial timber production. However, fishing, hunting, and the collection of nonwood forest products are permitted activities. Current Peruvian protected-areas legislation prohibits "destructive" activities such as agriculture and forestry production in the communal reserve, permitting only hunting by neighboring populations, both indigenous and peasant. At present, the boundaries of the communal reserve are not visibly delimited in the field and direct use by neighboring communities, and ladino colonists, has included hunting, fishing, and collection; isolated instances of land clearing for agriculture; and on a larger scale, illegal logging, for which communities receive a fee per square foot of lumber produced (Gaviría 1995).

The definition of use rights within the communal reserve has been modified slightly since its creation. The new definition of a communal reserve is an area reserved for the conservation of fauna and flora for the purpose of benefiting neighboring human populations, both indigenous and campesino. New human settlement, commercial timber extraction, and agriculture and livestock production are prohibited. In 1990, with the creation of the National System of Protected Areas (Sistema Nacional de Areas Naturales Protegidas por el Estado, or SINANPE), the category of Communal Reserve, which has legally existed since 1975, was included in the protected-areas system, so that the reserves have come to be considered areas for the protection of wild fauna and flora where traditional uses are permitted. Under the draft Strategy for the National System of Protected-areas of Peru, elaborated in 1995 with technical assistance from the GTZ, the latter definition of a communal reserve is in the process of becoming institutionalized. The draft strategy would require resource use to be conducted under a management plan implemented by user groups and approved by the competent authorities.

On the western border of the park, the granting of forestry concessions has been a problematic resource tenure issue in the past; such concessions were suspended nationwide in 1993 but reinitiated in mid 1995. Concessions are frequently granted without clear reference to existing boundaries. Although land titles in the region are clearly delimited, the concessions have in the past been issued somewhat haphazardly, creating a mosaic of overlapping claims. Indeed, a review by park personnel of the forestry concessions granted on the west of the park found that the total area covered by concessions exceeded the land area available (P. Aguilar, pers. comm., 1995). In some cases, concessions also extended within park boundaries, although the lack of roads generally prevented exploitation of these areas (*Plan maestro* 1987). It is not certain whether this situation will be repeated with the renewal of forestry concessions, but continued lack of capacity for administration and control by the Ministry of Agriculture suggests that it is likely.

Resource Use

Within Yanachaga-Chemillén National Park, permitted tourism and research activities have been minimal to date due to the unstable security situation. However, given the park's proximity to Lima and accessibility by highway, the area has the potential for rapid tourism growth. Pro Naturaleza is currently working to promote the research potential of the park among national and international academic institutions in an effort to attract additional resources.

Livestock raising, predominantly of cattle, and cash-crop production currently predominate in the western region of the park, with areas dedicated to cultivation of *rocoto* (*Capsicum pubescens*), a hot chile, extending to the park boundary. In the 1950s and 1960s, the province of Oxapampa was the country's largest timber producer, but this activity has declined dramatically due to overexploitation of the resource. No plans for sustainable forest management have yet been developed. All forestry concessions were suspended nationwide in 1993, further limiting the scale of production. Logging has continued along the western boundary of the park, and concessions were resumed in 1995. Prior to 1993, some forestry extraction contracts were granted for areas that extended within the park boundaries. In most cases, the absence of roads prevented logging from taking place within the park, but some incursions into the park have occurred along the Tunqui River (*Plan maestro* 1987).

To the east of YCNP, Yanesha communities residing in the Palcazú valley are oriented primarily toward a subsistence economy, with the use of currency as the basis of exchange introduced less than fifteen years ago. The Yanesha practice slash-and-burn subsistence agriculture, using their principal crops of yucca, corn, rice, plantain, and beans almost exclusively for household consumption. Population pressure has forced households in many areas, such as Oxapampa and Villa Rica, to reduce fallow periods, causing soil degradation and reducing productivity. There has been a shift from crops to livestock, and all communities now have domestic livestock, including cattle, black belly sheep introduced by the PEPP, pigs, and birds; recent estimates total fifteen hundred head of cattle, fifteen hundred head of pigs, and twenty-five hundred sheep (Gaviría 1995).

Only in the relatively isolated community of Alto Lagarto does subsistence agriculture continue to predominate. The communities of Alto Iscozacín, Shiringamazú, Loma Linda-Laguna, Santa Rosa de Pichanaz, San Pedro de Pichanaz, and 7 de Junio are highly dependent on the sale of standing timber to local dealers. Much of this activity is illegal, including timber extraction without permit, transfer of cutting and transport permits from communities to timber dealers, logging in the communal reserve and the San Matías Protection Forest, disregard for land-use zoning in the removal of trees from steep slopes, and failure to respect minimum cutting diameters. The communities 7 de Junio, Santa Rosa de Chuchurras, Buenos Aires, and Nueva Esperanza are located near the area of earliest colonization and therefore have been highly influenced by colonist livestock production. Livestock is heavily concentrated on colonist lands, totaling an estimated 14,800 head of cattle, 5,000 pigs, and 1,800 sheep. Members of Yanesha communities obtain cattle from colonists through the *al partir* system, in which community members breed and raise cattle in return for half of the calves obtained (Gaviría 1995).

Hunting, fishing, and collection remain important activities among native communities, providing an important source of protein that is shared reciprocally among family members. The Yanesha hunt at least once every fifteen days with shotgun and bow and arrow. In some instances, hunters penetrate a short distance into the national park. Important game species include deer (*Mazama americana*); big rodents, such as *majaz* or *zamaño* (*Agouti paca*) or *cutpe* or *misho* (*Dasyprocta* spp.); monkeys, including *mono coto* (*Alouatta seniculus*) and *mono choro* (*Lagotrix lagotricha*); and birds such as *paujil* (*Mitu mitu*) and partridges (Tinamidae). Fishing is a weekly activity, using nets, hook and line, explosives, and occasionally harpoons. Group fishing involving family units or communities is conducted using plant-based fish poisons, *barbasco* (*Lonchocarpus* sp.), or explosives. The species most commonly caught are *boquichico* (*Prochiltodus nigricans, Semaprochilodus* sp.), *palometa* (*Mylossoma* sp.), *zungaro* (*Zungaro zungaro*), *doncella* (*Pseudoplatystoma fasciatum*), *lisa* (*Schizodon fasciatum, Leporinus fridresii*), and *carachama* (*Plecostomus* spp.). Although hunting and fishing are primarily for household consumption, both game and fish are occasionally sold in nearby markets. Species collected most commonly for consumption are mollusks, crustaceans (shrimps and crabs), snails, larvae of *Rhichophorus* sp. (Curculionidae), *uvilla* (*Pouruma* sp.), *ungurahui* (*Yessenia batagua*), *"caimito" (Chrysophyllum sanguinolentum)*, *aguaje* (*Mauritia flexuosa*), and almonds (*"almendras," "Calmendras" Caryocar amygdaliferum, C. glabrum*) (Gaviría 1995; *Plan maestro* 1987).

Also collected in the zone of influence of the park are palm leaves (*"umiro,"* *Geonoma* spp.) and heart *"palmito"* (*Euterpe* spp.); medicinal plants such as *uña de gato* (*Uncaria* spp.), *chuchuhuasi* (*Heisteria* spp.), *copaiba* oil (*Copaifera officinalis*), *"sangre de grado"* (*Croton* spp.); and native fruits such as *pijuayo* (*Bactris gasipaes*), *guaraná* (*Paullinia cupana*), papayas (*Carica* spp.), and *granadillas* (*Passiflora* spp.) (INRENA 1995b).

Land-use studies from 1982 through 1985 in connection with the USAID-funded Palcazú Rural Development Program (PDR) indicated forestry production as one of the few feasible large-scale development activities. Although colonist landholdings had been largely cleared of forest for livestock grazing, large tracts of primary forest remained on communal lands, with a total area of some 11,400 hectares suitable for extensive forest production and 8,900 hectares suitable for intensive forest production. In 1986, the Yanesha Forestry Cooperative was established with seventy members under the auspices of the PEPP's PDR-Palcazú Forestry Development Unit, with assistance from USAID and the Tropical Science Center (TSC) of San José, Costa Rica. The objectives of the cooperative were to provide employment to members of native communities, to manage natural forests for sustained yield, and to protect the cultural integrity of the Yanesha people (Simeone 1990).

Participating communities contributed a portion of their communal forests, with the cooperative conducting resource inventories, developing management plans, and returning a portion of receipts to the community. The cooperative forestry management system was based on strip clear-cutting, in which strips of thirty to fifty meters in width and three hundred to five hundred meters in length were cut to simulate the disturbance regime associated with natural tree falls (Simeone 1990; Hartshorn and Pariona 1993). A variety of products were developed in order to achieve near-full utilization of all species removed from the strips. The cooperative constructed a sawmill, carpentry shop, and charcoal-producing facility. They produce charcoal, preserved posts, and sawn lumber—all for local sale (J. Arce, pers. comm., 1995).

By 1987, the project had identified and mapped forest production areas, and harvesting had been initiated in the first two experimental strips. However, USAID support for the project then ended abruptly. After the termination of USAID funding, work with the Yanesha Forestry Project was the only part of the Central Selva Resource Management project that continued, with progress considerably slowed by increasing terrorism in the region. In 1988, the Peruvian Foundation for Nature Conservation, with financing from the World Wildlife Fund, initiated a forestry extension program, training for members and nonmembers, technical and legal assistance, and a rotating fund for the purchase of machinery (Wells and Brandon 1992; J. Arce, pers. comm., 1995).

By 1990, the cooperative had begun to experience considerable difficulties. Charcoal produced by the enterprise proved too costly for local markets. In addition, although native tree species proved most suitable for the production of posts, the cooperative encountered little demand for these products in domestic markets. Sawn wood also proved difficult to market due to the predominance of nontraditional species, although some success was achieved in marketing it to "green" companies in the United States and Europe, which were able to offer above-market prices because of the product's association with a sustainable forestry project. As a result, only seven experimental strips were harvested during the history of the cooperative, with approximately half of the raw material being obtained instead from agricultural land clearing or selective logging (Linares 1991; J. Arce, pers. comm., 1995).

In 1992, the cooperative's activities ground to a halt, relations between it and involved NGOs having been severely strained by financial difficulties. Since that time, PEPP has attempted to assist the cooperative in developing a management plan for the reinitiation of forestry production and in locating funding assistance. Pro Naturaleza has begun examining options for production of nontimber forest products by some of the communities in the valley, principally orchids and uña de gato. The latter is a high-value, wild medicinal plant enjoying rapidly growing demand in domestic and export markets (J. Arce, pers. comm., 1995); the species has also been promoted as a viable alternative to coca following the collapse of world cocaine prices (Reforzarán sustitución de cultivos de la coca 1995).

Organizational Roles

The Fundación Peruana para la Conservación de la Naturaleza (FPCN), or Pro Naturaleza since 1995, was created in 1984 and in 1986 became involved in the conservation of YCNP with financial assistance from the Conservancy through a grant from USAID. In 1987, following termination of USAID funding for the Palcazú project (including YCNP), Pro Naturaleza stepped in to ensure the continued presence of the director and seven rangers contracted following the park's creation. Since that time, the NGO, with the partnership of the Conservancy, has assumed financial costs associated with contracting, training, and maintaining the director and guards. As established in the 1990 agreement signed by Pro Naturaleza and the Directorate General of Forestry and Fauna (DGFF, now the Directorate General of Protected-areas and Wildlife, located within INRENA), these personnel are granted official government recognition by INRENA, thus ensuring that they possess the authority required to fulfill their duties. INRENA also provides a small budget to support staff and operations.

Financing for construction and maintenance of park infrastructure, training, and operations is also provided by the Conservancy–Pro Naturaleza partnership. The park now has a central office and visitor center in Oxapampa and a biological station in Paujil. A substation is currently being constructed on the Pescado River on the eastern flank of the park to facilitate research and patrols. Park personnel are equipped with a four-wheel-drive pickup truck, motorbikes, boats, radios, generator, computer, and printer. The project has also supported boundary demarcation and trail construction, particularly along the critical western border.

PiP is the sole source of outside funding to Yanachaga-Chemillén

National Park and has undoubtedly played a critical role in park conservation in the virtual absence of government financial support for the project. Furthermore, despite severe economic and political difficulties in Peru during the project period, PiP has been remarkably effective in YCNP. A number of factors have contributed to this success. The Pro Naturaleza–DGFF agreement providing for Pro Naturaleza contracting of park staff has permitted close cooperation and coordination between the NGO and field personnel and has contributed to the importance of education and extension in the activities of the park authority. Both the park director and the rangers are well trained, experienced, and highly motivated. Although political and budgetary conditions have limited the direct involvement of the NGO in field activities, implementation solely by park staff may prove to be effective, given their permanent presence and commitment to conservation objectives.

Park personnel have also initiated a number of innovative strategies for community outreach, environmental conservation, and park conservation, including their successful efforts to delimit the western park boundary through direct agreement with local landowners. The absence of human settlements within park boundaries has undoubtedly facilitated this process by minimizing conflicts between the park and neighboring communities and by allowing park staff to focus their efforts on conservation and management activities within the park. Furthermore, the presence of the PEPP, which despite limited financial capacity is the most visible government presence in the region, has facilitated a division of labor between conservation and development activities. The establishment of good working relationships between the park and the PEPP presents a number of opportunities to strengthen linkages between park and buffer areas, as discussed below.

Linkages between Park and Buffer Areas

A number of efforts have been made to establish links between Yanachaga-Chemillén National Park and buffer areas, particularly on the western border of the park. The park staff are hired locally and have established productive working relationships with relevant official agencies, for example by coordinating with local authorities (PEPP, the National Program for the Management of Basins and Soil Conservation, and the Agrarian Office of Oxapampa) in delimiting and demarcating the western zone of the park. The park is also represented on the Multi-Sectorial Committee of Oxapampa, which is charged with coordinating provincial development activities, permitting frequent communication between the park authority and a variety of regional and local organizations. The park authority and PEPP are also working together to improve relationships between the park and native communities and to assist in the development of a management plan for the Yanesha Communal Reserve.

Park personnel have also initiated a program to delimit and demarcate the park's western boundary, when possible through formal agreement, with neighboring landowners expressing mutual agreement on the location of the boundary. The park office in Oxapampa has also developed a database on land ownership and land use that is being field-verified and updated by park guards for monitoring purposes. The information contained in the database includes location, names of owners and managers, lot information and contract numbers, land use and condition, agreements signed with the park, and observations and comments. Informal extension has been conducted with local landowners in connection with work to delimit the park's western boundary. Park staff have also conducted environmental education activities in neighboring communities, establishing tree nurseries in local schools, organizing visits to the park by schoolchildren, and giving public talks on the park and the importance of its conservation.

On the eastern boundary of the park, not only does the Yanesha Communal Reserve serve as a buffer area for YCNP, but the park also contributes to the protection of animal and plant resources available to local communities within the reserve. Existing legislation requires that the communal reserve be administered by its beneficiaries in accordance with local forestry regulations and in close coordination with the agrarian authorities of Iscozacín.

The creation of the communal reserve and its subsequent management have contributed to negative attitudes toward the reserve, despite considerable extension work among the communities following its creation and despite recent efforts by park staff and Pro Naturaleza to develop communication with local leaders. In 1988–89, an effort was made to incorporate Yanesha leaders as park personnel, but without success. The shortage of financial resources has hindered the implementation of other programs dcsigned to build community support for conservation. These difficulties have been aggravated by political tensions in the region. Local and regional politics are experiencing turmoil as a result of rising drug trafficking, and a military base has been established in Iscozacín to control drug trafficking. Earlier in the year, the region's principal trafficker was captured by Colombian police and jailed in Peru, and the mayor of Palcazú was charged with collusion in drug trafficking.

In early 1995, in an effort to improve administration of the reserve by local communities, coordinating meetings were held to create an administrative body, the Foundation for the Conservation of the Yanesha Communal Reserve, and a support group. Pro Naturaleza provided financial support for these meetings. The foundation's Directive Committee is formed of ten representatives of local communities and is headed by an administrator. The support group consists of seven representatives of community forestry enterprises, one representative each from the Ministry of Agriculture and the YCNP, and two representatives from PEPP. As a result of this process, a draft management plan for the reserve has been developed and as of 1995 was under review. As a result of representation by the park in the support group, relations between the park and community leaders appear to be improving.

The PEPP has maintained an active role in regional development efforts through PDRs in Pichis-Pachitea, Palcazú-Iscozacín, Oxapampa, and Chanchamayo-Satipo. Development activities include construction of secondary roads and bridges and maintenance of existing highways; introduction of and support for sheep production; extension to cattle ranches and dairy producers. In the Palcazú valley, the program has organized some twenty Yanesha enterprises dedicated primarily to agriculture, forestry, and livestock, but these enterprises are relatively new and are experiencing difficulty in acquiring working capital. The program has also supported local reforestation projects. The PEPP also provided technical and financial support for the creation of a yucca processing plant in one of the native communities, although it has experienced problems with internal organization and transportation of raw material to the plant (Gaviría 1995).

Although the PEPP has expressed increasing interest in providing assistance to the park in infrastructure development and public information and is working closely with Pro Naturaleza and the park authority regarding activities with Yanesha communities, there does not appear to be frequent communication between PEPP and the park authority regarding

PEPP development activities elsewhere. More frequent communication regarding projects such as road development and agricultural extension not only would allow the park to exert a greater influence over the development of the park's buffer zones, but also would be likely to highlight opportunities for productive interventions by Pro Naturaleza to support resource conservation activities.

Additional opportunities for local participation may arise as a result of recent efforts to revise protected-area policies by INRENA. The proposed Strategy for the National System of Protected Natural Areas of Peru would require the creation of Management Committees (Comités de Dirección) for each protected area in order to provide for participation by existing local organizations. The Strategy also considers the possibility of creating flexible participatory mechanisms in cases where local organization is weak, so that the Management Committees may encourage the creation of local organizations by offering participation in decision making as an incentive (INRENA 1995a).

Conflict Management and Resolution

At present, no formal mechanisms exist for conflict management and resolution. The principal source of conflict to date has been the lack of participation by Yanesha communities in the creation and management of protected areas. In 1988–89, Yanesha leaders were hired as park staff, but this effort ceased. The Yanesha leaders were busy trying to resolve problems related to the creation and management of the communal reserve. However, they wanted to receive payment to continue as park rangers, even though they were unable to carry out protection activities. Yanesha leaders reportedly told other Yanesha that their leaders' approval would be needed to join park staff. Tensions between Yanesha leaders and the park may be eased by increasing education and outreach activities by park personnel in the Palcazú, particularly if park staff are able to provide technical and other assistance in the management of the communal reserve.

Large-Scale Threats

Yanachaga-Chemillén National Park has benefited greatly from the research and planning activities and land titling programs associated with regional development programs of the 1980s. Human impacts within the park's boundaries are minimal, acute conflicts over land tenure have been prevented, and the Yanesha residing in the area of the park have been granted at least a limited measure of protection through titling and the establishment of the communal reserve. Nonetheless, new settlement along the western border of the park continues to be encouraged by official development projects and may exert growing impacts on the park itself. Furthermore, although incursions into the park have been minimized by the presence and energy of park staff, a stable source of long-term funding to support these controls has yet to be found.

On the eastern border of the park, efforts to encourage resource development and community participation have not achieved significant successes to date. Although development of a locally administered management plan for the Yanesha Communal Reserve is an important step toward natural resource conservation in the region, local cooperation in its implementation remains uncertain given native resentment of outside control over protected-areas in the past.

The population of the immediate area of the park appears to be stable at present, due in part to guerrilla activity and in part to the limited number of transportation routes. The small number of recent settlers may in fact have been offset by some outmigration as the result of the security situation. Relative peace has now been established in the area, accompanied by the return of some of the households that had fled the area, although the increasing importance of drug trafficking in the zone is bringing with it new problems.

Population growth and migration on the northern, western, and southern boundaries of the YCNP, and in the transition zone on the east between the protection forest and the park and communal reserve, may become problematic in the future. Human settlement in the area is currently concentrated west of the park along the highway extending from the town of Pozuzo south to Puente Paucartambo and east of the park along the Palcazú River and the highway running parallel to the river from Puente Paucartambo to Iscozacín. On the western highway is the town of Pozuzo to the north of the park and the towns of Huancabamba and Oxapampa immediately west of the park. Agricultural production extends to the western boundary of the park and has been prevented from encroaching on the park itself by the presence of park guards. In addition, electric energy is available in Oxapampa, coming from Yaupi, which is thirty kilometers away. Energy development is not likely to immediately affect park conservation but may serve to make the area increasingly attractive to settlers.

The region of the Central Selva as a whole is experiencing extremely high rates of population growth as it absorbs migration from more densely settled areas. Future settlement in the region may also be increasingly affected by recent changes in Peruvian legislation governing land and resource rights. The Peruvian Constitution of 1993 eliminates provisions of the 1979 Constitution that protected indigenous and peasant lands by prohibiting their division, sale, or transfer. The new constitution not only grants indigenous and peasant communities the right to freely dispose of their lands, but also provides that lands that are "abandoned" will revert to the state for their sale. Complementary land-use and forestry legislation is still being developed, making it difficult to predict the effects of this legislation with any confidence.

Also problematic for the long-term conservation of Yanachaga-Chemillén National Park, known to possess significant petroleum, natural gas, and mineral reserves, are official policies governing the development of hydrocarbons and other nonrenewable resources in protected areas, which have fluctuated greatly over the past two decades (INRENA 1995b). In August 1991, the Law of Investment Promotion in the Hydrocarbon Sector established that energy development and exploitation of nonrenewable resources in protected-areas required a previous environmental impact study and financing to restore affected areas. In November 1993, a new Regulation for Environmental Protection for Hydrocarbon Sector Activities provided that development of hydrocarbon resources in protected areas must be coordinated with relevant authorities to ensure fulfillment of the objectives for which the area was created. The Regulation for Environmental Protection for Electric Sector Activities similarly requires coordination with the Ministry of Agriculture. These laws and regulations establish uncertain protection at best and are clearly subject to change according to circumstance.

National Policy Framework

Peru's National System of Natural Protected-areas currently includes seven national parks, eight national reserves, seven national sanctuaries, and three historical sanctuaries, covering a total area of 5,513,426.1 hectares. The 1991 budget of INRENA totaled US$73,415, with regional governments providing additional funding in a few instances. INRENA has therefore relied heavily on resources provided by national and international donors; in 1990, these totaled US$389,758 for direct conservation activities, with an additional US$259,678 provided for indirect activities such as workshops and planning. Conservation NGOs currently provide sole funding for two national parks and two national sanctuaries and supplementary funding for one national sanctuary (INRENA 1995b).

Although financing of short-term conservation and management activities for YCNP has been achieved through the Parks in Peril program, long-term financing has been more difficult to obtain, and no source of revenue is available within the park. The Peruvian government has been active in developing long-term funding mechanisms for the protected-areas system and has created the National Fund for Natural Areas Protected by the Government (FONANPE) for the management of external funds supporting long-term protected-areas conservation. As of mid 1995, the fund had received commitments totaling US$16.3 million from a number of sources, including debt swaps with Switzerland, Canada, and Germany. The World Bank has also signed an agreement for a GEF donation of US$5 million to support the establishment of a National Protected-areas Trust Fund; the donation is accompanied by a grant from the GTZ of US$1.5 million to strengthen governmental institutional capacity to coordinate and implement conservation activities. However, although YCNP was included in a preliminary list of priority areas to receive support from FONANPE, most of the funding committed thus far has been earmarked by donors for more high-profile areas such as Machu Picchu (A. Camino, pers. comm., 1995).

Indigenous Peoples and Social Change

Several efforts have been made since 1980 to protect resident Yanesha from the effects of development and new settlement, through land titling, the creation of protected areas, and natural resource development and conservation projects. However, road construction, settlement and population growth, and development projects themselves have resulted in the increasing integration of native Yanesha into colonist-dominated market economies. Communal landholdings may not be of sufficient size to support internal population growth under the continuation of traditional slash-and-burn agriculture and small-scale livestock rearing. Although extraction of nonwood forestry products such as uña de gato represents an important alternative land use, it is also highly sensitive to fluctuations in international markets. An analysis of long-term factors affecting park management must therefore consider that land pressures are likely to intensify in the future due to growing land use for cattle grazing, a practice increasingly adopted from neighboring colonists and all the more attractive given the difficulties faced by forestry development and conservation projects.

Conclusion

The conditions surrounding the establishment of Yanachaga-Chemillén National Park differ from those of many other national parks; it was spawned by a large-scale development scheme. Yet unusual circumstances have combined to make this park relatively free from much of the complexity that faces other parks. The park is buffered by a number of other protected areas, which increases its ecological viability. The land tenure situation is relatively straightforward, since the boundaries were drawn to include unclaimed areas. Ironically, guerilla activity in the region has helped to keep pressures off the park and has reduced migration and settlement in the region. Relationships with neighboring Indian communities have been relatively good.

The challenges to the park's future security appear to be tied to a greater degree to large-scale and external threats, rather than those of local origin. The increasing importance of drug trafficking in the zone will bring new problems. Increased stability in the region is likely to lead to increased road construction and colonization of the zone. Exploration and development of petroleum, natural gas, and mineral reserves in the park are all potential threats, since the laws and regulations governing such activities in protected-areas are unclear and subject to change. The future of the park will largely depend on the levels and types of regional change and the government policies that either support or undermine such development.

Postscript

There have been no significant changes in the overall social context surrounding the park. The Master Plan for Yanachaga-Chemillén National Park was prepared in 1996. However, because it has not been officially approved by INRENA, it is still considered to be a draft document. In 1996 a new natural systems plan for Peruvian protected areas was developed by INRENA with the support of the German Agency for Technical Cooperation. In June of 1997 a new Law of Natural Protected Areas was passed. PiP money supported the work of Pro Naturaleza to provide technical assistance to the government to prepare the new law.

Consolidation funding (see chapter 1) for the site was secured in mid September of 1997 through a project entitled Conservation of the Forests of the High Amazon in the Central Selva of Peru. This project is supported by the government of the Netherlands and will provide $883,000 of financing for park management for a five-year period. In addition, $1,745,000 of financing over a five-year period is being provided to support rural development activities in the park's zone of infuence. INRENA will be implementing the park management component, and Pro Naturaleza will be implementing rural development activities in the buffer areas.

Note: Due to violence D. Rose was unable to travel to the site, so analysis was based on interviews in Lima and L. A. Yallico's knowledge of this site.

Glossary

DGFF—Directorate General of Forestry and Fauna

FECONAYA—Federation of Native Yanesha Communities

FONANPE—Fondo Nacional para Areas Naturales Protegidas del Estado (National Fund for Natural Areas Protected by the Government)

INRENA—Instituto Nacional de Recursos Naturales (National Institute of Natural Resources)

PDR—Palcazú Rural Development Program

PEPP—Pichis-Palcazú Special Project

Pro Naturaleza—Fundación Peruana para la Conservación de la Naturaleza or FPCN (Peruvian Foundation for Nature Conservation)

SINANPE—Sistema Nacional de Areas Naturales Protegidas por el Estado (National System of Natural Protected Areas)

TSC—Tropical Science Center

YCNP—Yanachaga-Chemillén National Park

PART III

Reality and Reaction

Saving Neotropical Parks

13. Comparing Cases: A Review of Findings

Katrina Brandon

Each of the case study sites has certain unique attributes that distinguishes it from other protected areas. The factors that contribute to this uniqueness vary widely—from particular species associated with those sites, to spectacular geographic features, to the particular social characteristics at the site. When one thinks of Rías Lagartos and Celestún, flocks of flamingos immediately come to mind, while quetzals are associated with Sierra de las Minas. Machalilla and Del Este feature beautiful beaches; Amboró, magnificent arborescent (tree) ferns and bold rocky outcrops. When conservationists familiar with Corcovado think of it, the long-term struggles between gold miners and the park and the changes on the peninsula spring to mind.

Identifying mechanisms to insure the long-term viability of parks and conservation of biodiversity requires moving beyond these first impressions and analyses of individual sites. Knowledge about the interaction of ecosystems, social context, and political environment at a single site is a necessary first step in analysis. But coming up with creative alternatives to current patterns of thinking requires a broader analysis of the common threats, the trends, and the challenges across the sites.

The special character of each site is captured in the preceding case study chapters. Rather than telling the story at each site, this chapter looks across the nine sites to see what characteristics they have in common, and which differ. The eleven themes that served as the basis for the case study analysis are the units of analysis presented in this chapter. In contrast to the case study chapters, which point out the special characteristics at sites, this chapter draws out the common threads in the case studies and the key features that are addressed in subsequent chapters.

Park Establishment and Management

Many biologists would consider the history of a park's creation to be esoteric and unrelated to the daily problems of management. Yet few factors are more important in determining how a park is managed, to what end, with what threats and social conflicts than the process of how it was established, where, and who was involved. While there is a vast literature on ideal reserve design, boundaries are, in the end, social and political markers. This section briefly summarizes the origin of the nine parks, the key features of park design and zoning, and marine components. The information presented in this section varies substantially in depth and detail; for some places, such as Corcovado, detailed historical information on how the park was founded exists and is presented in the case study. Such in-depth information is lacking for other sites. Information that should be precise, such as the park's size, is not always accurate.

At most of the sites there was little historical information on the actual process that led to the creation of the park and the selection and demarcation of boundaries. The origin of the nine case study parks varies dramatically, although all are of recent origin—none was a protected area prior to 1973. Five of the parks (Corcovado, Del Este, Machalilla, Podocarpus, and Yanachaga) were designated as national parks at the time of their establishment. Amboró and the Mexican sites were established under other national categories of protected areas, and their status was "elevated" to more rigorous protection status later. As Mexican laws governing protected areas changed, Ría Lagartos and Ría Celestún were upgraded to the Mexican category of special biosphere reserve. In the case of Amboró, the size was increased and the categorization changed from a natural reserve to a national park after eleven years—although in 1996, as described in the postscript to the case, the boundaries were reduced by 30 percent. Sierra de las Minas, the most recently created of the parks (1990) was established as a biosphere reserve and gained UN recognition as one in 1993.

Multinational corporations were influential in park creation—positively at Rio Bravo and Del Este, and negatively at Corcovado. At Rio Bravo and Del Este, land was given by corporate entities explicitly for park creation. At Rio Bravo, external pressure and lobbying by a variety of organizations to prevent Coca-Cola Foods Ltd. (CCFL) from converting rainforest into cropland, resulted in CCFL donating 92,000 acres of land,

and money for management, to create a sustainable-use area. The Nature Conservancy and other conservation organizations donated the support to purchase an additional 136,800 acres in several blocks. Most of the land for the creation of Del Este National Park was donated to the government by Gulf & Western Corporation with the condition that a park be established. The government then expropriated land near the Gulf & Western property, without compensation, so that the entire peninsula would be a park. The president of Costa Rica had dual objectives when he proposed a land swap with a U.S.-based company that had become a symbol of foreign exploitation of workers and resources: first, to create a park on the land that now forms Corcovado; and, second, to stop the national legislature from expropriating the company's land and thereby establishing what the president believed would appear as an unfavorable climate for foreign investment.

While many development projects are blamed for park destruction, it was such a project that led to the creation of Yanachaga National Park, where land titling and the creation of indigenous and forest reserves, and the park itself, were all explicit elements designed to protect residents and lands amid the Peruvian president's desire to develop Amazonian regions. USAID's funding of this project required assessment of environmental impacts. Anthropologists and biologists who conducted the assessments recommended that the government establish several contiguous protected areas to maintain forests and wildlife populations and to safeguard indigenous lands from colonization.

Ecological and social research laid the basis for the establishment of Machalilla and Sierra de las Minas. Biologists and archaeologists recognized the importance of the ecological and archaeological resources in Machalilla; five years later the park was created. At Sierra de las Minas, Defensores de la Naturaleza sought support from the World Wildlife Fund to conduct a multidisciplinary study of the area. This study led to a formal proposal to the government to create SMBR. Apart from Yanachaga, Machalilla, and Sierra de las Minas, the parks lacked good ecological information relevant to park establishment. One possible additional exception is Corcovado, where there was incomplete, although excellent, biological information on the Osa Peninsula. Little is known about the histories leading to the creation of Amboró or Rías Lagartos and Celestún. What is known about Amboró is that, once the park was established, a British biologist was hired as its first director. The single-minded pursuit of protection by this director and his staff, irrespective of long-standing residents there before the park was declared, led to hostility toward the park in the northern section, which has persisted to the present.

Design and Zoning

How boundaries are defined when parks are established and what features they include are of tremendous importance in biodiversity conservation within parks. Biodiversity conservation is more challenging in parks with roads through them, or with fragmented designs, or with significant gaps in vegetation. Zoning is a key park management tool; it allows for different kinds of uses in different areas, although those uses are not necessarily consumptive. The design of parks, and subsequent zoning inside parks, has been based on a variety of circumstances, as described below. At some sites ecological criteria have been used, while at others, pragmatic decisions about resource uses have formed the basis for zoning decisions.

Ría Lagartos and Ría Celestún are located at opposite ends of a long expanse of coastal wetlands that are critical to the life cycle of the American flamingo. Both reserves include sandy beach ridges with mangrove-fringed lagoons (*rías*) behind them. Ría Lagartos is highly saline, yielding high numbers of brine flies for flamingos and other wading birds, which use the area as a nesting and breeding ground. Ría Celestún has freshwater, which attracts the bird colonies during the winter. No special zoning has been defined within the reserve; a new management plan was under preparation in 1995.

Zoning within Rio Bravo is generally based on the donor's intent for the type of conservation method at the time of purchase or donation. For example, the Coca-Cola Foods Ltd. land donation for Rio Bravo required sustainable-use activities on the 92,000 acres donated. The majority of 110,000 acres acquired via donations from the Conservancy, the United Kingdom, and USAID are managed in the spirit of the donation—pure protection. Zoning at Sierra de las Minas is the result of a process that overlaid the results of substantial technical research with adjustments made for the location of communities. Four management zones (core 45 percent, sustainable-use 14 percent, buffer 39 percent, and recovery 2 percent) were identified, and Defensores de la Naturaleza has used this zoning to limit uses and promote local awareness concerning permissible uscs in each zone.

The design of Corcovado is based on what was easily acquired in the land swap. The entire park is managed as one uniformly protected unit, although ecotourism is permitted in the northwest corner of the park. To the extent that there is zoning, Corcovado may be thought of as the core zone, while the reserves that surround it act as sustainable-use zones. The Golfo Dulce Forest Reserve and the Guaymi Indigenous Reserve are contiguous to Corcovado, and the Sierpe Terraba Mangrove Forest Reserve begins near the northern end of the forest reserve. The park itself, at 41,788 ha., is too small to protect sizable populations of many of the species, such as jaguars and harpy eagles, that reside therein. The design of Yanachaga is similar to that of Corcovado; it is a national park of 122,000 ha. surrounded by other protected areas. The San Matías–San Carlos Protection Forest and the Yanesha Communal Reserve buffer the park.

Del Este comprises the entire end of a small peninsula in the extreme southeast of the Dominican Republic. The park also includes the offshore island of Saona. Recognizing the history of human use and communities within the reserve, the management plan includes seven zones: Intangible (mangroves and a nesting area open only to research), Primitive (forest areas to be managed for moderate public use), Extensive Use (buffer zones between communities and more strictly protected areas), Intensive Use (beaches for tourism), Historic-Cultural (primarily caverns), and Recovery, and Special Use Zones (Saona, park administration, and services). While the zoning guidelines have been observed, regulations governing human use have been only weakly enforced.

Machalilla was established in three fragmented sections in order to avoid including major towns. The approximate sizes of the blocks are: north 10,570 ha; south 29,580 ha; and Los Piqueros 900 ha. Grazing, fuelwood extraction and other resource uses throughout Machalilla have been so pervasive that there is no abrupt change in the habitat at the edge of the park. The only physical edges readily identified at MNP are edges between vegetation types created by natural physical conditions, and edges around moist forest remnants at upper altitudes created by recent logging activities. Park officials would like to connect the northern and southern blocks of the park via the purchase of a 2,000-ha corridor.

The same 1996 decree that reduced Amboró's boundaries created a Natural Area for Integrated Management, which is the area between the old and new boundaries. Uses are allowed in this zone, although it is unclear whether there will be strict internal control or zoning within this buffer zone.

Coastal and Marine Areas

Four of the sites, Ría Lagartos and Ría Celestún, Corcovado, Del Este, and Machalilla, have extensive coastal areas, although only Machalilla has legislation establishing offshore boundaries and protection of marine areas. Ría Lagartos is the only Mexican site listed by the RAMSAR Convention for the Protection of Wetlands of International Importance. The barrier beaches at both Ría Lagartos and Ría Celestún are protected; however, there is no protection for offshore waters. Likewise, at Corcovado, there is no offshore marine protection, although coastal areas of the park are patrolled. The Isla del Caño Biological Reserve was established in 1976 and is located twenty kilometers off the Pacific Coast. It has a land area of three hundred hectares and a maritime band of three kilometers surrounding the island; it is managed within the same regional conservation unit as Corcovado.

Del Este and Machalilla both protect offshore islands. Del Este protects Saona Island (eleven thousand hectares), which lies just off the coast from the mainland park area. While the park's management plan proposes the creation of a marine zone extending five hundred meters from the coasts of the peninsula and of Saona Island, these modifications have never been implemented. Therefore, the barrier reefs to the south of Saona Island, and coral formations located in the waters to the east and west of the park, are not protected in any way. Machalilla includes thirty-one miles of beach, La Plata and Salango islands, and twenty-thousand hectares of ocean in the form of a two-mile marine corridor that follows the coastline and encircles the two islands. The marine portion includes Ecuador's only continental coral reefs.

Analysis

One of the striking revelations that comes from a review of the history of park establishment is how recent most of the parks are—none is even twenty-five years old. The history of park establishment has not been well documented; with the exception of Corcovado, Rio Bravo, and Sierra de las Minas, there is little information on what led to decisions on location, boundary, demarcation, etc. Sierra de las Minas, the most recently established park, has combined scientific information with what is socially practical in the zoning process to a greater degree than other sites. While conservation texts abound with theoretical examples of ideal reserve design, few of the parks match some ideal. Rather, their location, design, and zoning tend to reflect what was possible at given sites. Coastal waters and marine life next to the four parks are under heavy pressure and are only legally protected at one of the four sites.

Land and Resource Tenure

Tenure is the form of rights or title under which property is held and that determines how an individual or group may use, share, sell, lease, inherit, or otherwise control property and resources. Tenure is most commonly used to describe land, but the systems of rights and rules that make up tenure also apply to natural resources such as water, trees, and wildlife. Tenure regimes and customary practices vary dramatically among countries and different groups of users. Understanding what is traditional, what is legal, and what are actual uses for both land and resources is necessary; all have an impact on biodiversity conservation.

Communities within Parks

Four of the sites have no communities within the park boundaries: Rio Bravo, Corcovado, Podocarpus, and Yanachaga. However, as described below in the section on claims within parks, there are other demands for resources within these parks. Ría Lagartos and Ría Celestún, Sierra de las Minas, Del Este, and Machalilla all include numerous communities within protected-area boundaries. Most of the settlements within Amboró are simply the result of a mistake, rectified as of 1996.

At most sites there was little analysis of land claims in the area prior to establishment of the parks. There are examples, such as Amboró, where such analysis and recognition of claims prior to park creation would have greatly simplified that process. When Amboró's boundaries were expanded in 1991, the total number of households within the park reached three thousand. Based on Bolivian law, both residence within a park and use of park resources is illegal; this created great conflict as people on both sides of the boundary argued that the boundary placement was unjust. In fact, while virtually all groups agreed that the expansion of the park with the decreed boundaries was a mistake, the challenge was to define the optimal size and boundaries both for the park and for communities. A process called the Red Line process was initiated to negotiate new 1996 boundaries. The process of how Amboró resolved this problem (chapter 11) is instructive in areas where capital city decision making has led to rural problems.

At Amboró, Del Este, and the two Mexican sites, parks also were superimposed over existing land uses and claims. At Del Este, three communities abandoned their settlements when the park was created. At Ría Celestún, the town of Celestún was incorporated into the reserve; at Lagartos, four communities were included. No consideration of tenure issues was included in the decree that established the areas.

Sierra de las Minas was developed as a biosphere reserve with full awareness of the communities that live there. There are seven indigenous communities within the core zone; as described in the section on resettlement, Defensores is exploring the voluntary resettlement of two of these communities. One hundred and ten communities are scattered throughout other zones in SMBR.

When Corcovado was established, existing settlements consisted of eighty to one hundred households, most of which had moved there during the 1975 dry season, just prior to the park's creation. They were resettled at the government's expense after the park was established. At Machalilla, the population was artificially low when the park was established, since a severe drought lasting for several years had induced people to migrate out of the area. When the drought ended, residents returned and found that their landholdings had been incorporated into a park. The park boundaries at Machalilla were specifically drawn to exclude large communities; nevertheless, a number of small communities throughout the park have claims on land and resources. Because of limits on resource use and agricultural expansion, there has been a net decline in the population in and around the park.

At Rio Bravo and Del Este the population levels were low in the areas that were privately held, since, in the donated portions, people who were there were illegally squatting on privately held lands. Among the case studies, Rio Bravo is unique, because most of the lands it incorporates were privately held, virtually eliminating tenure problems. At Del Este, however, additional land that had human settlements was included in the park; some of these communities have remained there. Yanachaga was superimposed over the traditional lands of indigenous groups; as compensation, the government formally gave them title to their own reserves adjacent to the park, and use rights within the park.

Both land and resource tenure issues remain formally unresolved at some sites, particularly Machalilla, Del Este, and Rías Lagartos and Celestún. For example, in Ecuador it is illegal for people to reside inside a national park; informally, however, authorities have overlooked this at Machalilla. While Corcovado is free from land tenure claims at present, a legacy of conflict remains over who was compensated when the park and adjacent forest reserves were created. However, a historian of the Osa notes that many of those who claimed land were in fact squatting on privately held land—that the "original" owners and the squatters had only recently claimed.

Claims on Land and Resources within Parks

Resource tenure claims were less common at most sites than were claims on land, but the only sites at which they are straightforward and well understood are Rio Bravo, Corcovado, and Yanachaga. In Rio Bravo, all land-based activities within the park are controlled by Programme for Belize. However, PfB does not control rights to rivers, subsurface deposits, or Mayan antiquities. No extractive or consumptive resource uses are permitted in Corcovado National Park, and there are no lands claimed within the park. Human settlement and natural resource use are prohibited in all Peruvian national parks, with the exception of ancestral indigenous populations—such as the Yanesha. Unless pressures increase dramatically, however, it is likely that most resource use will occur within the Yanesha forest reserves.

At Rías Lagartos and Celestún, there was no definition of tenure rights and there is still no clear delineation of property or of appropriate resource uses. At Sierra de las Minas, there is no reliable land tenure survey, and existing information indicates that there are overlapping claims and boundaries. While a variety of uses are formally recognized through the zoning process at Sierra de las Minas, there are still conflicts over the zones and the resource uses within them.

In Del Este Park, there are restrictions on land uses. Resource tenure is acknowledged by allowing honey and coconut production to remain on areas cleared before the park was established. However, official attitudes toward authorized productive activities have been ambiguous, leading to uncertainty regarding future rights of tenure and resource use. For example, following administrative and personnel changes within the Parks Directorate in 1995, the park administration began to remove ovens and other infrastructure located within the coconut plantations, storage sheds located within apiaries, and domestic livestock on Saona Island, as well as the mainland, as a means of limiting their impacts within the park.

At Machalilla, Podocarpus, and Amboró, land and resource tenure issues are conflictive, although it is still unclear what impact the park boundary change and the creation of a large buffer zone, with communities inside, around Amboró will have. Ecuadorian law states that land within national parks is national patrimony and the sole property of the state; private property within parks is illegal. Yet as a practical matter, Ecuadorian officials have looked the other way and ignored situations where people have long-standing land claims. This is certainly the case at Machalilla, where two studies have tried to clarify the tenure situation. Although the results of the two studies differ, jointly they suggest that: (1) at least five kinds of land tenure exist within the park, including individual and communal; (2) a high number of absentee residents retain their claims over land; and (3) 20 percent to 75 percent of the terrestrial lands are claimed, depending on which study one believes. Resource tenure within the park is even more complicated and less well understood, and, as described by Tello, Fiallo, and Naughton-Treves (chapter 9), "it depends on the nature of the resource and the organizational level and political power of its users." Because of the complexity of this case and the in-depth information it contains, readers are referred to chapter 9.

Although the same laws apply at Podocarpus as at Machalilla, the situation is not nearly as complex. Over seventy farms near new settlements are partially, or completely, within the borders of PNP. Whether these farms predate the park's establishment is subject to debate. Groups of gold miners and campesinos are also found within the park. However, there are no formal settlements within the park. So the uncertainty revolves around land ownership and use, the date of farm establishment, and resource uses.

Patterns of Land Use Surrounding Parks

Table 13-1 provides a summary of the land-use situation surrounding parks. Land uses are unstable at five sites: Rías Lagartos and Celestún, Corcovado, Del Este, Amboró, Machalilla, and Podocarpus. While the potential for land-use changes exists at the other sites, say, with the opening of new roads, at most of the other sites the tenure situation, even if unclear, appears to be relatively stable.

At Ría Lagartos cattle ranching is pervasive both within and outside of the southern part of the reserve. At Celestún, migration to the area has led to instability as agricultural and grazing lands are converted to vacation properties. The Osa Peninsula, where Corcovado Park is located, is undergoing a profound transformation—road improvement and electrification are leading to real estate speculation and significant land-use changes. New agroindustrial enterprises resulting from access and electricity have led to changes in what is planted. Subdivisions for "jungle" vacation homes are being platted out, and rapid land turnover and clearing are underway.

Much of the tenure instability at Amboró has been resolved with the reduction of the park's boundaries. However, there is substantial migration to the north side of the park with the potential for parkland to be claimed.

The land tenure situation around Machalilla is also in transition; the population surrounding the park is increasing. At Podocarpus, many of the farms along the western boundary are titled but are being converted to vacation homes. Ironically, two large private properties, totaling thirty-two hundred hectares, adjacent to the park want the protection that the park offers. Rapid colonization is underway along portions of the eastern and western boundaries.

At Del Este and Yanachaga there is potential for relatively rapid land-use changes in the future. Private lands on the coastline adjacent to the park could begin changing hands; especially since there is rapid development in other bordering areas. In Yanachaga, increased security in the region and infrastructure development could make areas attractive to settlers. The forest and indigenous reserves that have served as buffers for Yanachaga would then be more susceptible to logging and colonization. Additionally, the 1993 Peruvian Constitution eliminates earlier provisions that protect indigenous and peasant lands from division, transfer, or sale, leading to further instability.

Analysis

As demonstrated by the case study sites, land and resource tenure in and around parks is frequently unclear. The case studies reveal substantial differences in traditional and legal tenure and actual patterns of use. Tenure is not a significant issue at two sites: Rio Bravo and Yanachaga. Elsewhere, there are complicated tenurial issues over land or resources either in or adjacent to parks. These unresolved tenurial issues add to the social and

Table 13-1. Land and Resource Tenure

Site	Within Park	Adjacent to Park or Core Zone	Stable or Changing	Clear Information
Ría Largartos and Ría Celestún, Mexico	No definition of tenure in decree; private lands not expropriated; at Lagartos, mix of private, communal, and urban land. At Celestún, mix of private and ejido.	Lagartos: Cattle ranching to south of reserve continues. Demarcation and clear boundaries lacking. Celestún: Easier access and greater population increase pressure.	Lagartos: From 1993 to 1995 population doubled to 8,000, resulting in unstable tenure. Celestún: Seasonal changes but stable population; instability over private uses such as vacation home construction.	No clear information exists to delineate properties and clarify tenure and title.
Rio Bravo, Belize	All lands privately owned by Programme for Belize.	Well-studied tenure situation; 5 groups: large private lands; government reserves; Mennonite agricultural communities; mestizo agriculturists; and Creole communities.	Tenure changes are no threat within park. Adjacent lands generally stable; large private lands still forested; government reserves and Mennonites act as buffers; potential pressure from mestizo and Creole ag. expansion. Tenure threat from turnover and clearing of large private lands.	Clear within reserves; good overall studies in surrounding areas.
Sierra de las Minas, Guatemala	Land use and tenure differ on north and south. North still forested but being converted through slash-and-burn ag. South is large private lands. Approx. 45% of land is public, 50% private, 5% municipal.	Core zone has 45% of total area; most landholdings in core are large private tracts, which are inaccessible. Defensores owns 23,000 ha in core; 2,500 ha donated by private owners giving use rights for 30 years. Small amount used by indigenous groups.	Core zone relatively stable due to lack of access. In both northern and southern part of reserve, ag. frontier advances. Some community invasions of land to demonstrate possession. No legal mechanisms recognize traditional use rights. Strategy is to allow land ownership in reserve subject to zoning and appropriate management.	No reliable land survey for tenure and title; poor demarcation, overlapping boundaries for reserve, and zoning exist.

Corcovado, Costa Rica	No land use allowed or claimed within park.	Serious conflicts over land title in surrounding protected areas; conflict between govt. agencies with titles granted in forest reserve; then reversals leaving legal status of titles unclear. Real estate speculation and land subdivision for sale to foreigners on rise.	Extremely unstable land tenure situation has plagued the peninsula since the 1950s. Dependence of park on surrounding protected areas makes their conservation important, but rapid conversion of lands, new infrastructure development, weak government agencies make situation unstable.	Poor information exists about claims, tenure, and titles within the adjacent protected areas. No good info on foreign ownership.
Del Este, Dominican Republic	Land for park donated or acquired through expropriation without compensation. Communities in park subject to restrictions on use leading to conflict as population grows. Resource tenure acknowledged; coconut and honey production continue within park.	Communities adjacent to park are growing with some increased pressure. Some large private landholdings may be a buffer.	Private lands on the coastline adjacent to the park, coupled with rapid development in other bordering areas, are leading to a rapidly changing land tenure context. Presence of a resident community on Saona Island in the park and land and resource shortages in the region north of the park imply continued tenure instability.	Socioeconomic study in 1994 included tenure. Information on adjacent areas not formally documented.
Amboró, Bolivia	Dramatic example of tenure problems from park expansion. 1991 boundaries incorporated communities, but law made any resource use illegal. "Red Line" program and 1996 boundary modification intended to make boundaries acceptable to park and communities.	Different land tenure regimes on north and south sides. Land tenure clearer on south, and more people have legal title. North is less stable with recent colonization by indigenous people clearing land to demonstrate claim.	Creation of park has led to tenure instability on the south side, which the Red Line process has sought to correct. Tenure instability and migration on north still prevalent. Government nearly granted titles to colonists within park several times; instability, confusion, and conflict resulted.	As a result of the Red Line program, there has been an effort to map the titles and claims of all within park.

continues

Table 13-1. *(continued)*

Site	Within Park	Adjacent to Park or Core Zone	Stable or Changing	Clear Information
Machalilla, Ecuador	Private property is illegal within park, but this is ignored. Five tenure types exist within park, including private and communal with and without title.	Estimates of private landholdings within park range from 20% to 75%. Wide discrepancies in title and claims. Some evidence that up to half of private landowners favor land expropriation if given compensation.	Situation has been changing over the years. No resource uses are legally allowed, although they have been allowed. Policies concerning buyouts and community desires to defend land and resources will result in future changes.	Two studies have tried to get clear information on tenure; results vary dramatically, in part due to absentee owners and reluctance of residents to explain or prove legal status within park.
Podocarpus, Ecuador	Land tenure variation in and around park best summarized by three categories: long-term residents, recent colonists, and indigenous residents. Seventy farms of recent origin within park, although it is unclear whether these predate the park. No settlements in park, however.	Many farms along western boundary are titled, but they are being sold for vacation homes. Other adjacent farms have cattle that graze into parkland. Two large private properties (3,200 ha) in park want protection park offers. Rapid colonization along parts of eastern and western boundaries.	The land tenure situation is changing. Long-term residents, often absentee, are selling land. Gold, access to the region, and land markets are leading to increasing colonization around PNP and the potential for invasions of park. Shuar communities surrounding PNP are losing use of traditional yet untitled lands to colonists and to state.	A 1995 socioeconomic study included an overview of land tenure, but no specific information exists on tenure and titling within and around the park.

Yanachaga, Peru	No one lives in the park, and there are no claims. There is tenure security in the areas surrounding the park, including in other protected areas. Resource tenure adjacent to the park is problematic though, as Yanesha have traditionally used, and continue to use, these areas. Western boundary of park delimited with agreement of neighboring landowners.	USAID required that ten Yanesha communities near project area be given title—titles were given to Yanesha, European, and national settlers in the region. Forestry concessions have been granted on western boundary of the park. Land titles are clearly delimited, but concessions are overlapping and some even include parkland.	New Peruvian Constitution of 1993 eliminates earlier provisions that protect indigenous and peasant lands from division, transfer, or sale. Increased security in the region and infrastructure development could make areas attractive to settlers and protected areas that have served as buffers more susceptible to threats.	Information on tenure and titling was extremely well organized in the 1980s when such programs were underway, but accuracy may be declining with use changes. Information on concessions is less clear.

political complexity of park and ecosystem management. They also open the way for land-use changes—in areas where land- and resource-use rights are unclear, greater changes in land use are underway. These land-use changes, if significant in scale around protected areas, could significantly alter the ecological profile at some sites and lead to new patterns of threats or to greater stability. For example, the construction of vacation homes adjacent to some sites has the potential to secure greater areas for conservation, if forested areas are left intact, or could lead to greater pressure on parks as local peoples are displaced and search for land. Resource tenure varies substantially across sites as well and is not necessarily consistent with use patterns, as described in the following section.

Resource Use

Closely linked to land and resource tenure, and the regulations governing parks, are the resource uses that occur in and around parks. The uses vary dramatically across the sites, based on these factors: proximity of local populations, differing resource-use traditions, access, and the variety and availability of the resources themselves. Table 13-2 indicates the kinds of resources described in the case studies and whether the use is within or adjacent to the park. The description of commercial use in these tables refers to the sale of products, whether the market is local, regional, national, or international.

There are several caveats regarding the information in the table. First, there may be other resource uses, but authors were asked to review the primary kinds of uses at sites. With the exceptions of Sierra de las Minas and Rio Bravo, there is virtually no information at any of the sites on what resources are used, by whom, and why, and the impacts on wild populations.

Second, what is listed as "inside" or "adjacent to park" or "both" is dependent on the areas themselves. For example, at Sierra de las Minas, the whole area is a biosphere reserve—zoning allocates what is permissible within, but it is difficult to label particular resource uses as inside or adjacent to, given that context. Similarly, at other parks with communities inside them (Rías Lagartos and Celestún, Del Este, and Machalilla), it is difficult to define where the reserve begins and human activities end. Finally, it is difficult to compare what is legal and illegal at different sites, since what is illegal, and what prohibitions on uses are enforced, varies widely. What is illegal is clearer at the sites where no human extractive and consumptive activities are allowed, such as Rio Bravo, Corcovado, Amboró, and Podocarpus. At the Mexican sites and at Sierra de las Minas, Del Este, Amboró, and Machalilla, there is a high degree of dependence on resources for local livelihood. Many of the uses may not be legal at Del Este, Machalilla, or, until recently, parts of Amboró. However, authorities at those sites recognize the futility of fully restraining uses. At both Machalilla and Del Este, however, restrictions on resource use appear to have led to migration from the parks.

Some general patterns are evident. Fishing is mentioned as a local activity at seven of the sites. Yet there is virtually no management of marine areas or inland fishing at any of the sites. At three of the marine sites—Del Este, Machalilla, and Rías Lagartos and Celestún—fisheries appear to have significantly declined. Local hunting at most sites is for both subsistence and sale. Collection of plants and wildlife for local uses, such as medicines, parrots for pets, and ferns for decoration, is common at most sites. As with hunting, in many areas it is difficult to differentiate collection for household use from collection for commercial sale. Nonetheless, when a peasant is leaving a park with six parrots, or five monkeys, or ten orchids, park authorities assume it is for commercial trade in wild products, even if that trade is localized.

Logging is prevalent at the different parks, although commercial logging within parks has been controlled. Local logging and timber sales are common at most sites, with varying degrees of intensity. Livestock grazing, primarily of cattle, is a problem within many of the parks. At Machalilla, while the overall number of livestock is slowly declining, goats and cattle have wreaked havoc on local ecosystems. Commercial cattle grazing near protected areas exists at six sites; it poses a real or potential threat at Rías Lagartos and Celestún, Del Este, and Podocarpus.

Mining at three of the sites—Corcovado, Amboró, and Podocarpus—and salt extraction at Rías Lagartos and Celestún are extractive activities that often bring with them a variety of problems, as amply detailed in the case studies of Corcovado and Podocarpus. Not only can the mining processes cause environmental harm, but miners are likely to illegally use park resources. For example, at Corcovado, miners have been blamed for substantially reduced wildlife populations in certain sectors of the park. At Podocarpus, commercial and artisanal mining has been a serious problem within the park.

Table 13-2. Resource Uses at the Nine Case Study Sites

	Rías Lagartos & Celestún	Rio Bravo	Sierra de las Minas	Corcovado	Del Este	Amboró	Machalilla	Podocarpus	Yanachaga
Fishing, local	■	▒		□	□		■	▒	□
Fishing, commercial	■			□	□		■		
Hunting, local		□	■	▒	▒	■	▒	▒	□
Hunting, sport		▒						▒	
Hunting, commercial				■	▒		▒		
Local collection, plants & wildlife	■		■	■	▒	▒	▒	▒	□
Commercial collection	■			■	▒			▒	□
Agricultural expansion, subsistence	■	□	■	□	▒	■	▒		□
Agricultural expansion, commercial		□	■	□	□			□	□
Logging, local			■	□	▒	■	■	□	
Logging, commercial		□		□		■	■		

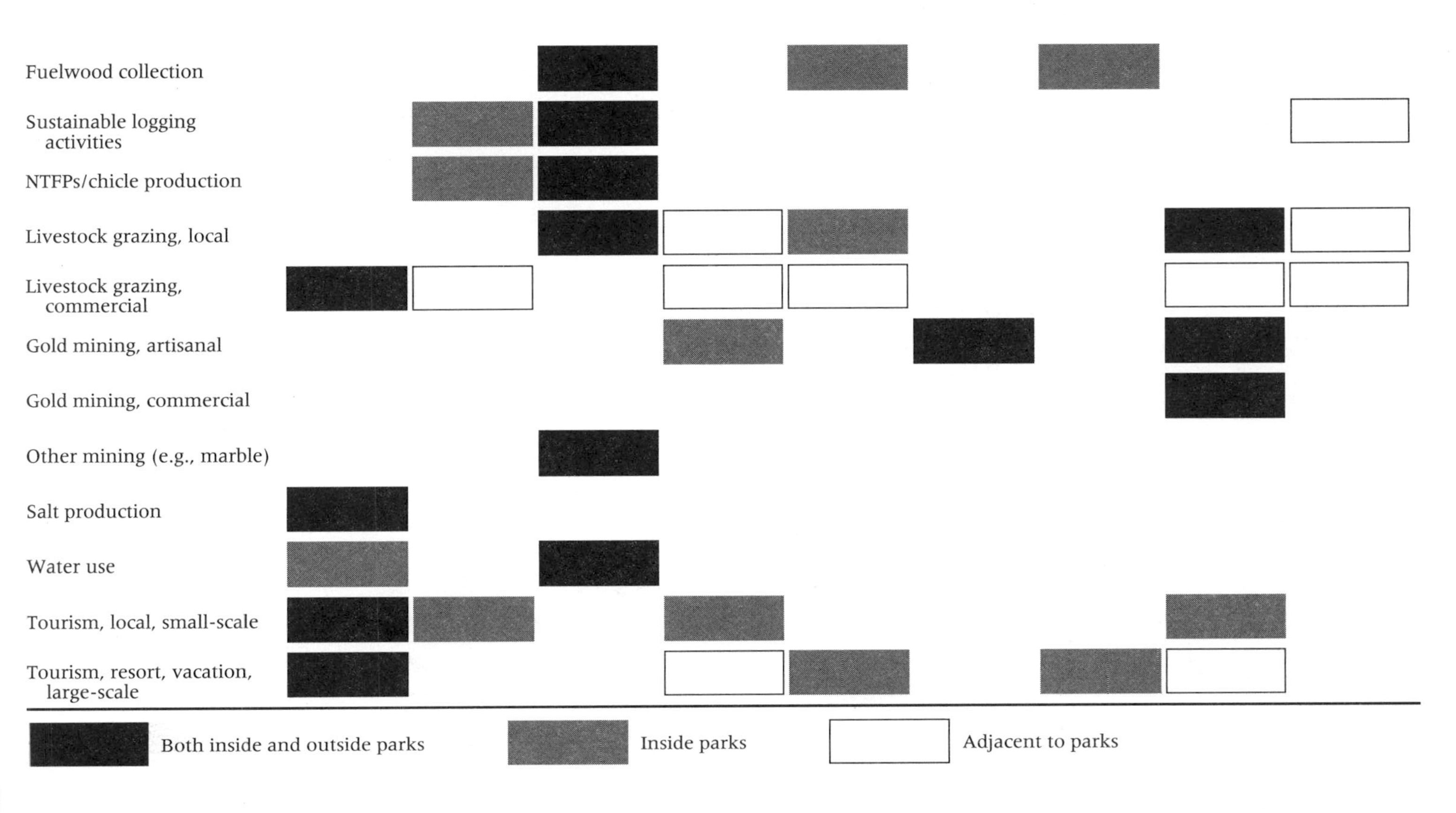

Fuelwood collection
Sustainable logging activities
NTFPs/chicle production
Livestock grazing, local
Livestock grazing, commercial
Gold mining, artisanal
Gold mining, commercial
Other mining (e.g., marble)
Salt production
Water use
Tourism, local, small-scale
Tourism, resort, vacation, large-scale
Both inside and outside parks
Inside parks
Adjacent to parks

Tourism, while nonconsumptive and nonextractive, constitutes a form of resource use. There has been small-scale tourism or ecotourism at some of the sites and large-scale or uncontrolled tourism at others (table 13-2).

How, and which, resources in and around a park are used may largely determine that site's potential for biodiversity conservation. Surprisingly, there is little information at most of the sites on resource use, with the exception of two sites: Rio Bravo, where uses are highly controlled and managed by Programme for Belize; and Sierra de las Minas, where such information formed an integral part of the zoning process. The extent to which there is resource use for subsistence needs versus market sales is also unknown at most sites. However, large-scale commercial logging has been controlled at all areas. One of the resource uses with the greatest impact in the future may be tourism.

Organizational Roles

This section briefly reviews the roles of The Nature Conservancy's partner organizations in the management and protection of sites. Traditionally, the NGOs in conservation have taken on a wide variety of roles. At most of the sites, many organizations, not just the Conservancy's partner organizations, are working on either conservation or development issues. Furthermore, at most of the sites there are many sources of funding. NGOs often offer one avenue for local participation in management planning and decision making. Such participation is most likely to occur if it is formally recognized as part of a process or is in some institutional form, such as an advisory committee.

Six of the Conservancy's partner organizations have agreements to support the work of the government agencies charged with park management, and all agreements allow NGOs to carry out work themselves. As shown in table 13-3, the role and type of support they provide vary from site to site, but include financial support, community outreach, environmental education, and the hiring of personnel, such as park guards.

The government agencies responsible for protected-area management are often weak institutions; many of them have gone through numerous changes in recent years. In some cases, their authority has been increased;

Table 13-3. Organizational Roles

Site	Partner and NGO Role	Jurisdiction and Government Role	Level of Local Participation	Other Key Organizations
Ría Celestún and Ría Lagartos, Mexico	PRONATURA: Provides support, esp. basic protection, research, environment-related education, and outreach.	Responsibility is within National Institute of Ecology in SEMARNAP; 2 state agencies, federal agencies, and other NGOs collaborate on reserve management.	Technical Advisory Committee formed in 1993 in Lagartos and Celestún offers potential vehicle.	CINESTAV is another key local NGO; also 2 state governments, a GEF-supported project, and other NGOs.
Rio Bravo, Belize	Programme for Belize, an NGO, owns Rio Bravo in trust for people of Belize.	Government retains subsurface or mineral rights.	No formal mechanisms for participation. PfB encourages local participation in development.	Many groups: Mass. Audubon, Save the Rainforest, USAID, National Fish and Wildlife Foundation.
Sierra de las Minas, Guatemala	Defensores de la Naturaleza acts as reserve administrator and purchases land in core. Violations referred to police or district attorney.	Defensores was granted management of SMBR under supervision of National Protected Areas Council (CONAP). Coordination with 5 departments and 13 municipalities and federal government.	Numerous participatory planning processes and evaluations occur yearly; participation in development activities.	High involvement, including: WWF, CARE, children's campaigns in U.S. and Sweden, MacArthur, Claiborne/Ortenberg, RARE, Peace Corps, Forest Service, and Guatemalan orgs.

continues

Table 13-3. (*continued*)

Site	Partner and NGO Role	Jurisdiction and Government Role	Level of Local Participation	Other Key Organizations
Corcovado, Costa Rica	Fundación Neotrópica: agreement with govt. to provide support and technical assistance in sustainable development and conservation throughout CR including Osa.	Management authority vested in ACOSA, the regional conservation authority responsible for managing and coordinating govt. activities that affect parks.	Participation by relevant groups in ACOSA technical committee and ACOSA's local committee.	High support for activities Osa-wide, including GTZ, USAID, EEC, TUBA, FINCA, OTS, TSC, Univ. of C.R., WWF-U.S., NORAD, and GEF funding.
Del Este, Dominican Republic	PRONATURA, a consortium of 17 conservation and development NGOs, assists with management and fund-raising.	Parks Directorate has signed yearly agreements with PRONATURA and the Conservancy. In 1994 a patronato was created as advisory body.	Very limited; no local involvement in development of management or operational plans.	PRONATURA consists of 17 NGOs in DR; Spanish Agency for Intl. Cooperation, govt. of Japan, 2 aquaria, U. of Miami, and others.
Amboró, Bolivia	Fundación Amigos de la Naturaleza awarded contract to manage the park, but govt. decided to manage park itself. FAN now works in buffer area.	Substantial changes in management authority over the past decade. National Directorate for Biodiversity Conservation is now authority for Amboró and other sites. FAN ceded vehicles, buildings, and equipment to govt. for park management.	No mechanisms for participation in management. FAN has assisted communities on south side with participatory development planning.	Many NGOs active in development around Amboró: PROBIOMA, Caritas, Aseo, Dominican Order, and labor unions. CARE funds these groups working in buffer zone. GEF funds park.

Machalilla, Ecuador	Fundación Natura (FN) supports INEFAN thru cooperation w/ PiP; active at national level, e.g., environment-related policy reform and policy reform.	INEFAN, a govt. agency, retains management authority; agreement with Fundación Natura in PiP program areas.	Local org. strong in some sectors, e.g. Peasant Committee to support MNP. Better relations offer potential to increase participation in park management.	Many local organizations—confederation of comunas and inter-institutional committees, govt. agencies, univs., CI, USAID, GEF, DeD, IAF, Italian govt. agency.
Podocarpus, Ecuador	Fundación Arcoiris a local NGO, and FN support INEFAN. FAI works locally; FN works on policy and helping local organizations.	INEFAN, a govt. agency, retains management authority; maintains yearly agreements with FAI, FN, and the Conservancy for assistance in PiP areas.	Some participatory rural appraisals are conducted to help communities engage in process of their own development.	Numerous local NGOs involved in and near park. WCS research.
Yanachaga, Peru	ProNaturaleza, since 1991 has had renewable agreement for cooperation in park management.	INRENA, state agency, has full authority for park management, though they welcome external collaboration.	Limited, but 2 groups with local representation created to provide input on management plans.	PiP is sole funding source; USAID funded park creation; WWF supported Yanesha when USAID $ stopped.

in other cases, there is a movement toward decreasing national authority and increasing local control. Management authority is clear at Rio Bravo—Programme for Belize, an NGO, manages the reserve in trust for Belizeans. At all of the other sites, management authority is retained by the government, although there is a wide spectrum of arrangements with NGOs for support. At the two sites in Ecuador, in the Dominican Republic, and in Peru, the government retains management authority but has agreements with the Conservancy's partner organizations for them to provide financial and technical support.

Defensores manages the SMBR under the supervision of a government agency, CONAP. Management of the SMBR requires coordination with five departments and thirteen municipalities, as well as several national agencies. When Defensores was granted management authority for Sierra de las Minas, it was the first example in Latin America and the Caribbean in which a government delegated management authority to an NGO. There are now numerous examples of this trend elsewhere in the region.

Mexico is perhaps the site where management authority has changed the most—numerous reorganizations have occurred within the Mexican government, frequently changing the agency responsible for park management. At Rías Lagartos and Celestún, two state agencies also have a say in the management of the areas. Costa Rica has experimented with a decentralized governmental management authority in a system known as regional conservation areas since 1987. These ARCs, as they are known, link governmental agencies with responsibilities in a region with local and regional authorities and community representatives.

At Amboró, numerous organizations have been responsible for managing the park. A joint government and NGO management committee was established in 1991. In 1995, the government requested interested NGOs to submit a proposal for managing Amboró and subsequently awarded the management to the Conservancy's partner organization, FAN. Shortly thereafter, the government reversed its own decision and now manages the park directly. As a consequence, FAN is no longer involved in park management and has turned over buildings, equipment, and vehicles to the government for its use in park management.

Local participation in park management is limited. At Sierra de las Minas, local participation is integrated directly into the management planning process. At two sites advisory bodies have been created to elicit participation in management: in Rías Lagartos and Celestún and in Costa

Rica. At other sites, there is no formal participation in park management (although there may be local participation in development activities or specific projects).

The various organizations working in a particular area do not necessarily share any common philosophy or objective, nor is there coordination among activities undertaken by different organizations. There is also a range of involvement by government agencies. How agreements and partnerships between international, national, and local NGOs and government are structured varies widely. The government of Guatemala has turned over management authority to Defensores, although policing functions are retained by the state. In Ecuador, in contrast, the state retains control of management, although assistance in all areas, including policing, is provided by NGOs. Finally, local participation is only formally recognized at three sites: Rías Lagartos and Celestún, Sierra de las Minas, and Corcovado. The level and quality of participation at these areas vary, but the formal mechanisms for participation have been established. At Machalilla, a peasant committee to support the park was established; for peasants involved with this committee, the park was viewed as a local resource likely to be exploited by state (outside) interests. Informal groups such as this often become formal stakeholders with a voice.

Linkages between Parks and Buffer Areas

Partners at most sites have undertaken a variety of activities to make protected areas more acceptable to and compatible with the lives of those living in and around them. These approaches are collectively known as integrated conservation-development projects (ICDPs) and are based on the premise that protected-area management must reach beyond traditional conservation activities *inside* park and reserve boundaries to address the needs of local communities *outside*. ICDP activities usually emphasize local development and alternative income sources for local people that do not threaten the flora and fauna of the protected area. Table 13-4 lists some of the most common activities underway at sites. One of the most crucial elements in undertaking development activities within the context of conservation programs is making sure that the development activities have necessary benefits for conservation. Without this linkage, such activities are often a series of unconnected project elements.

A wide variety of activities were undertaken within the case studies to provide compensation for loss of access to resources, substitute for destructive livelihood patterns, or focus on local development. The two areas with the most extensive efforts in both conservation *and* development are Sierra de las Minas and Machalilla. At both sites, the future of the reserve depends on balancing livelihood needs with conservation objectives. Other parks where local livelihood patterns matter, and where people reside within the park, are Rías Lagartos and Celestún, and Del Este. At Lagartos-Celestún, there has been a special emphasis on developing income-substituting activities that generate employment and income in ways that do not further deplete fisheries resources. These efforts, as described in chapter 4, have met with mixed success so far.

At Del Este, conservation activities are still in an early phase. Development activities have been focused on working to develop local institutions that are supportive of conservation. Few efforts have been made to work with communities inside the park, perhaps as a result of the uncertainty over how to deal with those settlements. At Machalilla, despite the illegality of residence within the park, government authorities and NGOs actively work with communities on an array of resource management activities and on developing local institutions.

Well-funded and high-profile development activities have taken place around Corcovado National Park. Some of these were initiated by the Boscosa project, one of the best-known and widely reported examples of an ICDP; its history is described in chapter 6. Initially, the project focused on influencing the overall patterns of development on the Osa and creating a variety of local organizations with the institutional capacity to promote sustainable activities peninsula wide. The project assumed that the creation of peninsula-wide organizations would be the best way to insure the long-term protection of Corcovado and the other protected areas that surround it. Many of the efforts currently underway around the park are supported by organizations other than the Conservancy's partner organization, the Fundación Neotrópica.

Programme for Belize has aggressively undertaken conservation activities at Rio Bravo; as a private conservation area, this emphasis is not surprising. Because of the low level of social pressure, the need to intensively pursue development activities outside of Rio Bravo is not pressing. Instead, Programme for Belize is undertaking strategic actions to work with communities outside the reserve to build support for Rio Bravo and stabilize potentially threatening forces. Despite the relatively low level of

Table 13-4. Linkages: Activities Underway at Sites

	Rías Lagartos & Celestún	Rio Bravo	Sierra de las Minas	Corcovado	Del Este	Amboró	Machalilla	Podocarpus	Yanachaga
CONSERVATION									
Park management plans	x	x	x	x	x	x	x	x	x
Guard training	x	x	x	x	x	x	x	x	
Improved enforcement	x	x	x		x	x	x	x	
Ecotourism promotion	x	x		x			x		x
Boundary demarcation	x	x	x		x	x	x	x	x
Conservation education	x	x	x	x		x	x	x	x
Direct employment		x	x	x					
Ecological monitoring		x	x				x	x	
Ecosystem restoration		x	x			x	x		
Resettlement		x	x						
DEVELOPMENT									
Community services				x		x		x	
Improved production	x	x	x	x		x	x	x	x
Improved marketing	x	x		x					x
Natural resource mgmt.	x		x	x			x	x	x
Irrigation and water control			x			x			
Land tenure analysis/titling	x		x	x		x	x		x
Local institutional devt.		x	x	x	x	x	x		x
Local participation	x		x	x		x	x	x	x
Conflict management			x	x			x	x	

threat from local peoples at Amboró and Yanachaga, numerous development activities are underway at both sites. At Amboró, activities in the buffer zone are carried out by a variety of NGOs, from local-level development organizations to international organizations such as CARE, which carry out large-scale agricultural activities. In contrast, the poor security at Yanachaga has meant that most efforts have been carried out by local NGOs, even when funds have come from international donors.

Most of the sites have a variety of activities underway in both conservation and development spheres. Conservation activities are underway at most sites and are the primary emphasis of most of the Conservancy's partner organizations. However, many of the partners have also begun a range of development activities at sites. How well the development actions are linked to actual threats is unanalyzed at most sites. Many of the development activities underway seem to be focused on increasing local participation in development and gaining local support for the implementing NGOs.

Conflict Management and Resolution

All kinds of conflict exist at sites, ranging from a low-level distrust of park guards to armed and violent conflict. However, the case studies make it clear that conflict should not be understood as simply people versus park. Many of the conflicts described in the case studies are rooted in changes in the overall social context surrounding parks, as discussed in greater detail in chapter 14. For example, at some levels, conflict is evident between local groups and outsiders, including: at Ría Lagartos and Celestún between recent migrants and long-term residents and between fishermen and tour-boat operators; at Sierra de las Minas between water-user groups; at Corcovado between different groups of residents and land-use proponents; and at Machalilla between those living inside and those living outside the park. Several sites are notably free of conflicts related to the park. There is no conflict around Rio Bravo. At Yanachaga, there has been guerrilla activity in the area around the park, but this has not been park based. However, there has been some resentment of the park by indigenous populations. At both of these sites the lack of conflict is due to the unique historical circumstances at the site and the particular way the parks were established.

At Sierra de las Minas, conflict has taken several forms. Perhaps the most dramatic case is that of a wealthy private logger who wanted to exercise his right to use a logging permit in the core area of the reserve. Defensores, as well as communities, opposed the logging—Defensores on grounds of ecological damage, and communities on a fairness principle: if they couldn't log, no wealthy logger should have the right, either. This issue was finally settled when the national press and an international letter-writing campaign led to government cancellation of the logging permit and the later acquisition of the land by Defensores. Other conflicts have been largely related to settlement in the core zone, which Defensores has resolved using a variety of tactics, from negotiation to voluntary resettlement, to legal measures.

Gold within Corcovado and Podocarpus has been at the root of violent conflict between park officials and miners, but also among the miners themselves. At both sites, dialogue has been important, as have the courts in providing guidance on what activities are legal and illegal within parks. Although both of these areas are temporarily stable, the potential for violent conflict could quickly reappear.

Zoning that pits local uses against park management objectives has been at the root of conflicts at Del Este and Machalilla and, until recently, at Amboró. At these three sites, attempts to control or limit resource use have led to hostility and conflict. At Amboró, the Red Line project served as a formal mechanism to resolve the boundary problems. At most other sites, no formal process exists to resolve disputes.

Conflict is pervasive among some groups and to some degree at all sites. It is important to note that not all of the conflict is park related, although parks have served as the "lightning rod" for clashes between hostile groups in some cases. There are few formal mechanisms to limit conflict at most sites, although the Red Line process and boundary redefinition at Amboró are significant. At Corcovado and Podocarpus, where there have been serious conflicts over gold, antagonistic stakeholders have been brought together by concerned park management authorities or NGOs. Opportunities exist to define conflicts and improve conflict mediation activities and mechanisms at most sites.

Large-Scale Threats

Virtually all of the parks are currently vulnerable to large-scale threats, which have their origins far from park boundaries. The exception is Rio Bravo, where no large-scale threats currently exist. As shown in table 13-5, there are more actual threats than potential threats. The category of threat most often noted by case study authors has its genesis in policies. The policy context can take many forms, such as weak government institutions, conflicting government policies, and changes in laws regarding tenure; and it is the area where park management agencies feel they have the least control.

The second-most-common threat, which is, in fact, closely related to policy, is infrastructure development in or near parks. Five of the sites have had infrastructure and access improved near them, leading to potential problems from logging, colonization, and resource use inside parks. On the Osa Peninsula, which surrounds Corcovado, several key roads have been paved, and phones and electricity have been installed in what was, until five years ago, a remote area with limited access. The regional changes that are swiftly taking place in the communities and forest reserves surrounding Corcovado cannot be understated. Not surprisingly, another commonly cited large-scale threat is colonization, brought about by increased access. At Yanachaga, government-sponsored development is the cause of colonization, rather than spontaneous colonization at

Table 13-5. Large-Scale Threats

Site	Logging	Mineral, Oil, & Gas Reserves	Roads, etc.	Colonization
Ría Celestún Ría Lagartos		Salt at both	At both	At both
Rio Bravo	Potential	Potential: oil		Potential
Sierra de las Minas				
Corcovado	Around park	Gold mining	Roads, phones, electricity	Colonization around park
Del Este				Tourism development outside park
Amboró			Access due to highway	
Machalilla		Pipeline—potential if rupture		
Podocarpus	Potential w/road	Gold mining	Road flanking park	Yes
Yanachaga	Potential problem	Potential—park has oil, natural gas, & minerals		Govt.-sponsored development

other sites. At Del Este in the Dominican Republic, tourism and resort development outside park boundaries are of concern.

One surprising finding when looking at this table is that large-scale logging, grazing, and agricultural development are of relatively minor concern at most sites. Logging is a potential threat at three sites—Corcovado is the only park where large-scale logging poses a current threat. Grazing is a problem at Machalilla, although it is from small-holders. In contrast, the threat from grazing at Ría Lagartos is from large-scale operations. Grazing is a potential threat at three other sites. Marine resource overuse is of concern at the four ocean sites; at all those sites, at least some of the overuse is the result of large-scale, commercial fishing. However, with the exception of Machalilla, the areas do not incorporate marine protection into park management.

Finally, exploration and mining for subsurface minerals, gas, or oil are potential problems at many of the sites. At Yanachaga and Rio Bravo, there are below-ground reserves of oil. Gold mining is a current problem at Podocarpus and Corcovado. The Podocarpus case offers an in-depth

Agricultural Frontier	Tourism	Marine Overuse	Grazing	Policy	Other
	At both	At both	At Lagartos	Conflicting policies	
Potential			Potential		
By communities in reserve				Tenure	Weak govt. institutions
	Vacation homes & tourism	Yes		Conflicting policies	Weak institutions & policy
	Potential threat if unmanaged	Yes	Outside	Inconsistent & unclear policies	
Potential threat				Management?	Potential from coca growing & processing
	Uncontrolled tourism	Yes	Within park	Potential policy conflicts	
Increasing threat			Potential	Potential policy conflicts	Potential transboundary development
Western boundary affected			Potential	Policy changes re land & resource rights	Potential drug trafficking problems

analysis of the claims on the park's gold, which is coveted by a range of individuals and corporations, from artisanal miners to regional merchants and international mining companies. Lobbying, environmental education, and research to build a constituency to defend Podocarpus and control the mining therein have been key elements of FAI's efforts. At Corcovado, most of the current threats are from artisanal miners, although the large mining equipment brought in in the past, suggests that there is significant money backing the operations. Organized invasions of the park by gold miners have been significant and are an ongoing source of threat to the park. At Machalilla, local interests were ignored with the construction of a coastal oil pipeline by Petroecuador. This pipeline has the potential for spills in the future. Salt extraction from large evaporation ponds is a problem at Ría Lagartos.

National Policy Framework

The national policy framework is arguably the single most important element within the case studies for the long-term survival of those sites. The problems in managing parks and conserving biodiversity that are rooted in national policies are similar across sites. All of the case studies mentioned the lack of commitment or political will or power to promote environmental objectives. In Mexico, national and local development objectives have proceeded without regard to protected areas. The environmental agencies have had only spasmodic commitment and have gone through frequent reorganizations. While the case study on Sierra de las Minas notes the tremendous steps taken to support conservation in the past few decades, the government agencies charged with conservation have been unable to implement policies, such as those favoring the creation of private nature reserves, that would benefit biodiversity conservation. In Costa Rica, despite rhetoric to the contrary, there has been a lack of political will to support parks and forest management. The government of the Dominican Republic has not had clear or consistent policies regarding park management, use of tourism revenues, or resource use in parks.

In Ecuador, environmental institutions are weak and subordinate to other sectors. This has made the enforcement of laws difficult. Similar problems exist in Bolivia, where numerous changes have occurred in the institutions charged with park protection and natural resource management. There is guarded optimism that the recent creation of the National Directorate for Biodiversity Conservation will help solve these problems, along with new laws and legislation on biodiversity and forests.

In many countries, the budgets of park management authorities have been seriously constrained; external donors, such as the Conservancy with USAID funding, provide most of the financing for parks. The case studies reveal that this shortage of funds, and of staff, has led park management agencies to rcly more closely on NGOs to carry out a variety of activities that previously would have been within their domain. The case study at Podocarpus, describes how policies to shrink the state sector have led to fewer park rangers. The same is true in Costa Rica and undoubtedly at other sites as well.

In some cases, agencies have mixed loyalties—they must support biodiversity conservation as well as other agendas. In Mexico, for example, the first environmental agency called SEDUE ushered in strict controls on or prohibition of human activities impacting wildlife and habitat. Several reorganizations later, a new agency called SEMARNAP, created in 1995, has a greater emphasis on community development programs. In Costa Rica, a strategy to improve governmental coordination and decentralize authority to regional conservation areas has effectively done neither at Corcovado. Problems such as controlling illegal logging, which go beyond a local scope, have been overlooked amid daily management concerns.

The site that is least dependent on national policies is Rio Bravo, as a result of its nongovernmental ownership, but this itself was the result of a national policy. Yet in Belize, the absence of a unifying national conservation policy, evident in the different actions of the three agencies involved in policy making relevant to biodiversity conservation, has had an impact on Rio Bravo's management. Government policies also influence what happens in the lands surrounding Rio Bravo.

A huge array of other policies can seriously affect how well parks are managed. These include everything from the local security situation and stability at particular sites (e.g., at Yanachaga) to encouragement of agrochemicals in agricultural production (Rio Bravo), to incentives for land clearing and cattle grazing and migration (Mexico). Many of these particular policies are outlined throughout the case studies, and they are discussed more generally in chapter 15.

Indigenous Peoples and Social Change

Many of the lands that are critical for biodiversity conservation in Latin America are inhabited by indigenous and traditional peoples. One of the critical elements affecting how traditional groups are likely to "manage" biodiversity is related to their ability to manage and control social change. The future of traditional lands and the biodiversity contained therein will depend on a complex set of national factors, local factors, and the interplay between the two. At three of the case study sites—Corcovado, Podocarpus, and Yanachaga—indigenous groups occupy legally declared indigenous reserves near or adjacent to the case study parks. At two other sites, the presence of indigenous populations also strongly affects resource use.

Guatemala's legal recognition of indigenous peoples is vastly different from the approach of many other areas in Latin America. Indigenous or tribal lands do not exist in Guatemala as they do in other countries in the Americas. Although Indian villages and lands were recognized by the Spanish Crown, an 1871 law declared that unoccupied lands, including traditional indigenous lands, were available to the government for sale. This resulted in abolishing traditional indigenous property. Within the current legal framework, the most effective strategy for ensuring traditional user rights is securing land ownership. Defensores has been assisting the twenty thousand Q'eqchi' and five thousand Poqomchi' living within the SMBR to do this.

The Guaymi moved to the lands that are now part of the Guaymi Indigenous Reserve, adjacent to Corcovado, in the 1960s. Approximately 120 people live on the 2,713-ha reserve created in 1985. When the reserve was created, non-Guaymi settlers had to be bought out, but few people received compensation until at least 1992. Since the Guaymi and their nonindigenous neighbors had colonized the area at the same time, the latter saw no reason the Guaymi should get special treatment in an area that was not their traditional homeland. Campesino tensions toward the Guaymi are further heightened by the relative perceived "lack of indigenousness" about the Guaymi. While the Guaymi have their own language, it is difficult to differentiate them from their nonindigenous neighbors. From a conservation perspective, however, one significant difference between the Guaymi and other recent migrant groups to the Osa is that the Guaymi have maintained most of the forest cover on their lands. Their reserve appears to be a stable and biologically significant buffer to Corcovado.

The Shuar are known as subsistence hunters and fishers who traditionally lived in small settlements dispersed throughout the Ecuadorian Amazon. Nine Shuar settlements, centros, near Podocarpus National Park occupy an area of 20,400 ha. Each centro holds a legal, collective land title. The Shuar plant subsistence crops (maize, cassava, yam, and plantains) on small plots. The Shuar are undergoing a rapid process of transformation: they use modern technology in hunting (shotguns) and in clearing the forest (chain saws). Colonist invasions of their Shuar lands has begun to limit the land available for hunting and shifting agriculture. There has been a rapid conversion to cattle raising, which may soon become the primary economic activity among Shuar settlements. Colonist invasions have also led to access to commercial markets, new human diseases, infrastructure, and the entry of Shuar youth into gold-mining activities. These changes in livelihood among the Shuar, particularly the increase in cattle raising, will diminish the forest cover around Podocarpus. It may also lead to conflicts between Shuar youth, other gold miners, and park authorities if Shuar increasingly seek entry to the park for gold mining.

The migrants on the northern side of Amboró are primarily Quechuas from Cochabamba and Potosí. While they represent traditional communities, they are not traditional to the zone. The Quechua migrating from the altiplanos and high valleys do not have practices that are well adapted to the zone's lowlands, resulting in a process of agricultural and social transformation. The community of El Carmen, also in the northern sector, is a small traditional community of Chiquitano origin, where residents practice palm weaving and subsistence agriculture. In the short term, pressure from indigenous populations migrating down from the highlands is unlikely to have a significant impact on Amboró. However, increases in land scarcity elsewhere, combined with substantial migration over the long term, would increase the potential for land invasions, particularly if the park is viewed as an open-access resource.

Finally, the Yanesha in Peru inhabit part of the area where they have lived for over a thousand years. The Yanesha Communal Reserve, the first of its kind in Peru, is a 34,745-hectare reserve, created in 1988, that forms a buffer zone between Yanachaga National Park and the Yanesha and colonist settlements east of the park. Most Yanesha communities in the

Palcazú valley practice slash-and-burn subsistence agriculture. Currency has been used as the basis of exchange only in the past fifteen years. One significant change is a shift from crops to livestock. Some communities have become highly dependent on logging, and lands, once cleared, are used for cattle and other livestock. Hunting, fishing, and collection remain important activities; the Yanesha hunt with both shotgun and bow and arrow and fish using nets, hook and line, and occasionally harpoons and plant-based fish poisons. Despite some traditional practices, road construction, settlement and population growth, and development projects themselves are increasingly integrating native Yanesha into colonist-dominated market economies, with the dramatic changes in resource use and management and social organization that those entail. Over time, if the Yanesha want more land than their current reserve, there could be pressure to return the park to them or open the park to greater use, since the lands that make up the park were traditionally theirs.

Transboundary Issues

While we initially conceived of transboundary issues in terms of international boundaries, the case studies point to boundary issues at state or provincial levels. Both positive and negative elements are associated with transboundary issues. For example, if one country builds a road to protect and build up its national border and a country on the other side has a park there, serious park management problems may arise due to colonization and incompatible land uses adjacent to one another. In numerous transboundary areas worldwide, an effort has been made to create parks on both sides of national boundaries—often called binational peace parks. In Central America, the creation of such peace parks has been promoted as one way of increasing regional peace while enhancing biodiversity objectives.

International transboundary issues arise at two of the parks: Rio Bravo and Podocarpus. Because Rio Bravo lies near the borders with Guatemala and Mexico, there are potential problems with both countries. Aguas Turbias National Park lies between PfB land and the Mexican border and is subject to frequent timber theft and illegal hunting. In 1993, approximately four hundred trees were logged from forest within RBCMA property along the Guatemalan boundary. Each country tends to identify its neighbors as being responsible for poaching timber and wildlife, leading to potential conflict. Despite the conflict between Guatemala and Belize over their shared boundary, the armed forces of both countries were called in to Rio Bravo to halt timber poaching. However, if the border dispute between the two countries dramatically worsened, Rio Bravo's vulnerability to illegal resource extraction would increase.

International transboundary issues indirectly influence Podocarpus National Park during certain times, such as the boundary conflict between Ecuador and Peru. At such times of international conflict, the park's proximity to Peru is a destabilizing factor and management activities at Podocarpus are constrained. A potentially more serious problem to be considered is the political strategy to promote settlement and development along the Peruvian boundary so as to lower the risk of losing forested (viewed as "empty") territory.

Transboundary issues arise across state, provincial, departmental, and municipal lines. At Podocarpus, Ría Lagartos, and Ría Celestún there is a need to coordinate across state and provincial lines. At Podocarpus, for example, it has been difficult getting representatives from the provinces of Loja and Zamora to be on the same committee; each province has wanted to create its own park committee. Regional antagonism between the two provinces has seriously affected park management. In Mexico, Celestún Reserve is located within the municipalities of Celestún in the state of Yucatán and Calkiní in the state of Campeche. Conflicts between the states of Yucatán and Campeche over borders and fisheries have made bistate management planning a real challenge.

At Corcovado, fishermen from Panama are blamed for overexploitation of marine resources. Finally, parks can be a staging area for another form of "transboundary" movement. Del Este serves as a staging area to collect immigrants for transport to Puerto Rico.

Resettlement

Resettlement, including both voluntary and involuntary resettlement, occurred at six of the parks, although the number of people involved varied greatly. There is no indication of resettlement at Ría Lagartos or Ría Celestún, Podocarpus or Yanachaga. At times, both voluntary and involuntary resettlement took place at the same site. When Rio Bravo was created, for example, four families were resettled voluntarily and one family involuntarily. All were given new land with a title and compensated for improvements to the land. Resettlement from the lands that constituted Corcovado Park and the surrounding forest reserves was extremely complex and conflictive and took up a vast amount of time and resources from numerous Costa Rican governmental agencies. Most of the land claimed by peasants was unsuitable for agriculture or logging, and it was already owned by a private corporation or the Costa Rican government, making the land claims illegal. In the absence of field visits to verify residence, it is likely that some squatters were compensated well beyond what was due them and that more people received compensation than actually lived there.

When Del Este National Park was established in 1975, twenty-two individual parcels, eight commercial parcels, and three communal landholdings were expropriated without compensation. The following year, some of the lands from three commercial parcels were exempted from expropriation, including those belonging to a private resort. Owners of other properties within the park were allowed to remain if they did not expand the areas in use. Due to those restrictions, some people simply abandoned their lands. The exemption of the private resort when communal and generally small-holder parcels were included in the park with no compensation has been a source of lasting hostility.

Records for Amboró indicate that combining families already residing in the park with families added during the 1991 expansion brought the total to nearly three thousand households within the park. Legally, residence within parks was not allowed, but there was never any intent on resettling anyone. The Red Line process was developed to prevent the need for resettlement while maintaining the strict protection category of a national park.

Finally, although there was no resettlement from Machalilla, plans existed throughout the 1980s to resettle people from the park. However, no money was ever allocated for this purpose. There is debate over local residents' attitude toward selling their land to the government when the park was created. Some sources say that many individuals initially wanted to sell their land, while other sources recall violent local opposition to land purchases and expropriations.

One of the most instructive examples of voluntary resettlement can be found at Sierra de las Minas, where Defensores has worked with two indigenous communities, assisting them with voluntary resettlement from the core zone. There has been a lengthy process of dialogue with the communities to find out what land and location they would prefer. One of the communities, Vega Larga, will begin relocating gradually in 1998 onto land that is much more suited for agriculture than the infertile land it occupied in the core zone. The Santiagüilá community relocation was hampered because the legally registered limits of the property purchased for its relocation in fact overlapped with neighboring lands, necessitating a land survey. The preliminary survey indicated that three other communities were already occupying part of the purchased land, so their lands will be titled as well.

Conclusion

This chapter reviews the collection of case studies using a set of eleven themes as the basis for analysis. Looking across the cases and themes, some tentative findings are possible. First, the process of how a park was established frequently sets the tone for the whole context in which future park management occurs. At sites such as Amboró and Rías Lagartos and Celestún, where legislation established parks with little awareness of the local context, there has been enduring conflict over boundaries, zoning, and uses. In contrast, even in densely populated and socially complex areas such as Sierra de las Minas, combining local consultation with technical information has produced a system of zoning that, while not completely accepted, is at least relatively well understood. Closely related to park establishment are tenurial issues—of who were the winners and losers when parks were established. The claims over land and resources, poorly known and unresolved at many sites, seriously complicate the park management context. Illegal resource uses at sites often stem from these unresolved issues of tenure. Other resource uses and users represent the winners in the tenure game—tourists and state mining companies come quickly to mind.

The variety of ways in which partnerships and alliances between governments, NGOs, and other organizations have developed is impressive. While the structure of the relationships varies between countries, one thing is clear: Government institutions responsible for park management tend to be politically weak and underfunded. These institutions have pragmatically responded to this situation by collaboration with NGOs, increasing the financial support for conservation. In turn, other organizations have often been drawn to these areas to support development activities.

Conflict is common at most sites—often among different resource-user groups. These conflicts are sometimes exacerbated by the large-scale resource uses and threats to areas. Government policies that either directly or indirectly affect park management were viewed by partner organizations as the greatest threat. These policies span everything from legislation governing parks to incentives for land clearing and cattle raising to transboundary policies. Indigenous and traditional populations were present at five of the case study sites. Within all groups, a process of social change is underway that could seriously affect their relationship with wildlands and wildlife and patterns of resource use. In most cases, these changes, if they occur, will reduce the areas outside of parks with forest cover. The impacts of all of these findings are explored in subsequent chapters.

14. Perils to Parks: The Social Context of Threats

Katrina Brandon

In reviewing the accomplishments and challenges facing the nine sites described in this book, it is important to remember that they represent the first group of Parks in Peril sites, which were chosen, in large part, because they represented the region's most imperiled ecosystems. The original idea of the PiP program, as described in chapter 1, was to take legally declared areas that were paper parks and institute the "emergency" measures necessary to safeguard them. The analogy of an emergency room is suitable, for the program has been intent on taking those actions necessary to get the patient stable, well enough to pass along to others who can provide a transition to the longer-term support that will be necessary.

The intent and actions of the program are relevant when looking at the social context of parks, precisely because many social issues won't necessarily be addressed in an emergency context—they are issues to be dealt with over the long term and may intentionally have been bypassed when PiP work plans were set. Yet they do form the underpinnings for setting priorities and are particularly important in defining the next steps as these parks undergo site consolidation. Site consolidation occurs when sites reach a predefined degree of functionality, when they no longer need emergency assistance (see chapter 1). In this way, as parks leave the emergency room successfully stabilized, other sites can enter the program.

So why does the social context matter? What can we say about it that is meaningful to park managers, beyond saying that the context is important yet site specific? The social context, from the perspective of a park, is usually the set of threats that faces a park. Virtually all threats to biodiversity result from human actions—usually from different types of uses by different social groups. The immediate sources of threats are often expanding agricultural frontiers, illegal hunting and logging, fuelwood collection and uncontrolled burning, colonization, cattle grazing, and large-scale development and infrastructure projects. These immediate threats are ultimately attributable to policies in a variety of different sectors such as placement of roads and other infrastructure, licensing and concessions for timber and mineral resources, and tax policies that encourage migration and certain types of land use. Park management must be planned with a clear understanding of the social and economic forces driving local actions.

Change around parks is often nested—local levels are influenced by regional change, which is in turn propelled by national policies, which are sometimes driven by international forces. The local social context that affects park management often includes: the types and patterns of resource use and production, such as farming and grazing systems; demographic factors, such as infant mortality, local rates of population increase, and settlement patterns; local tenure security; differences in gender roles in production, levels of local organization, and access to technical changes; or changes in consumption. Regional changes are often spurred by road construction, which leads to influxes of migrants, which in turn lead to changes in patterns of land use and local organization. Increased access to markets leads to changes in local agricultural production patterns and may increase local patterns of extractive resource uses. These regional forces will have their own dynamic yet will influence the decisions and resource management activities at local levels as well, and may cause profound changes in local-level systems. These regional forces are in turn shaped by a range of national policies, incentives, and economic adjustment programs, which are in turn affected by forces such as international debt, markets, and prices.

An analysis of the social context helps park managers identify and define both the proximate and root causes of threats, the different interest groups at all social levels, what these groups want or need, and what activities are likely to lead to intended outcomes for biodiversity maintenance (Stankey 1989). Each of the eleven themes used as the basis for analysis within the case studies (e.g., history of park establishment, land and resource tenure) represents one element of the social context surrounding parks. Understanding these elements helps to inform decisions concerning appropriate actions. But when? What should be done in an emergency phase, and what should be done at a later date when a park is more stable? The answers to these questions are only vaguely centered in ecology or biology. An illusion exists among conservationists that what they are doing is conservation—when the case studies make it clear that they are really doing large-scale social interventions in complicated settings.

This chapter reviews some of the "perils to parks" that are clearly rooted in the social context, in an attempt to explicitly guide park planners to recognize the social and political nature of their actions. While this list is by no means exhaustive, these perils emerged from the PiP case studies and are common to numerous other parks worldwide:

- unrealistic, externally imposed missions
- policy pitfalls
- circumstances of origin
- park definition
- basis for implementation
- resource management
- uncertainty and the decline of local institutions
- equity
- catalyzing development.

Unrealistic, Externally Imposed Missions

Upon taking the job, a park manager's first sensation has to be a tremendous sense of, What have I gotten myself into? Just trying to figure out how to deal with basic management is extraordinarily challenging at any park in need of emergency care. And as we will see in later sections, even something that sounds technical, such as putting up a sign or demarcating a boundary, is a political action that adds to the social complexity. But a park manager who decides to do some reading about park management and looks to the current wisdom in the field is likely to be even more frightened by the job requirements.

The 1982 World Parks Congress in Bali stressed that "protected areas in developing countries will survive only insofar as they address human concerns" (Western and Pearl 1989, 134). Since that time, rather than being separated from the local context, parks have been viewed as vehicles to balance natural resource management with local social and economic development. As a result, there has been a proliferation of projects, activities, and actions to link people with parks, and to have parks reach outside their boundaries to address local concerns. With each successive year and conference, the agenda for parks has become more crowded. In 1989, for example, there was a call for parks to move beyond meeting local development needs and become engines of regional growth, or "critical elements of regionally envisioned harmonious landscapes" (McNeely 1989, 156–157). At the 1992 World Parks Congress, the expectations of what parks should accomplish were ratcheted up even further to provide benefits at local, national, and international levels (Barzetti 1993).

Externally imposed expectations of the overall mission of parks evolved quickly. Parks used to be seen primarily as tools for conserving biodiversity. But as described in the introduction to this book, even the term *biodiversity* may substantially complicate park management if it is part of a park's mission, particularly if a definition of biodiversity is used that "is not limited strictly to plant and animal worlds [but] includes human cultural diversity as well" (Barzetti 1993). In short, the issues of what parks are supposed to do, who they are managed for, and the benefits they are supposed to provide have become remarkably complex. No longer are park managers just supposed to understand, and manage, large, spatially defined areas and the plant and animal life within them. They are supposed to use these areas to simultaneously benefit local populations, regions, and even entire countries.

In certain circumstances, it is possible to achieve some of these objectives. But we have lost sight of what we are really trying to do with parks. In trying to make them socially acceptable and "accepted," we are holding parks responsible for curing structural problems such as poverty, unequal land distribution and resource allocation, corruption, economic injustice, and market failures. Strong social organization and political will must serve as the basis to address these problems. Parks may provide one vehicle or mechanism to address these problems at local levels, but even within a regional context they cannot be expected to reach well outside their boundaries and solve broader problems. Such externally imposed visions of what a park *should do* dilute the potential for what parks *can do*. The mission of parks needs to be refocused, in practical management terms, with clarity and realism rather than rhetoric as the guiding principles.

Policy Pitfalls

As the case studies clearly point out, government policies or their inefficient application are at the root of most threats encountered by parks. Indeed, of all the themes, the one that was most frequently signaled as problematic by the case studies across all sites was the national policy context; next and closely related was the existence of large-scale threats.

First, the relationship of political will to conservation repeatedly was raised in the case studies. Most environmental institutions were politically weak and subordinate to other economic sectors (such as mining, tourism, forestry) and to other government agencies (such as those responsible for infrastructure and land titling). This seriously limits the potential for park managers and related authorities to have the resources, support, and authority to effectively manage their own parks, much less try and use them to shape the regional context.

Second, a constituency for conservation is missing within governments. As a result, government support and financing for parks and park management are low, and there is an extremely high dependence on external sources of funds. Ironically, even when parks are successful in capturing funds, through mechanisms such as tourism (for example, at Del Este and Machalilla), neither parks nor conservation sectors are able to retain the money (Brandon 1996). This inability of park authorities to retain revenue is often attributable to the political weakness of government agencies charged with conservation, which are unable to thwart tourism ministries or central banks. This means that pressure is exerted on park managers to maximize revenue and to distort the park's mission so that it is managed as a "beach area which brings in tourist money" rather than for biodiversity conservation. In such cases, neither biodiversity nor local people benefits much from tourism to the park. The problem of muddled missions is exacerbated by policy pitfalls such as viewing parks as a source of revenue.

Third, the subordination of environmental institutions to other sectors makes it difficult for park managers to effectively challenge other government agencies over actions that affect parks. Roads are built, mining and logging concessions are given, pipelines are constructed, and land titles are granted by other government agencies over the objections of park authorities. Less obvious but equally problematic are more subtle intrusions such as the placement of boundaries. Where boundaries should go is often determined by law; yet as we have seen in chapter 13, laws have been written with no field-based input on ecological or social issues. Park managers frequently lack control over the most basic elements of how parks are managed; for example, who can be hired as a guard, how many guards can be hired, and the function of guards should all be park-based management decisions. In reality, such decisions are usually determined by national policies and civil service requirements—another example of policy pitfalls.

These examples demonstrate how the broader policy environment exerts a defining influence over the maintenance of biodiversity and the management of parks. Even the very definition of biodiversity, plus what a park is expected to do, whose interests it represents, and how and where it is established, is colored by the national policy context. Unfortunately, as shown in chapter 13, park authorities often have little control over management within park boundaries. This makes the assertion that parks should be models of development that much more unreasonable.

Circumstances of Origin

Where to establish a park is a decision that is deeply rooted in the social and policy context of national and regional priorities. Parks can be divided into two general categories based on the reason for their creation: faraway parks and parks established to stop transforming forces. Faraway parks were usually established in remote areas, precisely because they were remote, either geographically or in terms of political power. These areas were perceived by government officials as far away—their people were poor and lacked political power; government services and infrastructure were limited; and lands and resources were viewed as having little productive value. As governments looked for places to establish parks, these faraway areas were the most promising in terms of their potential for biodiversity conservation and the perceived ease of establishing them with few vocal complaints. Examples of these sites are Rio Bravo, Amboró, Podocarpus, and Yanachaga.

The second category of parks was established to stop, or at least to control or manage, the effects of rapid regional changes. These regional-level changes are usually generated by forces external to the region, often rooted in government policies or programs, such as road construction and the consequent changes in land use. Within the case studies, these sites are represented by Rías Lagartos and Celestún, Corcovado, and Machalilla.

What is evident in trying to understand the situation at a given park is how strongly the circumstances of origin not only influenced what happened in the past, but also continue to influence the present-day management context. Parks formed with the intent of "stopping" transforming regional forces will successfully thwart these forces only if there is significant political will to identify and address the threats at both local levels and in the regional context. As indicated by the case studies, the parks established to save what is in remote areas are likely to remain relatively intact and subject primarily to local-level uses, as long as they remain remote.

This represents something of a contradiction in the selection of parks for emergency care. Parks in remote sites may be paper parks, and they clearly need a huge amount of help to be transformed from paper parks to well-managed entities that effectively conserve biodiversity. In contrast, parks intent on abruptly stopping social change represent such grave ecological emergencies that they need vastly more attention than a program such as PiP can provide. Clearly, parks will be created for reasons other than saving remote areas and stopping the effects of regional change. But the case studies make it evident that the parks established in remote areas are more likely to conserve biodiversity than those in areas undergoing rapid regional changes. *Unless* the political will and technical and financial resources to address regional-scale change are present, the parks trying to stop regional change will be much harder to effectively manage over the long term.

Park Definition

The perils of park definition lie in whether the appropriate type of park was established for the existing social context. The case studies highlight that certain kinds of parks are more effective than others in a given social context. Unfortunately, policy makers give little attention to which type of park will work best when they create parks. While there are a broad variety of parks and purposes (see chapter 2), for the sake of clarity, two very different kinds of parks are apparent in the case studies: *conventional parks* and those based on a *biosphere reserve* principle.

The management challenge at conventional parks is generally twofold: restricting uses inside the park, such as local residence and consumptive resource uses, and stabilizing threats outside the park that spill inside. This latter challenge of reaching outside park boundaries to stabilize uses is a relatively new twist on the role of conventional parks (Wells and Brandon 1992). In the case studies, conventional park models, with no livelihood activities allowed inside, include: Corcovado, Rio Bravo, Amboró, Podocarpus, and Yanachaga. (The 1997 legislation for Amboró created a conventional park with a sustainable unit around it, but the core is still a large, conventional park.)

Parks based on a variation on the biosphere concept, are mixed-use protected areas, with people living inside and systems of internal zoning to regulate uses (Batisse 1986). In these areas, livelihood needs must be specifically addressed in terms of their compatibility with biodiversity conservation objectives. To have a biosphere reserve function effectively means that different zones of land and resource use need to be managed by a central management authority. While other political units may be located within a biosphere reserve, they must willingly comply with the management objectives set out for the reserve as a whole. Representative case study sites include Rías Lagartos and Celestún, Sierra de las Minas, Del Este, and Machalilla.

Biosphere reserves have been adopted at many sites worldwide because, on the surface, they look like a politically expedient and socially acceptable way of meeting livelihood and conservation needs—a "win-win" proposition. In reality, management challenges at biosphere reserves are more socially complicated, at least over the long term, than at conventional parks. This is because the social relations between the park and the people are quite different from those of a conventional park. Three very specific types of agreements, whether formal or informal, are needed for biosphere reserves to be successful. First, local people, the residents, need to agree to stop using resources at some area or areas zoned as core areas. Second, residents have to agree to alter use patterns in other zones of the reserve in order to meet some specified level of intensity of use, usually reduced use. These areas are known as sustainable- or compatible-use zones or buffer zones. Finally, both local people and key regional actors, as well as national government officials and agencies, have to agree that some new kind of management authority, often with some political significance, will be superimposed over prevailing social and political systems to manage the whole area.

The case studies highlight two kinds of problems of creation: one, the wrong type of park for a given social context; and two, a park without the necessary policy context in which to fulfill its functions. Machalilla is an excellent example of the former problem. The law created it as a conventional park, but managing it as one was impossible, given the concentration and density of people and levels of resource use and the fragmented nature of what could be considered core areas. As a result, INEFAN, the government's park management authority, over time began managing the site as a biosphere reserve rather than a conventional park. However, much hostility would have been avoided, and management would have been somewhat simplified if Machalilla had been created as a biosphere reserve, and an emphasis had been placed on reaching agreements with local populations early in the management process. Corcovado is an example of the latter problem, where policy-based decisions (i.e., clarifying tenure, controlling logging) to control what is happening around the site are absent.

All parks are not created equal, and the management objectives of conventional and biosphere parks are quite different. Which kind of park works best depends on the social context in the area at the time of the park creation—and the circumstances of origin. Mismatches can hamper implementation until there are substantive changes—changes in mission, policies, or type of park.

Basis for Implementation

When management—the process of transforming a paper park into a real park—begins, the policy context is set, some physical space is allocated (even if just in legislation or on a map), and the type of park is decreed. PiP has to work with these constraints as given, at least in the short term, until appropriate changes are defined—e.g., laws are changed, boundaries are redrawn, or park categories are changed. If we create a matrix combining the two categories of park origins (the faraway parks versus those intent on stopping changes) with the two park types, (conventional or biosphere) we can get a quick idea of the management challenges at the time of the park's birth (which may really be the beginning of implementation) that endure into the future. Table 14-1 provides a brief example of how the different circumstances of a park's origin or definition affect management in these four types of parks.

As shown in table 14-1, the most straightforward process of park

Table 14-1. Comparison of Conservation Actions by Park Type and Site Type

	Remote-Stable	Rapid Change
CONVENTIONAL PARK	• Boundary demarcation with communities • Compensation for lost use or access to land or resources • Conflict identification and resolution	• Boundary demarcation with communities • Compensation for lost use or access to land and resources • Conflict identification and resolution • Stabilize threats • Increase enforcement • Identify strategies to stabilize land and resource uses outside parks, e.g. ICDPs • Strengthen existing land and resource claims adjacent to parks if this will stabilize change and reduce threats • Identify perverse policies or policies needing change to effectively manage inside park and stabilize land uses outside park
BIOSPHERE RESERVE	• Work with communities to define core areas (based on good ecological and social data) • Clarify jurisdictional issues—consolidate control over land and resource use by biosphere management authority • Identify mechanisms for compensation or substitution as appropriate • Secure agreements about the levels and types of uses permitted in other zones	• Work with communities to define core areas (based on good ecological and social data) • Secure agreements about the levels and types of uses permitted in other zones and what happens if agreements aren't met • Identify mechanisms for compensation or substitution as appropriate • Identify strategies to stabilize land and resource uses throughout biosphere reserve • Increase enforcement, particularly mechanisms to secure rights and use of those inside versus resource pirates from elsewhere • Clarify jurisdictional issues; consolidate control over land and resource use by biosphere management authority • Strengthen existing land and resource claims within biosphere reserve to stabilize change and reduce threats • Identify perverse policies or policies needing change to effectively manage inside park and stabilize land uses within biosphere reserve

establishment is at conventional parks with a relatively stable social context. The management emphasis at such sites should be to stress local participation in boundary demarcation, compensate people for their lost access to or use of land and resources, reduce conflict, and help stabilize threats to parks, which will usually be locally generated. A program such as PiP can accomplish a great deal on all of these fronts.

In contrast, initiating management at a conventional park in the midst of rapid social change presents a more complicated process of park management and is likely to be more intensive (in terms of technical and financial resources) and extensive (in terms of the geographic area and the time required to stabilize the park). Such sites must do everything a conventional park at a stable site should do, but must do so while also tightening up enforcement. Many parks also begin integrated conservation and development activities to try to stabilize the land and resource uses immediately surrounding the park (Wells and Brandon 1992). Working in this context is difficult, because migrants often overwhelm existing systems of local organization. In this context, parks may be seen positively as a vehicle to help reinforce claims to resources, particularly if mechanisms are included to deal with tenure security on lands adjacent to parks. Although it is reasonable to expect a program such as PiP to handle the park-centered issues, necessary regional interventions will often go well beyond the scope of a program such as PiP. Such interventions usually require policy coordination or change and political support.

Biosphere reserves implicitly assume a process of negotiation to get the appropriate parties whose lands or resources are involved to agree to three conditions mentioned above: a core area with no use, use restrictions in other zones, and the creation of a new authority for the entire area. At biosphere sites where the social context is relatively stable at the time of park creation, there is the luxury of time to build strong internal relationships and focus principally on the people and circumstances inside the reserve. In particular, mechanisms can be developed for local people to become real stakeholders and feel a vested interest in what happens within the reserve. In contrast, at biosphere reserves where rapid changes are underway, just as with conventional parks, it is necessary to identify and work with the local-level populations while simultaneously trying to control regional change, particularly those changes that affect the local context of resource use. However, this can be very difficult to do, as new actors and national and regional forces continually change what is possible, as illustrated in the Mexican case. Rapid shifts alter the tenurial relationship in areas, making it difficult to define who has which rights.

The typology in table 14-1 points out the minimum that conservationists can expect to undertake when beginning management at the four different kinds of sites. Often, the right kind of park will be in the right kind of place, and implementation, even if complex, is at least straightforward in terms of where to begin and what to do. Rio Bravo, Yanachaga, and Sierra de las Minas are examples of the right kind of park in the right kind of place—with no problems of origin or creation. In other places, the problems of origin and creation are so great that one of the main implementation challenges is to keep the basic elements of the park in place while lobbying hard for changes in government policy. Amboró is an example of the right kind of park, but its boundaries were in the wrong place. Machalilla is an example of the wrong kind of park, which has led to long-term hostility that probably could have been avoided if it had been decreed and implemented as a biosphere reserve from the beginning.

Park managers and programs such as PiP need to be sensitive to this context when they begin the process of implementation. Trying to promote the wrong kind of park in the wrong setting, or with inadequate political support, is likely to lead to substantial problems during implementation.

Resource Management

Managing resource uses is an integral function of both conventional parks and biosphere reserves. In the context of conventional parks, managing resource uses and managing threats are really two sides of the same coin. Conventional parks do not necessarily have to be concerned with resource uses outside of parks—while this concern is desirable, it only is relevant insofar as something can effectively be done outside of park boundaries. From the perspective of a biosphere reserve, resource uses are not necessarily threats, and the challenge is for park management authorities and biosphere residents to reach agreement on management goals. Common to both types of parks is the implicit assumption that core areas will remain intact only if uses in those areas are highly restricted.

If strong benchmarks are lacking on what resources are being used, by whom, whether they are needed, and what "need" means, then it is difficult, if not impossible, to design activities that promote sustainable use. To determine if extracting one product from an area is likely to be sustainable, it is essential to obtain information from a variety of disciplines, including: basic population biology and ecology (e.g., how many flowers can be removed before the gene stock weakens [Witkowski, Lamont, and Obbens 1994], or how many animals can be hunted sustainably [Robinson and Redford 1994]); sociology (what social institutions regulate use, will they be viable in the future); marketing (can this product be transported easily and efficiently, and who will buy it); and economics (what is the demand for a given product, and what will happen if costs change).

Particularly in the biosphere context, resident populations, whose uses are to be restricted, have a right to be involved in defining what sustainable means. Beyond their right to be involved is the necessity of their involvement. At biosphere areas, if the park management doesn't get local use modification, then the only management option left is defending the core. In such a case, the whole biosphere reserve concept is rendered meaningless.

Some parks have tried to achieve sustainable use by limiting certain types of technologies (see the Ría Lagartos case). Others have limited resource harvesting based on the season, the species, or even how the item is to be used—for example, subsistence is okay, but sales are not. Yet reaching agreement on what makes sense may be difficult, and the local populations may not always think that restrictions or mechanisms to limit use make sense. The Aché, an indigenous population in Paraguay, posed a series of questions when they were informed (rather than consulted) about their use rights. Their list of questions for park authorities highlights some of the difficulties of defining use rights, including:

- Who defines "traditional" methods of hunting, fishing, and gathering?
- Does the method (traditional vs. modern) that is used matter? Isn't the total amount that is extracted or consumed the key criterion?
- Can use rights be altered by outsiders or reserve managers?
- What levels of technical information are sufficient to inform the park managers' decisions?

- How are the terms "in danger" or "threatened" defined? Just within a part of the reserve, in the whole reserve, in Paraguay, or internationally? Over what time period? (Hill with Tikuarangi 1996).

These questions just scratch the surface of the issues regarding sustainable resource use. Coming up with measurable and verifiable rules is important, because the resource users—in this case, the park residents—have a right to be involved with and understand the implications of their actions. Yet coming up with such rules in the absence of baseline data may seem arbitrary and generate hostility. Biosphere reserves are likely to have substantially reduced biodiversity in the absence of clear social agreements when faced with mounting consumption pressures. Ironically, conventional parks have a much easier challenge in social terms—they need to understand the patterns and threats that resource uses represent, but they do not have to reach the complex set of agreements that biosphere reserves do.

Uncertainty and the Decline of Local Institutions

The clarity of resource ownership, including land, influences how resources are used. Ironically, the creation of parks has often led to uncertainty and lack of clarity over who owns and controls lands and resources. When an event, such as park creation or progressive implementation of rules triggers local uncertainty, people may begin extracting or claiming resources as fast as possible: for example, forest clearing may occur for resource extraction (sale of timber for profit) or to quickly plant a crop in order to expand land claims and/or demonstrate possession. This has happened worldwide near parks, such as in Royal Chitwan National Park in Nepal, where local residents who thought they would no longer be able to obtain firewood because of the park ignored traditional proscriptions regarding indiscriminate felling and began cutting trees and stockpiling fuelwood (Brower 1991). At Ría Lagartos, Machalilla, Amboró, and Corcovado, there was evidence that the creation of the park spurred increased activity for people to quickly exploit resources and/or expand their claims.

One assumption underlying many projects is that while people won't worry about management of areas viewed as commons (e.g., parks), they will manage privately owned land and resources effectively and take a longer-term time horizon into account, in part because uncertainty is diminished. This assumption is still being debated (Wachter 1992). What is clear from parks worldwide is that paper parks with no active management will often be viewed as "unclaimed" or open-access areas. Paper parks are particularly vulnerable if there is migration, new or improved access, or changes in production.

Evidence from the case studies, and elsewhere, indicates that strengthening tenure security, especially through park-based initiatives, may be an important way of increasing support for conservation and stabilizing changes in land and resource use (Richards 1996; Vandergeest 1996). Such initiatives may be a concrete way for conventional parks to reach outside of their boundaries, and for biosphere reserves to stabilize what is happening within boundaries. Parks in areas undergoing rapid social change gained special support when tenure was used to give long-term residents occupying lands adjacent to parks greater access to resources than was given to more recent migrants. For example, at Podocarpus, two large landowners sought to gain protected-area status for their lands as a way of protecting them against pressure from colonists, and Shuar leaders requested support to resist the invasion of their traditional lands by mining groups.

Another way to diminish uncertainty, particularly at biosphere reserves, has been to use tenure to differentiate "insiders" from "outsiders." For example, at Machalilla, local groups have used the park as a way of defending their interests against "outsiders," and park guards focus patrol efforts on defending timber from outsiders. In Sierra de las Minas, the NGO partner Defensores de la Naturaleza has helped indigenous communities that were displaced from their lands to secure land tenure. This has reduced uncertainty and gained goodwill.

Resource tenure is often overlooked until there is scarcity, and the case studies contain frequent references to scarcity, particularly of resources that are intensively used and are less available. When scarcity occurs, particularly when the resource is essential for livelihood (e.g., fish for fishermen, mammals for bush-meat hunters, medicinal plants for sale), there are often attempts by those who believe they have use or ownership rights to identify the reason for resource declines. Often, the immediate reaction is for one group to assert its rights and try and exclude other users. In fact, there are examples of lake reserves in Brazil created by "insiders" to guard fish from "outsiders" (McGarth et al. 1993). Those most likely to be excluded are recently arrived migrants, or other groups from outside the area.

Park managers can increase support for parks, and minimize local conflict, by helping to identify the reasons for scarcity; whether the scarcity is attributable to overuse; and, if so, by whom, as a result of which technologies, etc. In some cases, scarcity may be unrelated to use or may be exacerbated by use—but use won't be the primary cause of scarcity. Disease, changes in weather patterns, and cyclical population changes may be at the root of scarcity. Efforts to clarify resource scarcity and/or overuse need to be based on careful analysis, since the reasons are often much more complicated than they appear to be.

One reason for resource overuse is the decline of traditional institutions that have regulated resource use; apart from a brief discussion of the comunas system at Machalilla, none were mentioned. In many societies, local institutions have kept resource uses in balance and prevented overexploitation. These institutions vary from place to place but may be embodied in religious systems, regulated by traditional village leaders or a council of elders, or be part of an explicit management system specifically designed to govern resource use. Alternatively, they may be ingrained in the daily practices of certain indigenous groups, making it difficult to disentangle the institutions that regulate resources from other institutions that regulate people's lives. Much has been written elsewhere about the decline of these local institutions, and, as Berkes (1985) states, "Loss of community control over the resource base, commercialization, population growth, and technology change often occur simultaneously." For example, the cycles of migration and entry of displaced workers into fishing in Rías Lagartos and Celestún has vastly complicated the degree, type and extent of fishing in both lagoons and oceans, and all of the social changes cited by Berkes have taken place simultaneously.

The case studies point to the virtual absence of local organizations at many sites, a finding confirmed by studies elsewhere (Seymour 1994). In response, many of the Conservancy's partners have supported the evolution of new institutions, particularly at biosphere reserves, where local resource regulation needs are more urgent than at traditional parks. Such local institutions are more likely to arise in areas of rapid regional change, where scarcity may be more apparent and there is a greater need to define use rights, and also at biosphere reserves, where mechanisms to promote consensus at many levels—among use groups, within communities, and within the regional context—are needed.

The emergence of these new institutions offers a potential vehicle to regulate resource use, as described in the previous section, "Resource Management" (p. 426). Some key examples in the cases are the creation of:

- a federation of fishing cooperatives in four communities around Ría Lagartos that has become "the single most important organized force in fisheries in the state," creating a marine protected area that extends six miles seaward.
- the Confederation of Comunas from the South of Manabí, which unites some of the communities and comunas (see the Machalilla case study) around Machalilla National Park. (This organization grew out of one started by Fundación Natura but is now independent.)
- FECONAYA, the Foundation for the Conservation of the Yanesha Communal Reserve. While the Yanesha were initially hostile to the creation of Yanachaga-Chemillén National Park and Yanesha Communal Reserve, they now support the park and reserve; as the park and reserve become accessible to colonists, FECONAYA could be important in asserting local rights.

The emergence and short-term success of these new organizations and alliances do not mean that they will necessarily endure or that they will be supportive of park management objectives. Many organizations will be influenced by external factors (see the Corcovado case), such as the availability of outside financing and technical assistance. But they demonstrate that the potential for local institutions to emerge, manage, and control resource uses is greater when scarcity or pressure on resources is apparent and uncertainty threatens livelihoods. In such cases, the Conservancy's partners have increased their local base of support by helping certain groups reinforce local rights and claims.

Equity

Closely related to issues of management are issues of equity. Equity issues, or a sense of "fairness," can be extremely important in how local people respond to parks, perceived threats, and partners. Throughout the case studies, equity considerations, including insider-versus-outsider status, were underlying sources of conflict. Conflict over resources, who has what rights, and the levels and kinds of resource protection and enforcement that are desirable and sought from parks is often influenced by what is considered fair.

The answer at a conventional park is clear—everyone is considered an outsider. The context at biosphere reserves is frequently more complicated, with "levels" of insiders, depending on things like proximity to the core zones, relationships between scarcity and types of livelihood, and special rights for indigenous populations. These issues of equity are rarely considered in park management decisions, but they make a difference in how parks and partners are regarded by local populations.

Questions of equity embedded within the management context arise in defining protection, in defining threats, and in responses to threats. They also arise in perceptions of what uses are and aren't allowed. For example, from a management perspective, tourism may be regarded as a nonconsumptive use and therefore all right. But from an equity perspective, it is a use available to outsiders with leisure time and money. Within the context of resource protection, how a protection process is initiated, who it involves, and who it targets must either be equitable, or a justification for inequity should be explained to local or resident populations. For example, protection efforts that are initiated in one sector of a park but not another quickly seem unfair. Who is involved in protection efforts is important; it may be extremely difficult for the NGO that is promoting development activities and dialogue to also have a policing function. Actions to respond to threats also need to be fair; if a threat is rooted in multiple sources, targeting one group will be viewed as unfair. Unfortunately, in many areas the easy groups to target are local users. Their impact can often be significant, but they are rarely the sole source of threats. This doesn't mean that particular areas or groups shouldn't be targeted; what it suggests is that targeting must be strategic and the rationale should be shared as part of a process of local consultation.

Local perceptions of differential enforcement of rules can be at the root of intense conflict, particularly at a biosphere reserve, where all groups may expect to have equal access to and use of resources. A case from Sierra de las Minas Biosphere Reserve is illustrative. There was a strong process of dialogue with local communities and acceptance by local communities of the core area designation; however, when a wealthy logger with legal rights to logging in the core was not stopped by the reserve's NGO manager, Defensores de la Naturaleza, local communities quickly assumed that Defensores had lied to the community to protect resources for wealthy elites. Its resource guards were physically attacked and injured. Hostility ensued for five years while Defensores tried various approaches to have the legal basis of the permit withdrawn. In the end, cancellation of the logging permit and acquisition of the land by Defensores demonstrated to rural communities that the core zone was to be respected, and that unresolved legal issues, not wealth and influence, were at the root of the logging. This hostility and resentment took a significant toll on relationships and highlighted the perceived unfairness and inequality possible in access to and use of resources.

Even issues such as who guards are, what their role is, and where they are from have important equity considerations, although as noted under "Policy Pitfalls," such management decisions may be constrained by the policy context. At Corcovado National Park, park guards may be hired through the civil service or locally. Government financing cutbacks forced park managers to reduce staff, so the locally hired guards were fired, creating hostility based on the impression that the park was supporting outsiders and providing few local benefits.

Another source of conflict evident at sites was between large-scale extraction companies (e.g., mining, logging, fishing companies) and local users. For example, artisanal miners and employees of a large, multinational mining company clashed in the heart of Podocarpus National Park, at times even shooting at one another, over access to mining sites. At Machalilla, artisanal fishers blame the decline in fish on overfishing by industrial fishers, who frequently violate an eight-mile zone reserved for artisanal fishing. Equity concerns are apparent even within groups of illegal users, complicating management responses. At Corcovado, gold miners claimed that there were really two groups—those who mined there before the park, and more recent opportunistic migrants who started mining and hoped for a settlement from the government to stop. The "old-timers" felt that their legitimate rights had not been recognized but that "new" miners did not have legitimate rights and it would not have been fair if they were compensated. Yet the old-timers acknowledged that in the absence of a fair way to figure out who should be compensated and the method of compensation, there was strength in numbers. Therefore, they would not dissociate themselves from the other miners unless it was useful. Similar equity concerns arise for old and new fishermen in Rías Lagartos and Celestún.

Tourism at the case study sites led to problems of inequity and resentment at all areas except Rio Bravo, where tourism is tightly controlled. Similar problems from tourism and the distribution of benefits are found at other sites worldwide (Brandon 1996). Particularly when local people and communities do not receive tourism benefits such as employment, or tangible sources of revenue, tourism reinforces the insider-versus-outsider conflict. Even worse, tourism can negate the idea that parks are for biodiversity conservation, since tourism impacts are often in conflict with conservation objectives. When tourism or the needs of tourists or tour companies (e.g., a clean beach versus fishing and fish cleaning on the beach at Machalilla) are favored over local needs, this increases perceptions that parks are for foreigners or elites with disposable income, and for governments to collect revenue (Perreault 1996).

Other equity considerations revolve around the distribution of project benefits. Which communities or user groups should get projects, technical assistance, and/or financial resources? Both conventional parks and biosphere reserves target activities on a particular geographic area or set of communities, or target activities on particular user groups across many communities. Strategies that target one group or another without a process of consultation can backfire. While many equity considerations are site specific, general questions include whether benefits should be targeted for those who at the local level have the greatest potential role in degrading biodiversity. This is the strategy of many ecotourism projects, where the best hunters are given training on how to deal with tourists and turned into guides. From the standpoint of reducing hunting, this is a great strategy. But what signal is sent to others who respected rules and hunted only outside of parks or core zones? Should those who degrade the resources be "rewarded" by receiving project benefits and alternative employment (narrow benefit distribution)? Should benefits be distributed more broadly, which dilutes the overall level of benefit but increases the number of households with a stake in reducing threats? What if many of the threats to the resource base are caused by those individuals who are already better off, which is often the case?

At Machalilla National Park, development activities are underway both within and outside the park. Residents from communities outside the park perceive a preferential treatment compared to communities within the park (e.g., Salango). Should communities "inside" get different treatment from communities "outside"? Does the biosphere concept mean that these inside groups receive preferential treatment? Will this act as a magnet for migration or intensify activities outside in order to get benefits? Is it best to involve all key communities in some way, which complicates the logistics of projects tremendously (and increases the scope and scale while decreasing the intensity), or is it better to intensively work in a few areas? In the case studies, the way these issues have been dealt with varies substantially from place to place. How equity issues are addressed, and which groups and what actions are targeted, will largely depend on a thorough analysis of the social context and the ways in which park managers have chosen to respond to the aforementioned perils.

Catalyzing Development

Worldwide, parks have served as a catalyst for mobilizing political pressure on governments for reform or for increased access to services (Western and Wright 1994; Wells and Brandon 1992). Partners, and the local-level development work they sponsor, have acted as catalysts through their awareness of capital city politics and access to technical and financial support. The case studies and reviews of parks elsewhere indicate that the parks themselves are a magnet around which local interests coalesce—even if the local interest is figuring out how to change or get rid of park regulations. For example, unions increased their power base in the towns along the park's southern border by promising to take action against the 1991 boundary change at Amboró. Strengthened union power was instrumental in getting the government to renegotiate park boundaries, which attracted NGO programs. Without the existence of the park, CARE would have been unlikely to establish a $7 million development program in that particular area. The case studies show that the parks and the organizations working on their behalf are indeed attracting substantial technical and financial resources in areas adjacent to parks.

High levels of external technical and financial support do not always lead to local support for parks, or even development. In the absence of a unified coordinating authority, and when there is a high level of involvement from outside groups, a lack of coordination or of a common strategy among the institutions and projects working in those areas hinders conservation objectives. This appears to be of particular concern at areas where rapid social change is underway, such as at Corcovado, Rías Lagartos and Celestún, and Machalilla (see chapter 6, "Organizational Roles," p. 166).

While significant amounts of money and technical assistance are often used to promote development activities outside parks, there is evidence that they do not necessarily improve the odds for conservation of an area. One common assumption in the literature is that reductions in poverty around the periphery of a park will reduce pressures on the park, but experience from developed countries does not bear this out. The nature of the threats may change somewhat, but threats often persist.

From both social and ecological perspectives, the substantial investments in areas adjacent to traditional parks or within biosphere reserves may be a poor strategy, acting as magnets for migrants or creating growth poles (Brandon and Wells 1992). So what are the best land uses next to parks, and what kind of development should be encouraged? The answer to this is not as clear as most ecologists think. From an ecological perspective, the answer is to keep the areas adjacent to parks as similar to the parks as possible in order to extend ecosystem functions and stabilize habitats. There are several examples in the case studies, and many examples worldwide, where private game ranches, lodges, and ecotourism destinations are doing this. Where there is no land scarcity for agriculture, or if the land is inappropriate for agriculture, these land uses may be important in an overall conservation strategy. Whether the vacation homes springing up on the Osa, around Podocarpus, and at Rías Lagartos and Celestún are "good" for conservation will depend on how the land is maintained, who is displaced, and where displaced farmers go.

While an ecological perspective stresses keeping land adjacent to parks in similar uses, the case studies indicate that land uses that appear to conflict with parks may sometimes be best, particularly when those land uses discourage threats from logging, poaching, or encroachment. For example, at Rio Bravo, intensive agriculture by Mennonites surrounds part of the reserve, making it virtually impossible for anyone to gain access to the park from that side and virtually eliminating threats on that side. Other examples worldwide point to similar reductions of threats from golf courses, vacation homes, and intensive agriculture (assuming that the use of agricultural chemicals is controlled). While such uses may make parks into ecological islands (although one could imagine developing corridors), from a social perspective (depending on the site, of course) they provide clear demarcation of boundaries, stable land uses, and use of lands to their appropriate capacity. Socially, such clear and well-defined boundaries make sense and make enforcement virtually irrelevant. We need to acknowledge that there may be tradeoffs between what is best in ecological terms and expedient from a social perspective.

In promoting development adjacent to or near parks, it is important that a coherent strategy linking conservation and development objectives be specified (Brandon and Wells 1992). Figuring out what kinds of development make ecological, social, and political sense will be difficult and will require creative thinking. Sometimes uses that appear to be incompatible, such as intensive agriculture next to a park, may in fact be the best way to secure one section of a park. At times, the best kind of development may be promoting migration out of areas, by providing direct incentives such as voluntary resettlement (see the Sierra de las Minas case), or less directly, by providing education and health care and very little else. While this may sound unreasonable, it is a more honest position than promoting "sustainable use" without acknowledging that the low levels of use required for sustainability may not allow people to achieve the levels of development they desire. Ultimately, the levels of control over regional development, and the kind of incentives, infrastructure, and land uses promoted in a given area, will be highly dependent on the existing political will to support conservation.

Conclusion: Linking Conservation with Development

What parks are being expected to accomplish is extraordinary; and these case studies confirm that The Nature Conservancy and its partners are often doing the extraordinary. But should parks be expected to do all that is currenty being asked of them? For parks where the mission has become human use—or more vague but charismatic statements such as "parks are for people"—what will happen to resources as they are exploited? Who will balance biodiversity protection goals with revenue objectives? Who gets the benefits, and how will this be decided? Is it reasonable to expect parks to be the driving force in promoting regionally oriented sustainable development?

Two key challenges for conservationists are to change the expectation that parks are supposed to be the cornerstone of sustainable development activities and to refocus their attention and actions on biodiversity conservation. Meeting these challenges will occupy conservationists on a full-time basis. Meeting the challenges outside of protected areas is best left to professionals from other disciplines with expertise in rural development. Their challenges are also twofold: first, to intensively promote sustainable development initiatives outside of parks and, second, to focus renewed attention on addressing the fundamental economic and policy incentives that drive unsustainable land use and management.

This chapter offers lessons to conservationists in refocusing their attention on biodiversity conservation. In particular, it demonstrates that some key actions can help park management more effectively promote biodiversity conservation:

- reclassify parks and park boundaries, depending on circumstances of origin and park definition, to ensure appropriate parks given the existing social context
- review key management actions based on park type and social context (table 14.1) and implement corresponding actions
- promote policy changes (civil service, logging and mineral concessions) to give park managers authority within park boundaries
- initiate local participation appropriate to the type of park—particularly at biosphere reserves, this means high levels of local involvement in zoning, monitoring, and management
- clarify land and resource tenure in or adjacent to parks to minimize uncertainty that leads to unsustainable use
- be explicit with local populations over management decisions that may have implications for equity concerns
- don't assume that rural development will lead to conservation or, conversely, that more intensive land uses adjacent to parks are always "bad"—conservation opportunities arise in unexpected ways.

The institutional capacity to coordinate the separate and integrated programs necessary to manage biodiversity, or natural resource, policies is lacking in most countries. Traditional sectoral approaches to development inhibit the formulation of policies and programs necessary for sustainable resource utilization. Ecological consequences of economic mismanagement are overlooked because they fall between ministries' portfolios.

Since biodiversity is only a subset of natural resource management, including land management, it is generally disregarded or accorded low governmental priority.

Parks are one vital part of the solution to save biodiversity, but greater attention must be given to improving land uses elsewhere within countries and to conserving biodiversity on lands outside of parks. Broader initiatives to encourage conservation on lands outside parks are essential if parks are to endure. Such initiatives must start with approaches that are realistic in the context of the socioeconomic development aspirations of developing countries. This development context will ultimately determine whether biodiversity is conserved or lost.

15. The New Politics of Protected Areas

Steven Sanderson with Shawn Bird

Unless the debt crisis, privatization, economic liberalization, domestic structural adjustment, and political democratization turn out to be just routine elements of social life, the macropolitical and economic ground is trembling underneath conservationists as we face the twenty-first century. The traditions of preservation and conservation, steeped in internationalist NGOs, guided by paternalist states, and regulated by the increasing presence of market and development, are all in question. So are the principles of economic development, now dubbed "sustainable development."

The many strains of thought in global conservation show that tectonic shift. From the IUCN Sustainable Use Initiative to *Caring for the Earth,* (IUCN/UNEP/WWF 1991), "protecting nature" has taken on an increasingly human form. And appropriately so, since research indicates that for the first time in the history of the biosphere, terrestrial ecosystems are practically all "human managed," directly or indirectly. Whether we take Vitousek's (1994) stylized fact that about 40 percent of net primary productivity is under human management, or even more global inferences suggesting that local environmental change is *global* environmental change, the earth is being shaped by human hands.

Parks are evocative examples of this human management. Despite popular conceptions to the contrary, parks are far from natural. They are constructed—politically, geographically, even ecologically. Human populations inside and out turn biota into resources (Sanderson and Redford 1997), regulate the natural events that shape and structure ecosystems (for instance, by regulating fire outbreaks or spread, or flooding), and turn politically convenient spaces into ecologically important sites—all by human hand.

Not surprisingly, those shaping hands are wearing political gloves. This chapter deals with the ways those forces make their presence known in protected areas; for the most part, specific examples of the Parks in Peril program can be found in the case studies and synthesis chapters found elsewhere in this volume. Hopefully, some of the observations are relevant for conservation politics writ large.

Parks As Islands

Images of global integration notwithstanding, and the loss of pristine wild areas (Denevan 1992; Redford 1992) acknowledged, the social constructions we call parks and protected areas are rather like islands (Freemuth 1991). They are mapped as such, and in the process the mapmakers zone the world away by erecting boundaries between parks and the outside world. In large measure, that is what protection is about. In that process of creating boundaries, several secondary political questions arise, which are strikingly familiar to students of nation building since the breakup of European empires after World War II.

The first, and most obvious, is who deserves—or receives—citizenship on these islands? The days of expelling resident populations from parks without compensation are thankfully on the wane, though not yet gone entirely. A central constitutive principle of parks today has to do with who has standing as an islander/park resident. The habits, rules, and entitlements of the island are different from those of the mainland, and islanders have both the prerogatives and burdens of being political curiosities. Islanders may also include nonhuman species, but they influence action only indirectly, as they are objects, not subjects of protection.

The intriguing possibility of assigning "citizenship" to nonhuman species is weakened by their lack of organized action or voice (political volition), which is a central requirement of political participation. But in a variety of ways they demonstrate their functional importance to protected areas, so that humans can represent them in ways familiar to the old politics of guardianship over indigenous peoples, remnants of which still survive in Latin America. A more fulfilling idea is to represent certain keystone elements and structuring processes in a protected area (Redford et al. 1995).

Usually, citizenship is conferred reluctantly; in fact there is a long and mistaken tradition of expelling residents from parks. When residents do win their place in parks, it is usually—but not always—on the basis of their social status. Obvious examples include indigenous peoples, a designation that itself can be created or modified, as was the case in the petition of Afro-Colombian communities for indigenous status following the promulgation of the 1991 Colombian Constitution, which gave indigenous peoples preferential claims to reserves (*resguardos*).

Likewise, some long-standing residents, such as rubber tappers in Brazilian extractive reserves, or *caboclo* or *ribereño* peoples in flooded forest, base their claims on tradition, not indigenous status. Developmentally "desirable" (or unavoidable) populations of more recent vintage fill out the category of the potentially entitled. Interestingly, these are often "problem populations," created by the state itself through colonization efforts or land reforms, which then turn out to be problematic for conservation goals. The era of Latin American land reform in the 1960s and the more spectacular and well-known Amazonian colonization schemes left such people scattered throughout the countryside. Once there, they became "development clients" for various state and international development agencies.

How human islanders are distinguished from human outsiders is a deeply political process. The rights citizens have vary dramatically from country to country and park to park. In Del Este, residents can live in the park, but there are restrictions on all kinds of rights, such as new housing construction. Residents of Saona were permitted to remain on the island but were similarly faced with severe restrictions on land use. The construction of permanent buildings and infrastructure on the island has been prohibited since the park's creation, leading to a shortage of housing to accommodate intrinsic population growth or tourism development. Expansion of lands dedicated to agricultural and livestock production is also prohibited. The rights at Sierra de las Minas are more complete but are conferred through a participatory zoning process.

The zones in Sierra de las Minas are typical of parks and biosphere reserves. The *core* must be used for the preservation of the natural environment, biological diversity, water sources, scientific research, and ecotourism. Extractive activities and human settlements are not allowed, with some exceptions. The *sustainable use zone* forms a narrow ring around the core. "Sustainable use" of the forest and other natural resources is allowed, but the establishment of permanent human settlements is not. The outer ring of the reserve is called the *buffer zone*; it contains about 110 human settlements, which include villages, ranches, and co-ops. The objectives of this zone are to achieve the sustainable use of natural resources, especially through traditional activities that benefit both the local communities and natural resources.

In fact, there is nothing more fundamentally political than conferring these zones and the political status of "local community" on people. For the state agency, environmental zoning determines social status. In turn, special claims, such as indigenous status, determine zoning. Sometimes, NGOs operating under state charter zone people in or out. Legal access to natural resources is the result of overall zoning norms. Sometimes, the results are most unhappy on the fringes of the park, as when border residents suffer depredations from park animals but have no rights to park resources. Other times, populations on the fringe act as intermediaries between park and nonpark. In the end, it is less about "use" than it is the legality of use, which puts the outsiders in charge of life inside protected areas.

It is not as if these processes are "objective." Status confers privilege, and both are constructed and argued over in a socially defined space. Winners in one time will be losers in another. Indigenous peoples were viewed as "ecologically noble," "primitive," "pre-human," or "pre-modern" within the political lexicon of, say, conquering imperial Spain. This view did not require consultation with the peoples of the new world, of course, though a discourse with them would have been unthinkable without a common language (Pagden 1987). Despite the renaissance of pro-indigenous politics in modern Latin America, in one way little has changed: Outsiders still determine to a significant extent who insiders are (Spooner 1987), how they are viewed by the world (Adams 1996), and whether they are viewed positively (as are indigenous peoples these days—and it is well to remember that this is a recent reinvention of the image of indigenous peoples) or negatively (as are colonists—again, recently) in the discourse of conservation.

Importantly, and sadly, even in the countries most committed to indigenous rights and participatory development strategies—Colombia again comes to mind—local and indigenous peoples are virtually dictated to as far as the concepts, strategies, and goals of sustainability are concerned. Agenda 21 is a top-down document and an international discussion played out in countries and regions but virtually without representation by local peoples and nowhere questioning the virtues of the UNCED vision of sustainability.

This shortcoming makes more important the question about how a progressive park policy would link the island with the outside world. We might think of the link as a bridge or ferry, offering limited access, relatively tight control, and a clear distinction between islanders and visitors. Literally, in some cases, outside conservation agents invite insiders to protect gateways to parks, to arm themselves against "poachers," "marauders," and other interlopers. In Central African preservation, often it seems that one population of local people is in fact much like another, but some have been diverted to one side or another of an ever changing game of good and bad, whose rules have been made up and modified mainly by Europeans, Asians, and Americans. In the PiP program, this game is evident, too; for example, the Mexican government has changed land tenure rules in now protected areas first with the national agrarian reform, then with parks and protected-areas legislation, and, finally, with a change in overall property rights legislation at the national level. To say that residents inside or around protected areas are "participants" in this process is to elevate democracy to levels unheard of in most countries.

Slightly less intrusive than the bridge is the ferry, in which the remoteness of a park is preserved by a lack of intrusive infrastructure. So park insiders are marooned like so many islands in Lake Titicaca, with tourists coming to visit the islands and their denizens, often viewing them as artifacts of a lost world. The ferry is controlled by the ferry master, not by the communities he visits. The price of isolation is lost opportunity, a bargain easier for an outsider to make than an insider. Sometimes the islander becomes or mimics what the visitor wants to see: Amazonian indigenous peoples put on shows for visitors, and locals parrot the international conservationist discourse in ways that are amazingly adaptive. Of course, communities are increasingly using parks as a way of keeping outsiders out, if that is possible. Unfortunately, it is not often the case that the park is isolated enough or clearly enough separated from the general population to permit such quarantines.

Variability in the System, Consistency in Goals

Suspending for a moment the infelicities of being on either the inside or the outside of such a park, what do "we" want from a park? This, too, is a political calculation, with no real objective function, especially in specific histories of parks in the worlds we know. Resident populations want to "use" the park to live in—though that use may be dressed up in different ways on different occasions. How could one imagine otherwise? If such a question were posed to any resident of any place, the answer at some level would have to be the same: resident populations have to use the resource in ways and degrees that sustain their livelihoods. Such uses change, too, as demands change within the community; small, low-output communities are never immune from social change, and their "consumer preferences" vary over time and with external influences.

The utilitarian aspect of land use does not *necessarily* connect local populations' visions of need to outsider notions of development, but the terms to describe residents' needs have been close enough to invite the developmentalist community in, where they have stayed now for some time. Outsiders have determined that all individuals are rational actors at heart, and that within each resident beats the heart of a consumer, whose preferences can be modified in ways that suit the development agenda. The incommensurability of this "dialogue" between residents and the outside world is never addressed, and surveys of the goals and hopes of resident populations rarely make up the foundation of park or regional planning.

So, in a short span of a decade or two, the answer to "what we want from a park" boils down to a reprise of the battle between preservation and "wise use" that took place in the United States at the turn of the twentieth century (Worster 1993). But aside from the obvious use values from a resource, and the aesthetic values of a precious environment (Norton 1992), the antagonism between preservation and use issue, has not been addressed in park protection.

So far, we have not discussed who the "we" is in the above discussion. In fact, parks have multiple partners, multiple clients. Sometimes "we" means local residents; other times, outsiders. "We" can be personified by local government, NGOs, or local communities. It can even be the U.S. government, which through USAID has allocated millions of dollars to parks in Latin America.

Of course, the goals of a park are related to the timing and motivation that inspired its creation in the first place. As Brandon shows elsewhere in this volume, parks in the PiP program were created because they were either relatively untouched, in danger of being transformed, or taken up by innovative actors. This "organizational ecology" should not be surprising. First conditions help determine future paths. So, when in the early 1970s a military leader who liked to fish for the big predator *paiche*

(Arapaima gigas) began to worry about the preservation of part of the Peruvian Amazon, a process was begun to create the Pacaya-Samiria conservation area, with legal status, zoning, rules, and—a park to be defended. In Costa Rica, when an unclaimed stretch of closed-canopy tropical forest has offered an opportunity to create a protected area without opposition, NGOs or government have reached out to establish parks. So the land becomes ecologically valuable because the protected area has been established.

In fact, the political geography of parks in Latin America displays interesting characteristics that make parks look more political than natural in their occurrence: Argentina's parks are mainly on its borders; in Brazil, older parks are near consumer populations, while new parks and protected areas are in the remotest parts of the Amazon; in Costa Rica, any patch of "natural landscape" still intact might find a bidder interested in making it a park. Patches that elsewhere are ecologically trivial are "significant" in Costa Rica.

So what "we" want from a park is a statement of intent to preserve or protect. But what is it that we are trying to protect, anyway, and what makes us think a park will protect it? These questions lead directly to park design, politics, and policy: what is to be protected?

This central question has been answered in the most general and tautological ways, by saying that we protect what is in the park. By zoning an area for protection, we protect what is in the zone, and maybe some collateral areas around the zone.

Many—perhaps most—parks have no initial ecological assessments worth the name, much less a serious appreciation of the dynamics of the ecosystems within the park boundaries. In some cases, as in Yellowstone Park, one of the putative protection criteria is a minimum viable population of grizzlies (Grumbine 1992). In this and many other cases—savannah Africa, for example—charismatic megafauna define park boundaries and politics, even if only in a symbolic way.

Sometimes use of keystone or indicator species has special merit—for example, when the objective of a park is to protect frugivores that are critical forces in seed dispersal. According to recent research, by focusing on primates in the Kibale forest of Uganda, one can protect animals that represent a disproportionate influence on seed dispersal, and whose decline represents a clear threat to the long-term diversity and resilience of the forest (Chapman and Chapman 1995). But what of a protected area that is created because of (as distinct from "for") a large predator, and no follow-up research determines the role of the park in protecting that species; or the relative importance of the species to the park, compared with, say, other prevalent species; or the importance of the park relative to other areas that might "compete" for protection?

The simple conclusion is that without clear ecological assessments and strong scientific understanding of the dynamics of a park landscape, it is nearly impossible to tell how the park is doing, or how important the park is, much less how policies are affecting the state of the park in its various facets. Yet, in the coarsest way this "anticipatory protection" is vital, since the information needed to do the serious ecological assessments is very hard to come by and will be a long time coming. Ironically, if conservation were to wait for such assessments, protected areas might be less valuable than they are today. In the cases found in this volume, only in Sierra de las Minas and Yanachaga were sound ecological assessments a part of park establishment. That Sierra de las Minas is the most recently established park offers the hopeful sign that we are learning.

Nevertheless, the scientific foundation of conservation should not be lost in politics, nor should the political value of science be underestimated. To understand the long term values of conservation in protected areas, proper ecological assessment is essential. In fact, of course, ecosystem processes and the goods and services they produce are widely different across systems, and within systems across key variables. The meaning to the overall condition of an ecosystem of a life history, a local extinction, a fire event or a flood, prey–predator relationships, forest age, or patch mosaic, among many other variables, differs according to the resilience, the phase, and the impact of human disturbances. To a floodplain, a flood is necessary, but its incidence, interannual variability, nutrient load, and other factors provoke different responses among human populations equipped with a "tool kit" of adaptations cued by the flood itself (Padoch and deJong 1987, Coomes 1996). In turn, human populations affect the flood, its reach, its intrusion into their lives, creating landesque capital that can even change the course of rivers for human purposes. The human–nature interaction defines the adaptive context of park conservation or degradation. To ignore this interaction in favor of political expediency is dangerous.

This in itself is not novel, but human policy responses beyond the local scale and beyond the ambitions of "low-output" or subsistence populations tend to be insensitive to ecological variability in parks. Simply put, park management seeks a stable state. As the seemingly inexorable rules of Weberian organization work their way, human policies tend to gravitate toward single equilibria, not multiple equilibria or multiple stability domains (Holling and Sanderson 1996). Parks in the context of modern society and hierarchical organizations are objects of a "one size fits all," or template, approach. Extractive reserves in the Western Amazon quickly became the darlings of international sustainability advocates, and efforts were quickly made to put together extractive reserves all over the Amazon, in Mexico and Central America, and in Southeast Asia. Many of these efforts were without appreciation for the complex historical, institutional, political, and landscape contexts important to the first reserves. Needless to say, studies were not done to assess the viability of even the most successful reserves over the long term. This same "rush to adoption" of certain policy packages now haunts micro-lending, which based on some early successes in South Asia is now quickly and rather uncritically adopted by development agencies and grassroots development NGOs.

The point is that the likelihood that a park can accommodate what "nature dictates" with all its variability is unlikely, for two reasons: first, the park is not what nature dictates in the first place; and, second, modern human management devices tend toward the blunt instruments of policy we know today. Nowhere is that bluntness more evident than in the quest to provide secure land tenure as a path to park policy.

The Magic of Tenure

Along with extractive reserves, micro-lending, and a long list of other design "winners" in conservation and development, tenure security has been seen as a magical intervention to guarantee better environmental outcomes. Most generally, tenure security is recommended with the idea that people will invest more in resources that are "theirs" and degrade them less. Some students of sustainability have taken the tenure issue further and said that privatization of property rights is preferable (Southgate and Whitaker 1994); much foreign and national development assistance has pushed the privatization agenda. One counter to this assumption relies on the idea that high levels of uncertainty (rather than certainty) generate predictable behavior by limiting actors' flexibility (Heiner 1983, 573). Of course, this says nothing about the merit of that predictable behavior, so it could as easily imply deforestation as forest stewardship.

The magic of tenure security is based on a false premise of tenure homogeneity—i.e., that everyone's tenure is the same—which is revealed most clearly in parks and protected areas. In virtually all of the Parks in Peril cases in this volume, and in many other parks throughout the world, tenure forms are highly variable (land tenure, resource tenure, and wildlife tenure are all different) and nested (resource tenure does depend on land rights to some extent). Tenure has different goals for different actors, too. In Colombia, a national protected area may overlap an indigenous resguardo, which itself assigns tenure rights differently within the indigenous community, depending on whether the resource is agricultural, hunting and fishing, or other extractive (e.g., minerals and forest products). Seasonality also determines tenure to some extent, so that one (harvest) year's claim on land for agriculture may yield another year to other purposes, or to none at all.

Some resource uses have little clear tenure, as is the case with clearing back swamps of *aguaje* (*Mauritia flexuosa*) for fruit and wood for grub ranching. Some forms of tenure are tied to land; others less so (Naughton-Treves and Sanderson 1995). Interestingly, where any tenure is established, however unclear, it tends to push unregulated-use practices outside the tenure system's reach. So, where wildlife or forest resources are regulated, users may diminish their use in favor of unregulated resources, such as fisheries. Such is the case in Del Este.

Tenure is not homogeneous, it does not revolve around purely economic models of transactions, and it does not necessarily yield better environmental outcomes under private regimes. Or at least the evidence is not strong or systematic enough to reach such categorical conclusions. Tenure varies according to the nature of the resource (land, wildlife, minerals, genetic material), the nature of the economic good (private or public), the culture of community rules (as in the resguardo case, indigenous vs. modern state regimes), and the heterogeneity of the park itself.

Tenure is not only quite heterogeneous within patches and communities, it is historically and hierarchically "nested." For example, in Amboró, a paper park had one set of rules, affected later by a new environmental law, in an area divided by two different land tenure systems—all being pushed about by new colonization efforts and external NGO pressures for sustainable-use regimes. In Rías Lagartos and Celestún, ejidos were the social (and tenure) invention of the Mexican revolution, based loosely on precolonial forms of communal landholding. Later, the national government dropped a park on top of the ejidos, with different rules—somewhat akin to national political elites declaring a set of environmental standards and anointing indigenous peoples or long-standing residents as their stewards. (This begs for a comparison with the current debate on "traditional" forms of landscape management vs. state management vs. new community-based management in savannah Africa.)

In Guatemala, land tenure is insecure, made more so by years of population movement related to the guerrilla insurrection, the settlement of the Petén, and changing economic opportunities, which translate into changing labor markets. There, and in relation to Sierra de las Minas, the discussion revolves around "controlling" population movements and "securing" land tenure or making it more "regular." In this case, though, land tenure is the dependent variable of such macroeconomic and macropolitical forces as war, peace, and economic change. While Costa Rica has not suffered the direct effects of war, its landscape and rural population have been transformed by economic change over the post–World War II era. Its people, too, are "tenure insecure" and mobile. The idea that park policy will drive the fundamental forces behind land tenure is strikingly optimistic against such backgrounds of differing political histories.

Wherever tenure is a consideration, the ad hoc, "on the fly" quality of tenure claims around a park would test any model professing to claim that one or another tenure system produces a given outcome. Tenure security as a variable leading to better park protection should continue—not as an article of faith but as an interesting hypothesis, treated skeptically by policy makers.

Who's a Partner?

In the last decade or two, governments, markets, and NGOs have all lined up with different solutions to the institutional requirements for conservation. As perhaps the most prominent theorist of such institutions has argued, advocates try to resolve the problems of management robustness and salience by getting prices, policies, or institutions right (Ostrom 1988). Advocates of government-based decisions have criticized the market; market advocates have declared public intervention to be "perverse"; and advocates of local institution building have tended to concentrate on the local scale as if it were relatively autonomous—or insular.

Lost in the shuffle are two central concerns of the politics of conservation: first, making the increasingly evident forces of national and international integration work for, rather than against, sustainability; and second, looking for strange bedfellows as a matter of everyday conservation practice. Let's take the latter first.

It cannot be surprising that in a politically and socially constructed park dialogue, friends and enemies are declared. The state is declared to be friend or enemy, but mainly enemy in the era of NGO ascendance. The NGO is friend, by disposition, not evidence. The jury is still out on the market, because it does well at allocating resources by exchange, but it says nothing about distribution, and that is what access is about. It is easy to find conservationists and conservation analysts who will offer up stylized statements about government, market, and NGOs in talking about "their" landscapes or species or systems.

Such stereotypes are fatal to conservation, and thereby to parks, in a period of rapid change. In an era of state decline, such as the period since 1982 in Latin America, little has been done to build capacity in the state apparatus, so the state in general suffers the upside-down problem of its own agencies exercising more autonomy than the state itself (e.g., Petroecuador vs. the Ministry of Mines and Energy, or vs. the government's general incapacity). Little fiscal or political authority has offered room for the state to reflect anything other than traditional power constellations and the inequities of society (viz., expropriations in the Dominican Republic to create parks). Multi-ethnic societies do not necessarily yield multi-ethnic politics—compare Bolivia, Peru, Mexico, Guatemala, and Ecuador on indigenous rights.

What is remarkable about the PiP program is that the discourse is about not enemies but alliances. The partner NGOs are more interested in roles than in blame. Sometimes NGOs take over government roles in the absence of strong government presence. Once again, the "we" in the equation varies by origin, interest, and level of organization. The interesting point is that PiP has created a mechanism whereby different constituents with sometimes conflicting goals and interests come together around conservation.

This is one of the most critical political qualities of parks programs and is often overlooked. One of the central political questions for parks management has to do with the creation of novel alliances for conservation. Yet the standard literature treats markets, states, and NGOs as if they were intractable enemies. PiP shows that, in fact, often governments, NGOs, and commercial considerations can go hand in hand—not smoothly, not without contestation, but forward in the service of conservation and development.

This happens best when autonomy over the parks does not reside at only one level. The national government is not close enough to the ground, and it has too many interests at play to be faithful to the park. Park peoples have their own interests, which are not always consistent with conservation; and they are hard pressed by external forces, including population and development pressures. Parks in Peril shows us that external support is important: USAID and The Nature Conservancy are fundamental to the success of the program, both organizationally and financially. In a time of government retrenchment, we cannot lose sight of the importance of external support. Yet that support can go for naught, as so many small-scale development experiments have shown through the years. Strong local organization is equally important.

The truly novel element in the 1990s is the growing strength of "devolution," or the return of political initiative to decentralized units of governance and community initiative. This movement exists from South Africa to India and certainly pervades the PiP experience. But to frame devolution with a spirit of conservation and development is a difficult task. It has succeeded most in the PiP cases where it has invited new coalitions of partners, from all sectors. Even so, this recognition of new possibilities is more of a promise of future politics than a description of today's conditions.

It is surprisingly unclear whether democratic governments at the national level make a positive difference in local-level resource management. Much, much more needs to be written about the connection between national-level politics and local-level sustainability, and the cases in this book do not speak to this issue directly. What is important is to find the cross-scale linkages between government at the national level and local-scale resource use, whatever the national regime. And, one would hope, those linkages over time would sustain a strong tendency toward democracy from the local level up, finally establishing a robust connection between democracy, sustainability, and conservation.

Conclusion

In the end, parks may be islands, but they are, in fact, under siege, as Freemuth (1991) has said. The forces of internal change, lack of clear ecological assessments, tenure problems, population pressure, and equivocal state policy all converge on parks to make their survival more problematic and perhaps less important in terms of their impact on biodiversity. The thesis in the introduction has been borne out in the political realm: Parks are the repository of too much environmental attention, without the corresponding effort outside the park. So, when they fail, parks can be the excuse for political systems that have given up on conservation—often without a proper fight.

PiP tells us that for parks to "win," local organizations must be strong, national policy upright, and intentions virtuous. The international economic system must be held at bay, and decision makers must understand the dynamics of ecosystems over the longer term. And politicians under pressure must avoid the sirens' call of easy tenure solutions and "one size fits all" remedies for protected areas.

The politics of protected areas could not be more complex. Each of the cases in this volume is nested within a very complicated domestic political system that has changed greatly over the past two decades. So summary assessments are difficult and necessarily overly general. But if programs such as PiP do not exist, there is little reason to hope that the concatenation of local, national, and international politics can take place with a single-minded focus on a park and its people.

16. Holding Ground

Kent Redford, Katrina Brandon, and Steven Sanderson

People should have no illusions about the severity of the problems protected areas will face in the coming years. The conflicts of tomorrow will be even more difficult than those of today, as resource scarcity, climate change, economic imbalance, population growth, expanding consumption, and continuing use of inappropriate technology form a witch's brew of challenges to protected areas, and to sustainable use of the environment as a whole. But such challenges mean that protected areas have an even more important part to play in securing a productive future for the people of our planet.

—Jeffrey A. McNeely, Jeremy Harrison, and Paul Dingwall

This book arises from the tension between two sets of conservation practitioners. One group is well represented in the published literature. Often contentious in their tone, these writers are only remotely connected to on-the-ground implementation. The other group is a mirror opposite: frequently mute in the published literature, intimately involved in trying to achieve on-the-ground park protection, and little aware of the debate swirling around it. Our volume is designed to bridge the gap between the two groups by sharing the experiences of practitioners working in a large multinational effort to conserve biodiversity in Latin America and the Caribbean, in the hope that this will: (1) improve the chances for conservation success at other sites; and (2) illuminate a debate largely based on opinion and anecdote. We are convinced that the debates surrounding park-based biodiversity conservation must move beyond general rhetoric to case-based experiences and robust hypotheses. The fragmentation of conservation described above must end.

This book also stands squarely in defense of protected areas, embracing the premise that protected areas are extremely important for the protection of biodiversity. Defense of protected areas must include a recognition that requiring them to carry the entire burden for biodiversity conservation is a recipe for ecological and social failure. Yet burdening parks with an overwhelming set of social goals has become all too commonplace. The 1992 World Parks Congress set the achievement bar at an impossible height for parks by stating that "protected areas must be managed so that local communities, the nations involved, and the world community all benefit," and, further, that parks should become "demonstrations of how an entire country must be managed" (Barzetti 1993). Ghimire and Pimbert (1997) continue in this vein, crippling parks by assigning them a Sisyphean task: "Conservation programmes are *only valid* and sustainable when they have the dual objective of protecting and improving local livelihoods and ecological conditions" (italics added). This is clearly not what parks were established to do—it is only after the fact that parks have been assigned the role of achieving a whole range of social objectives and then been judged harshly for failing to achieve those objectives.

Parks were established to protect nature in one of its many guises. Creating and maintaining parks is a social value; the observation that parks are a "social space" (Ghimire and Pimbert 1997) is a vital recognition of this. Indeed, "parks are political creations that are physically bound to greater ecosystems and emotionally tied to human events" (National Parks and Conservation Association 1994). Parks and protected areas have their advocates who defend the social values underlying their creation. Over 25,000 protected areas have now been established, covering over 5 percent of the globe. Only 1,470 of these are national parks; the others include game reserves, watershed-protection forests, indigenous reserves, and recreational forests (McNeely, Harrison, and Dingwall 1994). In the Latin American and Caribbean region the most dramatic growth in protected-area systems took place in the 1970s and 1980s. Seventy-six percent of Central American protected areas were declared in the 1980s; 65 percent of the Caribbean areas were declared in the same decade; while in South America, 38 percent were declared in the 1970s and 37 percent in the 1980s (McNeely, Harrison, and Dingwall 1994). Worldwide, parks are still being actively created: Russia has actively increased its parks system (Stepanitsky 1996); Germany recently created several new parks (Bruggemann 1997); and there is talk of expanding several protected-area systems in South America.

Not only are new parks being created, but new kinds of parks are being created as well. World Heritage Sites, one type of protected area, are now being designated not only to protect significant natural areas and historical sites, but also to protect cultural landscapes and heritage, such as rice terraces of the Philippines or historical pilgrimage routes in Spain (IUCN 1997). Protected-area systems are being expanded to protect everything that humans want to shield from change, a daunting task in an increasingly changing world. Using parks in this way further dilutes their mission of nature conservation.

Despite the outpouring of interest in creating parks of all sorts, many people loudly condemn them as a failure—but a failure measured in a currency that was not part of the reason for their creation (cf., McCloskey 1996). Parks were designed to preserve nature, not to "cure structural problems such as poverty, unequal land distribution and resource allocation, corruption, economic injustice, and market failure" (Brandon, chapter 14). These competing demands have predictably led to wild confusion, with park managers and advocates trying to promise that their sites will be all things to all people. This cannot yield a stable long-term solution. Many of the criticisms against parks have become embedded in clichés whose use has begun the process of "discourse truncation" (Kuran 1993), limiting the ability of all parties to address the roots of disagreement and conflict. Below we list several of the most commonly heard of these clichés.

We follow with a set of generalizations that emerge from the cases considered in this volume and from The Nature Conservancy's work more broadly considered (cf., Weeks 1997). We offer these as the beginnings of what we hope will be a robust discussion about the value of parks, which when combined with the statements found in Ghimire and Pimbert (1997), Kramer, van Schaik, and Johnson (1997), Brandon (1997), and McNeely (1995), outline a broad ground for discussion. It is worth emphasizing that many of the general conclusions reached by Ghimire and Pimbert (1997) are contradicted by the results of these case studies. In these contradictions lies much fascinating analysis waiting to be done. If we are to pursue an open dialogue about the values of parks, both good and bad, then we must openly represent the full range of opinions.

Conservation Clichés

"The parks frontier is closed." According to the logic that produced this cliché, empty spaces are gone, so there can be no more parks created. But, increasingly, we realize that there was very little empty space to start with, and that parks and other types of protected areas have almost always been created on top of existing populations or areas used by someone. When this cliché is used, it is often in a hopeful sense—hopeful that the political will does not exist to generate new parks in areas occupied or claimed by people. Yet recent statistics show that the number and extent of protected areas created in 1990–94 exceeded that of any previous five-year period (WCMC, cited in *Oryx* 1997).

"Empowerment of local communities will save more biodiversity than will parks." This cliché is based on the assumption that there is such a thing as local people who operate in a cohesive community fashion. All too often this is not the case (Agarwal, in press). As Borrini-Feyerabend (1996) states "Communities are complex entities, within which differences of ethnic origin, class, caste, age, gender, religion, profession, and economic and social status can create profound differences in interests, capacities and willingness to invest in the management of natural resources." It is clearly not that communities are "bad" but rather that they must not be stereotyped. Some will actively work to conserve some components of biodiversity; others will not, and have not.

"People have created biodiversity, so they are essential to its survival." As with many of these clichés, this one contains a grain of truth. Biodiversity is a social invention; people are its inventors as a meaningful concept. However, that does not mean that manipulation of biodiversity leads to its conservation. Furthermore, this cliché erroneously assumes that human influence in the selection of certain species and the structure of certain ecosystems has resulted in changes that would not be maintained in the absence of humans. It further incorrectly assumes that the sort of selection practiced by earlier human generations continues to be practiced by contemporary peoples.

"Biodiversity per se can be both used and conserved." The term *biodiversity* has very frequently been appropriated from its biological roots by political actors less interested in conserving the biosphere than in who gets to use the biosphere, under what property rules, and with what allocation of the losses and gains from use (Sanderson and Redford 1997). As a result, it is

used as a monolithic term in phrases such as this one, which ignore the fact that biodiversity has different components (genetic, population–species, community–ecosystem) and different attributes (structure, function, composition). Each one of these components and attributes is differentially affected by different types and intensities of human use (Redford and Richter, submitted). Ignoring the complexity of the term allows the politically expedient conclusion that humans can both use and save "biodiversity." The power (and danger) of this cliché in the parks arena is demonstrated in a document produced from a meeting of representatives of the park systems of fifteen Latin American and Caribbean countries, which contains the statement, "Little by little it is being recognized that biological diversity must be simultaneously protected and used" (FAO 1994). This logic, from park authorities themselves, belongs in a looking-glass world, where use and conservation are the same. Its simplicity is betrayed by its evident denial of the need not to consume.

"Parks must be viewed as resources." The previous cliché is echoed in this closely related one that directly addresses parks. This expression comes from a belief that the social value of protecting nature is not important in and of itself, and that parks must justify their existence in strictly economic terms. As Reid (1996) states, "The very name 'protected area' is a throwback to early conservation philosophy that viewed conservation as an alternative to development, not a component of development. . . . The term conveys the message that barriers exist between the resource and society." But it is exactly these barriers that were created by the society to maintain parks and their socially derived "non-resource" values.

"Local people hate parks," or *"You have to choose between local people and parks."* Ghimire and Pimbert (1997) state that "a growing body of empirical evidence now indicates that the transfer of 'Western' conservation approaches to the developing countries has had adverse effects on the food, security, and livelihoods of people living in and around protected areas." Despite this broad claim, the cases in this book and others (e.g., MacKinnon 1997) illustrate that parks and the organizations that support parks can bring strong benefits to local people, benefits that would not otherwise be made available to these people.

"Because of use of the 'Yellowstone model,' parks are imperialistic impositions on third world countries." The argument can be made that land and the animals and plants it contains has been set aside from use by interested groups for many centuries in many parts of the world; from the Chinese

and Persian hunting gardens to the sacred groves of India and West Africa. The New Forest in England has been a "protected area" since the twelfth century, although what it was designed to protect has changed from game through ship timber to wild nature (Heathcote 1994). The claim that national parks are a "rich-country institution" (Southgate and Clark 1993) is to deny inhabitants of other than rich countries the right to choose what options they would like to use in developing their own ways of life.

Conservation Generalizations

"Parks may be ecological islands, but they are part of the social and political mainland." Parks are islands in some respects but clearly not in others. By generalizing their insular qualities, it is easier to use isolation as an excuse for economic integration. Acknowledging parks as part of a set of societal values allows them to be supported for what they are and not condemned for what they are not.

"Ignore history at your own peril." Understanding the biological and social history of a given site, together with the political circumstances surrounding its creation, is essential in creating feasible conservation programs. As Brandon (chapter 14) points out, the circumstances of origin create significant phylogenetic or design constraints that can strongly influence the success or failure of conservation actions at a given site. Standardized approaches must be used as the raw material from which to tailor locally appropriate, enduring conservation solutions.

"Ignore scale at your peril." Each site is linked to regional, national, and international scales through agricultural, trade, and colonization policies and the politics of conservation, development, and local peoples (Sanderson and Bird, chapter 15). These connections can interact with one another and create conditions that impact threats, partnerships, and policies. Moreover, there is no "right scale," but a set of cross-scale dynamics important to biodiversity. When crafting local approaches, it is vital to understand the proximate and ultimate driving forces that have influenced and will continue to influence conservation actions.

"Work at protected areas needs to concentrate on alleviating threats to the biodiversity components that the site is designed to protect." Much work has been done at sites that is not directed specifically at ensuring the long-term conservation of those things that the site was established to conserve.

Much of the work at integrated conservation and development projects has not clearly linked development activities to specified conservation objectives and has therefore not guaranteed conservation outcomes (Wells and Brandon 1992). In fact, some inappropriately focused development activities have resulted in "death by friendly fire"—the destruction of that which they were designed to preserve. Without being precise about the purposes of a given conservation area, it is difficult to develop appropriate conservation actions (Weeks 1997).

"NGOs can be effective agents for conservation." These case studies and others in the PiP program show that NGOs can navigate the constantly shifting terrain between nature, local people, nonlocal people, national governments, multilateral organizations, and other NGOs. They can bring attention and resources to help protect a given site and to help ensure that people living near the site receive government services. Though the terrain is slippery, they can fulfill functions of national governments in ensuring the long-term survival of national patrimony.

"Parks cannot be conserved without national governments." All too often the role of national government is neglected, yet it is within the network of national policy and politics that parks must exist. Neglecting this fact can only risk failure. All too frequently the rhetoric surrounding parks has focused on local people and international actors, failing to focus on the vital role, good and bad, played by national governments.

"Be prepared for creative partnerships," and *"Look for the charismatic leader."* Common goals can make for uncommon partnerships. The PiP program has created a means for different constituencies with sometimes conflicting agendas to find common ground. This common ground and the desire to locate it has frequently been catalyzed by self-selecting individuals who can emerge to play vital roles in crafting enduring solutions.

"Conflicts are not constant, but parks must be." These case studies demonstrate that conflict concerning a given protected area shifts over time, involving different threats, different interest groups, and different social values. When developing ways of resolving these conflicts, it is vital to understand these shifting contexts and not compromise the long-term viability of the park itself under the belief that resolving a given conflict will provide an eternal solution.

"Stereotypes are fatal to new solutions." Nonconformity and the possibility of unexpected solutions are frequent surprises. These may arise from unexpected people, unexpected coalitions, unexpected agencies, and novel

circumstances. The case studies have in common the unexpected solution and the openness to explore the unexpected. Stereotypes and clichés serve only to prevent recognition of novelty.

Conclusion

The biodiversity that parks are designed to protect is a social good. Many of the parks in Latin America and the Caribbean were created in the 1980s, before the decade of biodiversity—the 1990s. The anomalous nature of the term *biodiversity* has contributed to the criticism that parks are not achieving their mission, and its increasing adoption worldwide has led to an expectation that parks were designed to save "biodiversity." Yet this term is essentially a political one whose appropriation by politically interested actors has led to a significant critique of national parks (Sanderson and Redford 1997).

As these case studies illustrate, the choices that a society must make are often quite stark: beef or Yucatán flamingos; gold or Andean cloud forests; timber or Amazonian jaguars; coffee or Costa Rican quetzals; tour boats or Great Maya Reef corals. Beef, coffee, and the other resources are fungible, they can be produced in many places, but Yucatán flamingos and Andean cloud forest cannot. Their survival depends on protecting them where they occur. This choice between fungible resources and nonfungible species and ecosystems will be made by stakeholders. But which stakeholders will prevail—those interested in commodities or those committed to noncommodity nature? Are there true stakeholders in both? This public struggle must be informed by history; between 1945 and 1990 45 percent of the agricultural land in South America was degraded and 74 percent of the land in Central America was degraded (Gardner 1997). The search for sustainability all too often is confined to the unplowed and uncut frontier, with ravenous claims being made on wildlands, as the production machine that runs our modern world feeds on nature and expels degraded land. Human societies in the late twentieth century have sought to create sustainability on the small percentage of the earth on which we have not already failed to achieve it.

Yet the pressure remains inexorable on parks, a meager 5–10 percent of the earth's surface. Parks have become the stage on which many demand action to redress rural poverty, social justice, gender inequity, and the plight of indigenous peoples. Parks are also supposed to be the testing ground for sustainable development and compatible resource use. The strident voices of critics the world over condemn parks for not solving many of the ills accumulated over centuries of capitalist excesses. Why are these critics focusing on parks and not on the 90–95 percent of the rest of Earth's land surface? Is it because they are unable or unwilling to make demands of the powerful groups that control the destiny of this vast majority of the earth?

Continuing to serve as handmaidens to the forces of industry would allow these forces both to continue turning Earth's entire surface into a failed experiment in sustainability and to condemn to extinction the last remaining bits of relatively undestroyed forest, grassland, and tundra. This destruction would take with it the remaining traditional peoples who rely on these areas for continued cultural survival (Redford and Mansour 1996).

The Parks in Peril program is a feisty, creative middle ground. It is true that parks may have been created by "top-down" forces, but that is the only way they could have been created. "Bottom-up" in situ efforts have created systems of sacred groves and sacred forests but nothing of a scale sufficient to preserve large portions of ecosystems. But top-down efforts will never ensure the conservation of a place that they have succeeded in creating. For this, the good will and enthusiasm of local forces are essential.

We stress that parks are necessary, but not sufficient, for biodiversity conservation. They must be seen as part of a national, regional, or ecoregional scheme that will comprehensively and effectively address biodiversity conservation issues in parks as well as outside of parks. Park-based conservation must be integrated with conservation efforts focused on agriculture, forestry, grazing, pollution, water diversion, and urban areas. Parks may be jewels in the crown, but they will not survive in isolation. Parks aren't a failure any more than they are a success. They are a hope, a hope to be realized at single sites where a scientific understanding of biodiversity is married to the management of human progress and dignity. They are a reflection of the human desire to not completely destroy that which sustains us. Park advocates and park managers must work in close alliance with others trying to ensure a compatible future for humans and their societies, along with the myriad other species and systems inhabiting the earth.

Appendix: PiP Site Scorecard Criteria

Minimum Protection Activities

Physical Infrastructure

5 = All physical infrastructure necessary for reserve management (as defined by PiP partner and including boundary demarcation) in place

4 = Most physical infrastructure for reserve management in place; one or more of components listed in 5 (above) missing or inadequate

3 = Some physical infrastructure for reserve management in place, but significant gaps exist

2 = Little physical infrastructure for reserve management in place

1 = No physical infrastructure for reserve management in place

On-Site Personnel

5 = Number of on-site personnel sufficient to perform all planned protection activities

4 = Number of on-site personnel adequate to perform most planned protection activities

3 = On-site personnel able to perform some protection activities

2 = Some on-site personnel, not enough to adequately perform protection activities

1 = No on-site personnel

Training

5 = Training needs identified; training program begun

4 = Training needs identified; some basic courses provided

3 = Training needs identified; no training yet initiated

2 = Training needs being identified

1 = No indication of personnel training needs

Land Tenure Issues

5 = Land tenure information mapped and in use by site manager
4 = Some land tenure information available and being used by site manager
3 = Land tenure information available but not being used by site manager
2 = Inadequate access to land tenure information
1 = Land tenure information not available from any source

Threats Analysis

5 = Threats identified, ranked, and being addressed through management actions
4 = Threats identified and ranked; specific strategy drafted to address specific threats
3 = Threats analysis done; no specific strategy yet drafted to address threats
2 = Threats analysis under way
1 = No analysis of threats

Protected-Area Status

5 = Official declaration of protected area obtained at appropriate level with reserve boundaries correctly demarcated
4 = Proposal for official declaration, with reserve boundaries correctly demarcated, submitted to proper authorities; no declaration yet obtained
3 = Proposal for declaration being prepared with reserve boundaries correctly demarcated
2 = Protected-area decree exists; boundaries incorrectly demarcated
1 = No protected-area decree exists

Long-Term Management

Reserve Zoning and Buffer Zone Management

5 = Reserve zones defined; land-use patterns conform to usage standards established for zones
4 = Reserve zones defined; land-use patterns mostly conform to standards established for zones
3 = Participatory process underway to develop use zones and allowable uses within buffer zones
2 = Studies underway to determine appropriate use zones
1 = No division of usage zones within the reserve

Site-Based Long-Term Management Plans

5 = Long-term management plan completed, landscape-scale management plan guiding actions

4 = Long-term management plan completed, guiding reserve management

3 = Long-term management plan completed but not yet implemented

2 = Long-term management plan in preparation

1 = Long-term management plan not yet begun

Science and Information Needs Assessment

5 = Local/international scientific/research organizations and individuals coordinating with reserve management to address reserve information needs

4 = Conservation science needs identified, ranked, and distributed; contact made with science/research organizations and funding sources to address those needs

3 = Conservation science needs identified and ranked

2 = Needs generally known

1 = Needs essentially unknown

Monitoring Plan Development and Implementation

5 = Timely monitoring information and analysis in site manager's hands; being used for management purposes

4 = Accurate, threat-related monitoring variables being monitored

3 = Accurate, threat-related monitoring variables identified; baseline information being collected and classified

2 = Some baseline information being gathered, but with no clear relation to monitoring needs

1 = No environmental monitoring of any significance underway

Long-Term Financing

NGO Self-Sufficiency Plan

5 = NGO implementing strategy for achieving operational self-sufficiency

4 = NGO has completed strategy for operational sustainability and has begun implementation

3 = NGO completing strategy for operational sustainability

2 = NGO beginning strategy for operational sustainability

1 = NGO has no strategy for achieving operational sustainability

Long-Term Financial Plan

5 = Long-term financial plan completed; diversified portfolio of funding sources available to cover operational expenses

4 = Long-term financial plan near completion; recurrent/sustainable sources and mechanisms identified

3 = Draft financial plan completed; recurrent/sustainable sources and mechanisms for park operations identified

2 = Financial planning under way

1 = No financial planning or diversification of funding sources in evidence

Site Constituency

Management Committee or Technical Advisory Committee

5 = MC/TAC includes reserve-area communities, actively participates in reserve management decisions

4 = MC/TAC includes some community representation and occasionally participates in management decisions

3 = MC/TAC exists but doesn't participate in management decisions

2 = MC/TAC being formed

1 = MC/TAC nonexistent

Community Involvement in Sustainable Resource Use

5 = Well-documented pilot projects for sustainable resource use undertaken in cooperation with major community organizations

4 = Well-documented pilot projects for sustainable resource use involve community organizations

3 = Well-documented pilot projects for sustainable resource use involve individual communities or residents

2 = Pilot projects for sustainable resource underway but don't involve communities

1 = No pilot projects for sustainable resource use underway

Policy Agenda Development

5 = Conservation policies that promote park security in place at all appropriate levels

4 = Plan for conservation policies that promote park security completed; policies being actively pursued at some levels

3 = Plan for securing appropriate conservation policies completed

2 = No formal plan developed for promoting appropriate conservation policies; however, action being taken on an as-needed basis to develop policies that promote park security

1 = No action being taken to develop or promote conservation policies for park security

Environmental Education Programs

5 = Measurable positive impact of environmental education programs contributing to local support and conservation of reserve

4 = Environmental education programs well established but formal assessment of impact not completed

3 = Environmental education programs being conducted

2 = Environmental education programs being developed by PiP partner

1 = No environmental education programs under development by PiP partner

Glossary

CARE—A U.S.-based development organization
CITES—Convention on International Trade of Endangered Species
DANIDA—Danish Agency for International Cooperation
EEC—European Economic Community
GEF—Global Environmental Facility
GTZ—German Agency for Technical Cooperation
ICDP—Integrated Conservation Development Project
NGO—Nongovernment organization
PiP—Parks in Peril program
UNDP—United Nations Development Program
UNEP—United Nations Environment Program
USAID—United States Agency for International Development
UNESCO—United Nations Educational, Scientific, and Cultural Organization
WWF—World Wildlife Fund

References

Note: All case studies were informed by Parks in Peril evaluations and quarterly reports, and the following publication: Mansour, J., ed. 1995. *Parks in Peril Source Book.* Arlington, Va.: America Verde Publications, The Nature Conservancy.

Introduction

Ascher, W., and R. Healy. 1990. *Natural Resource Policymaking in Developing Countries: Environment, Economic Growth and Income Distribution.* Durham, N.C.: Duke University Press.

Brandon, K. 1996. *Ecotourism and Conservation: A Review of Key Issues.* Environment Department Paper #33. Washington, D.C.: World Bank (April).

Brandon, K. 1997. "Policy and Practical Considerations in Land-Use Strategies for Biodiversity Conservation." In *Last Stand: Protected Areas and the Defense of Tropical Biodiversity,* R. Kramer, C. von Schaik, and J. Johnson, eds. Oxford: Oxford University Press.

Burley, F. W. 1984. The Conservation of Biological Diversity: A Report on United States Government Activities in International Wildlife Resources Conservation, with Recommendations for Expanding U.S. Efforts. Washington, D.C.: World Resources Institute.

CEQ (Council on Environmental Quality). 1980. *Environmental Quality 1980: The Eleventh Annual Report of the Council on Environmental Quality.* Washington, D.C.: Executive Office of the President, Council on Environmental Quality.

Cornelius, Stephen E. 1991. *Wildlife Conservation in Central America: Will It Survive the '90s*? Transactions of the Fifty-sixth North American Wildlife and Natural Resource Conference, pp. 40–49.

Ehrlich, P., and E. O. Wilson. 1991. Biodiversity studies: Science and policy. *Science,* 253:758–762.

Ghimire, K. B., and M. P. Pimbert, eds. 1997. *Social Change and Conservation: Environmental Politics and Impacts of National Parks and Protected Areas.* London: Earthscan Press.

IUCN. 1998. *1997 United Nations List of Protected Areas.* Gland, Switzerland: IUCN.

Janzen, D. H. 1994. Wildland biodiversity management in the tropics: Where are we now and where are we going? *Vida Silvestre Neotropical* 3:3–15.

Kramer, R., C. von Schaik, and J. Johnson, eds. 1997. *Last Stand: Protected Areas and the Defense of Tropical Biodiversity.* Oxford: Oxford University Press.

McNeely, J. 1995. "Foreword," in *Parks in Peril Source Book.* J. Mansour, ed. Arlington, VA: America Verde Publications, The Nature Conservancy.

Nietschmann, B. 1992. *The Interdependence of Biological and Cultural Diversity.* Kenmore, Wash.: Center for World Indigenous Studies.

Pimbert, Michel P. 1993. *Protected Areas, Species of Special Concern, and WWF.* Gland, Switzerland: World Wildlife Fund (November).

Redclift, M.R. 1989. The environmental consequences of Latin America's agricultural development: Some thoughts on the Brundtland Commission report. *World Development* 17, no. 3, pp. 356–377.

Redford, K. 1990. The ecologically noble savage. *Orion* 9:24–29.

Redford, K., and J. Mansour. 1996. *Traditional Peoples and Biodiversity Conservation in Large Tropical Landscapes*. Rosslyn, Va.: The Nature Conservancy.

Redford, K., and B. Richter. 1998. *Conservation of Biodiversity in a World of Use*. Submitted for publication.

Robinson, J. 1993. The limits to caring: Sustainable living and the loss of biodiversity. *Conservation Biology* 7, no. 1 (March):20–28.

Robinson, J., and E. Bennett, eds. In press. *Hunting for Sustainability*. New York: Columbia University Press.

Robinson, J., and K. H. Redford, eds. 1991. *Neotropical Wildlife Use and Conservation.* Chicago: University of Chicago Press.

Sanderson, S. E., and K. H. Redford. 1997. "Biodiversity Politics and the Contest for Ownership of the World's Biota." In *Last Stand: Protected Areas and the Defense of Tropical Biodiversity,* R. Kramer, C. von Schaik, and J. Johnson, eds., pp. 115–132. Oxford: Oxford University Press.

Stevens, S., ed. 1997. *Conservation through Cultural Survival: Indigenous Peoples and Protected Areas.* Washington, D.C.: Island Press.

WCED (World Commission on Environment and Development). 1987. *Our Common Future*. Oxford: Oxford University Press. (This publication is known as the Brundtland report.)

WCMC (World Conservation Monitoring Centre), comp. 1992. *Global Biodiversity: Status of the Earth's Living Resources*. London: Chapman and Hall.

———. 1996. Unpublished data. Cambridge, United Kingdom.

Wells, M., and K. Brandon. 1992. *People and Parks: Linking Protected Area Management with Local Communities.* Washington, D.C.: World Bank.

West, P. C., and S. R. Brechin, eds. 1991. *Resident Peoples and National Parks: Social Dilemmas and Strategies in International Conservation*. Tucson: University of Arizona Press.

Western, D., and R. M. Wright. 1994. *Natural Connections: Perspectives in Community-Based Conservation*. Washington, D.C.: Island Press.

Wilson, E. O., ed. 1988. *Biodiversity*. Washington, D.C.: National Academy of Sciences Press.

Wood, D. 1995. Conserved to death: Are tropical forests being overprotected from people? *Land Use Policy* 12, no. 2:115–135.

WRI (World Resources Institute) et al. 1992. *Global Biodiversity Strategy.* Washington, D.C.: World Resources Institute.

Chapter 2

Biodiversity Support Program (BSP), Conservation International, The Nature Conservancy, Wildlife Conservation Society, World Resources Institute, and World Wildlife Fund. 1995. *A Regional Analysis of Geographic Priorities for Biodi-*

versity Conservation in Latin America and the Caribbean. Washington, D.C.: Biodiversity Support Program.

Dinerstein, E., D. M. Olson, D. J. Graham, A. L. Webster, S. A. Primm, M. P. Bookbinder, and G. Ledec. 1995. *A Conservation Assessment of the Terrestrial Ecoregions of Latin America and the Caribbean.* Washington, D.C.: World Bank.

Green, M. J., M. G. Murray, G. C. Bunting, and J. R. Paine. 1997. *Priorities for Biodiversity Conservation in the Tropics.* World Conservation Monitoring Centre (WCMC), Biodiversity Bulletin No. 1.

IUCN (World Conservation Union). 1978. *Categories, Criteria, and Objectives for Protected Areas.* Morges, Switzerland: IUCN.

———. 1992. *Protected Areas of the World: A Review of National Systems.* Volume 4: *Nearctic and Neotropical.* Gland, Switzerland: IUCN.

McNeely, J. A., J. Harrison, and P. Dingwall. 1994. "Protected Areas in the Modern World." In *Protecting Nature: Regional Reviews of Protected Areas,* J. A. McNeely, J. Harrison, and P. Dingwall, eds. Morges, Switzerland: IUCN.

McNeely, J. A., and K. R. Miller. 1984. *National Parks, Conservation, and Development: The Role of Protected Areas in Sustaining Society.* Washington, D.C.: Smithsonian Institution.

The Nature Conservancy. 1997. *Designing a Geography of Hope: Ecoregion-Based Conservation in The Nature Conservancy.* Washington, D.C.: The Nature Conservancy, 68 pp.

Pressey, R. L., C. J. Humphries, C. R. Margules, R. I. Vane-Wright, and P. H. Williams. 1993. Beyond opportunism: Key principles for systematic reserve selection. *Trends in Ecology and Evolution* 8:124–128.

Redford, K. H., A. Taber, and J. A. Simonetti. 1990. There is more to biodiversity than the tropical rainforests. *Conservation Biology* 4, no. 3:328–30.

Scott, J. M., T. H. Tear, and F. W. Davis. 1996. *Gap Analysis: A Landscape Approach to Biodiversity Planning.* Bethesda, Md.: American Society of Photogrammetry Remote Sensing.

Vitousek, P. M., J. Aber, R. W. Howarth, G. E. Likens, P. A. Matson, D. W. Schindler, W. H. Schlesinger, and G. D. Tilman. 1997a. Human alteration of the global nitrogen cycle: Causes and consequences. *Issues in Ecology,* no. 1. Ecological Society of America.

Vitousek, P. M., H. A. Rooney, J. Lubchenco, and J. M. Melillo. 1997b. Human domination of Earth's ecosystems. *Science* 277:494–499.

Wells, M., and K. Brandon. 1992. *People and Parks: Linking Protected Area Management with Local Communities.* Washington, D.C.: World Bank.

WCMC (World Conservation Monitoring Center). 1992. *Global Biodiversity: Status of the Earth's Living Resources.* London: Chapman and Hall.

———. 1996. *Tropical Moist Forests and Protected Areas.* Digital Files. CD-ROM.

Chapter 4

Abundes, M. E., and M. C. Gamiño. 1991. *Programa de educación ambiental en los refugios faunísticos de Ría Lagartos y Ría Celestún, Yucatán.* Unpublished report. Mérida, Yucatán: Pronatura Península de Yucatán, A.C.

Arellano, G. A. 1994. *Programa de uso turístico para la Reserva Especial de la Biósfera de Ría Celestún, Yucatán, Mexico*. Mérida, Yucatán: Pronatura Península de Yucatán.

Arengo, F., and G. A. Baldassarre. 1995. "American Flamingos and Ecotourism on the Yucatán Península, Mexico." In *Integrating People and Wildlife for a Sustainable Future*, J. A. Bissonette and P. R. Krausman, eds., pp. 207–210. Proceedings of the First International Wildlife Management Congress. Bethesda, Md.: Wildlife Society.

Batllori, S. E. 1990a. *Celestún: Caracterización ecológica*. Unpublished report.

———. 1990b. *Celestún: Problemas de manejo*. Unpublished report.

Demayo, A., World Bank, 1995, personal communication. Washington, D.C.

Global Environmental Facility (GEF). 1995. *Quarterly Operational Report* (August). Washington, D.C.: GEF Secretariat.

Instituto Nacional de Ecología (INE). 1993. *Programa de manejo de la Reserva Especial de la Biósfera de Ría Lagartos, Yucatán*. Yucatán, Mexico: INE.

———. 1995. Reserva Especial de la Biósfera Ría Lagartos: Programa operativo anual 1995 para el manejo de la Reserva.

Leslie, H. 1997. *La distribución espacial y temporal del turismo en la Ría Celestún: Una propuesta para manejar el turismo*. Mérida, Mexico: PRONATURA.

Moan, S. A. 1992. *Ecotourism on the Yucatan Peninsula: Ecotourism Potentials in the Rio [sic] Lagartos Wildlife Preserve*. M.L.A. Thesis, SUNY College of Environmental Science and Forestry, Syracuse, New York.

Programa de desarrollo social y ordenamiento ecológico del territorio costero del estado de Yucatán: Fase descriptiva (versión preliminar). 1995. University of Colorado at Boulder, CO: SEMARNAP, NAWCC, USFWS; Mérida, Mexico: CINVESTAV-Mérida y PRONATURA.

Chapter 5

Brown, M., et al. 1996. *Un análisis del valor del bosque nuboso en la protección de cuencas, Reserva de Biósfera Sierra de las Minas (Guatemala) y Parque Nacional Cusuco (Honduras)*. Guatemala City, Guatemala: Fundación Defensores de la Naturaleza/RARE Center for Tropical Conservation.

Campbell, J. A. 1982. *The Biogeography of the Cloud Forest Herpetofauna of Middle America, with Special Reference to the Sierra de las Minas of Guatemala*. Doctoral dissertation, University of Kansas.

Congreso de la República de Guatemala. 1990. Ley de Areas Protegidas, Decreto 4-89. *Diario de Centro América*, Guatemala, February 11.

———. 1989. Reserva de la Biósfera Sierra de las Minas, Decreto 49-90. *Diario de Centro América*, Guatemala, October 4.

Conservation Data Center (CDC). 1993. *Evaluación ecológica rápida de la "Reserva de la Biósfera Sierra de las Minas."* Universidad de San Carlos, Guatemala: CDC/CECON.

Dix, M., Universidad del Valle de Guatemala, 1991, personal communication.

Duarte, A., Guatemala Mission to the Organization of American States, Washington D.C., 1996, personal communication.

Flores, I.,1997. *Análisis del potencial artesanal de la cestería en Chilascó, BV, como una alternativa a la oferta turística de la RBSM*. Guatemala: Universidad del Valle de Guatemala. Guatemala: Defensores de la Naturaleza, Guatemala City, Unpublished document.

Fundación Defensores de la Naturaleza. 1992. *Plan maestro de la Reserva de la Biósfera Sierra de las Minas 1992–1997*. Guatemala: Defensores de la Naturaleza.

———. 1995. *Plan operativo 1995 de la Reserva de la Biósfera Sierra de las Minas*. Guatemala City, Guatemala. Unpublished document.

Fundación Defensores de la Naturaleza and WWF. 1990. *Estudio técnico para dar a Sierra de las Minas la categoría de la biósfera*. Guatemala City, Guatemala: Unpublished document.

Greenberg, R., P. Bichier, A. Cruz Angón, and R. Rietsma. 1995. *Bird Populations in Shade and Sun Coffee Plantations in Central Guatemala*. Washington, D.C.: Smithsonian Migratory Bird Center, National Zoological Park.

Margoluis, R. and Gálvez, E. 1993. *Diagnóstico para la integración humana a la Reserva de la Biósfera Sierra de las Minas*. Guatemala City, Guatemala: Defensores de la Naturaleza.

Méndez, C., Defensores de la Naturaleza, Guatemala, 1994, personal communication.

Presidencia de la República de Guatemala. 1990. Acuerdo Gubernativo No. 739-90, *Diario de Centro América*, Guatemala.

Chapter 6

ACOSA. 1993. *Plan anual operativo*. San José, Costa Rica: Government of Costa Rica.

Alfaro, M., et al. 1995. *Selección y ubicación de zonas prioritarias para programas de reforestación y manejo de bosque natural con pequeños y medianos productores en el área de conservación de Osa (ACOSA)*, Tercer Borrador, (February). San José, Costa Rica: Unpublished report.

Alvarez, H., and L. Márquez. 1992. *Plan de manejo de la Reserva Forestal Golfo Dulce*. San José, Costa Rica: Fundación Neotrópica.

Brenes, L.G., et al. 1990. *Informe sobre la explotación de oro en la Península de Osa*. Puntarenas, Costa Rica: Unpublished report.

Burnie, D. 1994. Ecotourists to paradise. *New Scientist* (April 16):23–27.

Camacho, M. A. 1993. *Regional Planning and People's Participation in Costa Rica: A Case Study at the Natural Protected Area of the Osa Peninsula, Brunca Region*. Doctoral dissertation, School of Development Studies, University of East Anglia, March.

Christen, C. A. 1994. *Development and Conservation on Costa Rica's Osa Peninsula, 1937–1977: A Regional Case Study of Historical Land Use Policy and Practice in a Small Neotropical Country*. Ph.D. dissertation, Johns Hopkins University, Baltimore, Maryland.

Cuello, C. 1997. *Sustainable Development in Theory and Practice: A Costa Rican Case Study*. Ph.D. dissertation, University of Delaware, Newark, Delaware.

Cuello, C., and A. Piedra-Santa. 1995. *Diagnóstico de la Península de Osa: Principales tendencias socioeconómicas, ambientales, y organizativo-institutionales*. San José, Costa Rica: Fundación Neotrópica.

Donovan, R. 1994. "BOSCOSA: Forest Conservation and Management through Local Institutions." In *Natural Connections: Perspectives in Community-Based Conservation*, David Western and R. Michael Wright, eds., pp. 215–233. Washington, D.C.: Island Press.

Fundación Neotrópica. 1985–1995. Internal Project Documents on Costa Rica. San José, Costa Rica.

García, R. 1994. *Taller de revisión de propuesta: Programa de investigación del área de conservación y desarrollo sostenible Osa*. Agua Buena, Península de Osa, Costa Rica: Centro BOSCOSA.

González, G., and C. Rocío. 1992. *Impacto ambiental de la explotación de oro artesanal, Península de Osa, Puntarenas, Costa Rica*. Thesis, Department of Geography, School of History and Geography, Faculty of Social Sciences, University of Costa Rica, San José, Costa Rica.

Hitz, W., E. Alpizar, and F. Montoya. 1995. *Evaluación proyecto BOSCOSA, Fundación Neotrópica y recomendaciones para la transición*. Draft Final Report. San José, Costa Rica: USAID.

Jiménez, J. J. 1996. *Diagnóstico del mercado de madera en la Península de Osa*. San José, Costa Rica: Fundación Neotrópica.

Maldonado, T. 1997. *Uso de la tierra y pérdida de bosques en la Península de Osa. Algunas áreas críticas y repercusiones en la conservación del área de conservación Osa.* Costa Rica (Borrador) San José, Costa Rica: Fundación Neotrópica.

Naughton, L. 1993. Conservation versus Artisanal Gold Mining in Corcovado National Park, Costa Rica: Land Use Conflicts at Neotropical Wilderness Frontiers. *Yearbook, Conference of Latin Americanist Geographers*, 19:47–55.

Redford, K. H. 1992. The empty forest. *Bioscience* 42:412–22.

Silk, N. 1993. An ecological perspective on property rights in Costa Rica. *Land Use Policy* (October):312–21.

Soto, R., ed. 1992. *Evaluación ecológica rápida, Península de Osa*. San José, Costa Rica: Fundación Neotrópica.

Thomsen, K. 1997. *Potential of Non-timber Forest Products in a Tropical Rain Forest in Costa Rica*. Ph.D. dissertation, Faculty of Natural Science, University of Copenhagen, Copenhagen, Denmark.

Ugalde, A. 1997. Carta al Presidente José María Figueres Olsen, sobre la situación del Parque Nacional Corcovado, San José, Costa Rica.

Umaña, A., and K. Brandon. 1992. "Inventing Institutions for Conservation: Lessons from Costa Rica." In *Poverty, Natural Resources, and Public Policy in Central America*, S. Annis, ed. Washington, D.C.: Overseas Development Council.

USAID. 1995. *Costa Rica Project Paper: Regulation for Forest Management* (REFORMA), AID/LAC/P-82, Washington, D.C.

Wells, M., and K. Brandon. 1992. *People and Parks: Linking Protected Area Management with Local Communities*. Washington, D.C.: World Bank.

Wessels, A. 1992. *Conservation without Evictions: The Search for Social and Ecological Justice on Costa Rica's Osa Peninsula.* Undergraduate paper, Department of Environmental Studies, Brown University, Providence, Rhode Island.

World Wildlife Fund. 1985–1995. Internal Project Documents on Costa Rica, Washington, D.C.

Chapter 7

Abreu, D., and K. A. Guerrero, eds. 1997. *Evaluación ecológica integrada del Parque Nacional del Este, República Dominicana. Tomo 1: Recursos terrestres.* Santo Domingo, Dominican Republic: The Nature Conservancy, United States Agency for International Development, Banco Interamericano para el Desarollo (BID), Dirección Nacional de Parques (DNP), and Fondo Integrado Pronaturaleza (PRONATURA).

Cano, C. C. 1993. *Proyecto uso público, protección y recuperación de vida silvestre del Parque Nacional del Este: Documento técnico de proyecto, Volumen I.* Santo Domingo, Dominican Republic: Dirección Nacional de Parques y Agencia Española de Cooperación Internacional (AECI). (November.)

Congreso Nacional. 1974. "Ley No. 67 que crea la Dirección Nacional de Parques." *Gaceta Oficial* No. 9349 (November 20).

DNP (Dirección Nacional de Parques). 1980. *Plan de manejo Parque Nacional del Este.* Santo Domingo, Dominican Republic.

Lladó, J. A. 1994. *Tourism and the Sustainable Development Prospects of "Parque Nacional del Este" (PNE).* Report to The Nature Conservancy/PRONATURA Rapid Ecological Assessment Project. Santo Domingo, Dominican Republic. Unpublished report.

Marte, D. The Nature Conservancy, Santo Domingo, Dominican Republic, 1995, personal communication.

Pugibet, B. E. E. 1995. *Resultados del inventario de las pesquerías (CREEL Survey) realizado en el Parque Nacional Del Este,* Draft paper on file at PRONATURA, Santo Domingo, Dominican Republic, March 12–18.

Troncoso, B. 1995. *Propuesta de plan de manejo ecoturístico para el Parque Nacional del Este (PNE).* Discussion paper prepared for The Nature Conservancy, draft, May 5.

Unión Dominicana de Voluntarios Incorporados (Union of Dominican Volunteers, Incorporated). (UNIDOS). 1995. *Consulta para el mejoramiento de los modelos de colaboración entre la Dirección Nacional de Parques (DNP) y las organizaciones no gubernamentales (ONGs) de la República Dominicana: Informe final.* Santo Domingo, Dominican Republic. Unpublished.

Chapter 8

Bolland, N., and A. Shoman. 1977. *Land in Belize, 1765–1871, Law and Society in the Caribbean,* No. 6. Kingston, Jamaica: Institute of Social and Economic Research.

Cowo, P., chiclero, 1995, personal communication.

Cruz, Benjamin, head ranger, PfB, 1995, personal communication.

Cullerton, Bridget, managing director, Best Enterprise for Sustainable Technology (BEST), 1995, personal communication.

Dominguez, Gloria, domestic worker, San Felipe Village, Belize, 1995, personal communication.

Fabro, Ismael, chief environmental officer, Ministry of Natural Resources, Belize, 1995, personal communication.

Government of Belize. 1990. *Department of the Environment Policy and Strategy Statement*. Belmopan: Government of Belize.

International Travel Maps. 1995. *An International Travel Map of Belize*. Vancouver, B.C., Canada: ITMP Publishing.

Ku, Gregorio, ranger, PfB, 1995, personal communication.

"The Land Deal of the Century." 1987. *Spearhead* (periodical, April/May): pp. 1, 16.

Likes, George, USAID, Belize Program, 1995, personal communication.

Loskot, Joseph, owner and general manager, New River Enterprises, Belize, 1995, personal communication.

Memorandum of Understanding between the Government of Belize and Programme for Belize. 1988. Belize City, February 22.

Nicolait, Lou, executive director, Belize Center for Environmental Studies, 1995, personal communication.

PfB. 1995. *Programme for Belize Overview*. Belize City: Programme for Belize.

———. 1996. *RBCMA Management Plan*, 3rd ed. Belize City: Programme for Belize.

PfB/IDB. 1995. *Towards a National Protected Area Systems Plan for Belize*. Belize City: NARMAP.

Romero, Bart, station manager, La Milpa Field Station, Belize, 1995, personal communication.

Salas, Osmany, planning and development coordinator, Belize Audubon Society, 1995, personal communication.

Vernon, Dylan, general coordinator, Society for the Promotion of Education and Research (SPEAR), Belize, 1995, personal communication.

Chapter 9

Arnal, H. 1993. *Parks in Peril Program 1993 Evaluation Report—Machalilla National Park*. Arlington, Va.: The Nature Conservancy.

Arriaga, L. 1987. "Manejo de recursos costeros en el Ecuador." In Proceedings of the *Seminario sobre la pesca artesanal y su problemática de desarrollo en el Ecuador*. pp. 3–10. Quito, Ecuador: Escuela Superior Politécnica del Litoral, Centro de Planificación y Estudios Sociales, Instituto Latino Americano de Investigaciones Sociales.

Becker, C. D., and C. Gibson. 1996. "The lack of institutional supply: Why a strong local community in western Ecuador fails to protect its forest." International Forestry Resources and Institutions (IFRI) Research Program Working Paper, Bloomington, Illinois.

Black, J., The Nature Conservancy, 1995, personal communication. Quito, Ecuador (deceased).

Bucheli, F., INEFAN-Quito, 1995, personal communication.

Bustamante, T., Fundación Natura, 1995, personal communication, Quito, Ecuador.

CAAM (Environmental Advisory Committee to the Presidency). 1995. *Lineamientos para la estrategia de conservación y uso de la biodiversidad en el Ecuador.* Quito, Ecuador: CAAM.

Coello, S. 1993. *Diagnóstico de la actividad pesquera en la zona de influencia del Parque Nacional Machalilla.* Studies in Protected Areas no. 5. Guayaquil, Ecuador: Fundación Natura.

Cuellar, J. C., et al. 1992. *Estudio de las poblaciones del área interna y de la zona de influencia del Parque Nacional Machalilla.* Estudios en Areas Protegidas no. 3. Quito, Ecuador: Fundación Natura.

Dodson, C. H. and A. Gentry. 1991. Biological extinction in western Ecuador. *Annals of the Missouri Botanical Gardens* 78:273–95.

Dove, M. 1982. "The Myth of the 'Communal' Longhouse in Rural Development. The Kantu of Kalimantan." In *Too Rapid Rural Development,* C. MacAndrews and L.S. Chin, eds., pp. 14–78. Athens: Ohio University Press.

Emmons, L. and L. Albuja. 1992. (Mammals Section) In *Status of Forest Remnants in the Cordillera de la Costa and Adjacent Areas of Southwestern Ecuador,* T. A. Parker III and J. L. Carr, eds. RAP Working Papers 2. Washington, D.C.: Conservation International.

Fiallo, E. 1994. *Local Communities and Protected Areas: People's Attitudes toward Machalilla National Park, Ecuador.* M.A. thesis, University of Florida, Gainesville.

Forster, N. 1992. Protecting fragile lands: New reasons to tackle old problems. *World Development* 20, no. 4:571–85.

FN (Fundación Natura). 1991. *Conservación y desarrollo.* Quito, Ecuador: FN.

———. 1992. *Propuesta para un plan de acción sobre áreas protegidas en el Ecuador.* Quito, Ecuador: FN.

———. 1994. *Reflexiones en torno al tema de la adquisisción de tierras para la conservación de las áreas protegidas,* Position Document no. 5. Quito, Ecuador: FN.

FN-ESTADE. 1993. "Proyecto de ley de áreas protegidas y vida silvestre." Draft. Quito, Ecuador: FN.

Gallardo, J., INEFAN-Portoviejo, 1995, personal communication.

González, S., Comuna El Pital, 1995, personal communication.

IUCN (World Conservation Union). 1982. *IUCN Directory of Neotropical Protected Areas.* Dublin: Tycooly International.

Josse, C., PUCE, Pontificia Universidad Católica del Ecuador, 1995, personal communication, Quito, Ecuador.

Josse, C., and H. Balshev. 1993. The composition and structure of a dry, semideciduous forest in western Ecuador. *Nordic Journal of Botany* 14:425–34.

Light, S. S., L. H. Gunderson, and C. S. Holling. 1995. "The Everglades: Evolution of Management in a Turbulent Ecosystem." In *Barriers and Bridges to the*

Renewal of Ecosystems and Institutions, L. H. Gunderson, C. S. Holling, and S. S. Light, eds. New York: Columbia University Press.

Martínez, A., Comuna Agua Blanca, Confederación de Comunas del Sur de Manabi, 1995, personal communication.

Martínez, S., INEFAN-MNP and Comuna Agua Blanca, 1995, personal communication.

Medina, L., INEFAN-MNP, 1995, personal communication.

Ortiz, J., D. Parra, J. Cárdenas, and F. Cepeda. 1995. *Plan de manejo turístico del Parque Nacional Machalilla de la República del Ecuador*. Quito, Ecuador: INEFAN-Fundación Natura.

Parker III, T. A. and J. L. Carr, eds. 1992. *Status of Forest Remnants in the Cordillera de la Costa and Adjacent Areas of Southwest Ecuador*. RAP Working Paper no. 2. Washington, D.C.: Conservation International.

Paucar, A. M., et al. 1987. *Diagnóstico del Parque Nacional Machalilla de la República del Ecuador*. Quito, Ecuador: Dirección Nacional Forestal.

Pincay, A., Comuna Salango, 1995, personal communication.

Platt, D., environmental educator, 1995, personal communication, Quito, Ecuador.

Ponce, A., Fundación Natura, 1995, personal communication, Quito, Ecuador.

Sigle, M., DeD (German Development Assistance Organization), 1995, personal communication, Quito, Ecuador.

Silva, M. I., and C. McEwan. n.d. Machalilla: El camino de la integración. *Colibrí*, pp. 71–75.

Sylva, P., and S. León. 1992. *Acciones de desarrollo y áreas naturales protegidas en el Ecuador. Parque Nacional Machalilla*. Quito, Ecuador: Fundación Natura.

Vizcarra, J., INEFAN-Portoviejo, 1995, personal communication.

Zambrano, C., INEFAN-MNP, 1995, personal communication.

Zambrano, C. 1995. *Programas del Parque Nacional Machalilla actividades realizadas 1993–95*. Informal document, Puerto López, Ecuador: INEFAN-MNP.

Chapter 10

Apolo, W. 1984. *Plan de manejo: Parque Nacional Podocarpus*. Quito, Ecuador: Ministry of Agriculture.

Arnal, Hugo. 1993. *Parks in Peril Program 1993 Evaluation Report—Podocarpus National Park*. Arlington, Va.: The Nature Conservancy.

Arnal, H., The Nature Conservancy, 1995, personal communication. Arlington, VA.

Black, J., The Nature Conservancy, 1995, personal communication. Quito, Ecuador (deceased).

Cabrera, P., affiliation unknown, 1995, personal communication. Cajanuma, Ecuador.

Calderón, S., INEFAN/PNP, 1995, personal communication. Loja, Ecuador.

Chiriapo, M., Federación de Nacionalidades Shuar/Achuar, 1995, personal communication. Loja, Ecuador.

Cordovez, S., Subsecretary of Mines, INEMIN, 1995, personal communication. Quito, Ecuador.

FAI (Fundación Arcoiris). 1992. *El Parque Nacional Podocarpus: Area natural protegida ecuatoriana amenazada por minería.* Loja, Ecuador: Fundación Arcoiris.

Feinsinger, P., University of Arizona, 1995, personal communication. Tucson, Arizona.

FN (Fundación Natura). 1992. *Parque Nacional Podocarpus: Acciones de desarrollo y áreas naturales protegidas en el Ecuador,* no. 5. Quito, Ecuador: Fundación Natura.

Haworth, J., F. Jiggins, A. Jiménez, and J. Willin. 1996. *Mercury Levels in the San Luis Mining Area, Podocarpus National Park.* Unpublished manuscript available from F. Jiggins, Goulds Farm, Rayne, Braintree, Essex, CM75DF England.

Hernández, F. H., et al. 1995. *Estudio socioeconómico del área de influencia del Parque Nacional Podocarpus.* Quito, Ecuador: Fundación Natura.

Hoffman, S., Fundación Colinas Verdes, 1995, personal communication. San Pedro de Vilcabamba, Ecuador.

INEFAN (Ecuadorian Institute of Forestry, Natural Areas, and Wildlife). 1995. "Documento de posición institucional INEFAN frente al problema minero del Parque Nacional Podocarpus." Quito, Ecuador: INEFAN.

Kapila, S. 1993. *Application of Participatory Rapid Appraisal Techniques to National Park Management—a Case Study: Podocarpus National Park.* London: Parrots in Peril.

León, B., and K. Young. n.d. "The Podocarpus." Paper on file with Fundación Arcoiris, Loja, Ecuador.

Light, Stephen S., Lance H. Gunderson, and C. S. Holling. 1995. "The Everglades: Evolution of Management in a Turbulent Ecosystem." In *Barriers and Bridges to the Renewal of Ecosystems and Institutions,* Lance H. Gunderson, C. S. Holling, and Stephen S. Light, eds. New York: Columbia University Press.

Machlis, G. E., and D. L. Tichnell. 1985. *The State of the World's Parks: An International Assessment for Resource Management, Policy and Research.* Boulder, Colo.: Westview Press.

Naughton, L. 1993. Conservation versus artisanal gold mining in Corcovado National Park, Costa Rica. *Yearbook, Conference of Latin Americanist Geographers* 19:47–55.

PREDESUR. 1978. Características socio-económicas del grupo Shuar: Valles de Nangaritza, Zamora, Yyacuamui. Quito, Ecuador: PREDESUR Report.

Schulenberg, T. S., and K. Awbrey, eds. 1997. *The Cordillera del Cóndor Region of Ecuador and Peru: A Biological Assessment.* Rapid Assessment Program, Working Papers no. 7. Washington, D.C.: Conservation International.

Silva, D., Fundación Ecociencia, 1995, personal communication. Quito, Ecuador.

Suárez, L., and P. Mena. 1992. *Estudios ecológicos, entrenamiento y conservación en los bosques nublados del Ecuador.* Quito, Ecuador: Ecociencia.

Tambo, L., INEFAN/PNP, 1995, personal communication. Loja, Ecuador.

Tapia, R., Fundación Arcoiris, 1995, personal communication. Loja, Ecuador.

Vallée, D. 1992. *Environmental Impacts of Gold Mining in the Podocarpus National Park in Southern Ecuador.* M.Sc. thesis, Imperial College of Science, Technology and Medicine, University of London.

Vallée, D., S. Kapila, and E. Toyne. 1992. "Gold Mining." Paper on file with Fundación Arcoiris. Unpublished report.

Wells, M., and K. Brandon. 1992. *People and Parks: Linking Protected Area Management with Local Communities.* Washington, D.C.: World Bank.

Chapter 11

FAN (Fundación Amigos de la Naturaleza). 1994. *Límites del Parque Nacional Amboró-Propuesta institucional.* Unpublished proposal. Santa Cruz, Bolivia. (November.)

CARITAS Arquidiocesana. M. C. Ayreyu, Coordinación. 1994. *Estudio socioeconómico y de manejo de recursos naturales en la zona norte de amortiguamiento del Parque Nacional Amboró.* Santa Cruz, Bolivia (July).

CORDECRUZ (Corporación Regional de Desarrollo de Santa Cruz)—Consorcio IP/CES/KWC. 1994, 1995. *Proyecto de protección de los recursos naturales del departamento de Santa Cruz. Plan de uso de suelo (PLUS): Una propuesta para el aprovechamiento sostenible de nuestros recursos naturales.* Santa Cruz, Bolivia: Cooperación Boliviano-Alemán.

El Deber, El Día. 1994–95. Miscellaneous newspaper articles.

Federación Unica de Trabajadores Campesinos de Santa Cruz. n.d. *Estrategia global para la protección del Parque Nacional Amboró, integrado al desarrollo rural en la zona de influencia (Uso múltiple).* Prepared for PROBIOMA and ALAS. Santa Cruz, Bolivia.

Holdridge, L. R. 1967. *Life Zone Ecology.* San José, Costa Rica: Tropical Science Center.

PROBIOMA and USCC (Productividad, Biosfera medio Ambiente) and Catholic Relief Services, Bolivia Program. 1994. *Diagnóstico socioeconómico y de manejo de recursos naturales en comunidades colindantes al Parque Nacional Amboró, sector sur.* Santa Cruz, Bolivia. (August.)

Chapter 12

Aguilar, D. P. 1990. Environmental Conservation and the Palcazu Project. *Tebiwa* (December): 13–16.

Aguilar, P., director, Yanachaga National Park, 1995, personal communication.

Arce, J., ProNaturaleza, Lima, Peru, 1995, personal communication.

Camino, A. FONANPE, Lima Peru, 1995, personal communication.

Gaviría, A. 1995. *Estudio aproximación a un plan de manejo para la Reserva Comunal Yanesha.* Unpublished report. Lima, Peru: Pro Naturaleza.

Hartshorn, G. S., and Pariona, A. 1993. "Ecologically Sustainable Forest Management in the Peruvian Amazon." In *Perspectives on Biodiversity: Case Studies of Genetic Resource Conservation in Development,* C. S. Polter, Joel I. Cohen, and D. Jonezewski, eds. Washington, D.C.: American Association for the Advancement of Science.

INRENA (Instituto Nacional de Recursos Naturales). 1995a. *Estrategia del sistema nacional de áreas naturales protegidas del Perú: Plan director.* Unpublished draft (June).

———. 1995b. *Plan director del sistema nacional de áreas naturales protegidas por el estado, proyecto ayuda en la planificación de una estrategia nacional para la conservación de las áreas protegidas.* Unpublished report.

———. 1995c. Plan de manejo, Parque Nacional Yanachaga-Chemillén. Unpublished document. Lima, Peru.

Japón muestra interés en financiar la construcción del Central Paucartambo II. 1995. *El Peruano,* August 10, B-2.

Lazaro, M., M. Pariona, and R. Simeone. 1993. A natural harvest. *Cultural Survival Quarterly* (Spring): 48–51.

Linares, B. C. 1991. Análisis de sostenibilidad de un plan de manejo forestal: Caso Palcazu, Perú. *Revista Forestal del Perú* 18, no. 2: 83–199.

Plan maestro Parque Nacional Yanachaga-Chemillén. 1987. Lima, Perú: Dirección General Forestal y de Fauna, Instituto Nacional de Desarrollo, PEPP/PDR/Palcazu, USAID, Ronco Consulting Corporation, The Nature Conservancy.

Reforzarán sustitución de cultivos de la coca. 1995. *El Peruano,* August 10, A-2.

Simeone, R. 1990. Land use planning and forestry-based economy: The case of the Amuesha Forestry Cooperative. *Tebiwa* (December): 7–12.

Stocks, A., ed. 1990. Resource management and tribal development: Issues from Peru's Palcazu Project. Special issue of *Tebiwa* (December).

Stocks, A., and G. Hartshorn. 1989. "The Palcazu Project: Forest Management and Native Amuesha Communities." In *The Social Dynamics of Deforestation in Latin America: Processes and Alternatives,* S. Hecht, and J. Nations, eds. Berkeley, Calif.: University of California.

Wells, M., and K. Brandon. 1992. *People and Parks: Linking Protected Area Management with Local Communities.* Washington, D.C.: World Bank.

Chapter 14

Barzetti, V., ed. 1993. *Parks and Progress: Protected Areas and Economic Development in Latin America and the Caribbean.* Washington, D.C.: IUCN (The World Conservation Union).

Batisse, M. 1986. Developing and focusing the biosphere reserve concept. *Nature and Resources* 22:1–10

Berkes, F. 1985. "Fishermen and the 'tragedy of the commons.'" *Environmental Conservation* 12, no. 13:199.

Brandon, K. 1996. *Ecotourism and Conservation: A Review of Key Issues.* Environment Department Paper no. 33. Washington, D.C.: World Bank (April).

Brandon, K., and M. Wells. 1992. Planning for people and parks: Design dilemmas. *World Development* 20, no. 4:557–70.

Brower, B. 1991. Crisis and conservation in Sagarmartha National Park, Nepal. *Society and Natural Resources* 4:151.

Hill, K., with T. Tikuarangi. 1996. "The Mbaracayu Reserve and the Ache of Paraguay." In *Traditional Peoples and Biodiversity Conservation in Large Tropical Landscapes,* K. Redford and J. Mansour, eds., pp. 159–196. Rosslyn, Va.: The Nature Conservancy.

McGarth, D. E. et al. 1993. Fisheries and the evolution of resource management on the lower Amazon floodplain. *Human Ecology* 21, no. 2:167–195.

McNeely, J. 1989. "Protected Areas and Human Ecology: How National Parks Can Contribute to Sustaining Societies to the Twenty-first Century." In *Conservation for the Twenty-first Century,* D. Western and M. Pearl, eds. Oxford: Oxford University Press.

Perreault, T. 1996. Nature preserves and community conflict: A case study in highland Peru. *Mountain Research and Development* 16, no 2:167–175.

Richards, M. 1996. Protected areas, people and incentives in the search for sustainable forest conservation in Honduras. *Environmental Conservation* 23, no. 3:207–217.

Robinson, J., and K. H. Redford. 1994. Measuring the sustainability of hunting in tropical forests. *Oryx* 28, no. 4 (October): 249–256.

Seymour, F. J. 1994. "Are Successful Community-Based Conservation Projects Designed or Discovered?" In *Natural Connections: Perspectives in Community-based Conservation,* D. Western and R. M. Wright, eds., pp. 472–496. Washington, D.C.: Island Press.

Stankey, G. 1989. Linking parks to people: The key to effective management. *Society and Natural Resources* 2:245–50.

Vandergeest, P. 1996. Property rights in protected areas: Obstacles to community involvement as a solution in Thailand. *Environmental Conservation* 23, no. 3: 259–68.

Wachter, D. 1992. "Land Titling for Conservation in Developing Countries?" The World Bank Environment Department Divisional Working Paper no. 1992-28. Washington, D.C.: World Bank.

Wells, M., and K. Brandon. 1992. *People and Parks: Linking Protected Area Management with Local Communities.* Washington, D.C.: World Bank.

Western, D., and M. Pearl, eds. 1989. *Conservation for the Twenty-first Century.* Oxford: Oxford University Press.

Western, D. and R. M. Wright, eds. 1994. *Natural Connections: Perspectives in Community-Based Conservation*. Washington, D.C.: Island Press.

Witkowski, E. T. F., B. L. Lamont, and F. J. Obbens. 1994. Commercial picking of *Banksia hookeriana* in the wild reduces subsequent shoot, flower and seed production. *Journal of Applied Ecology* 31:508–20.

Chapter 15

Adams, J. 1996. Cost–benefit analysis: The problem, not the solution. *The Ecologist* 26, no. 1 (January–February):3.

Brandon, K. 1994. *Resource Tenure and Its Links to Conservation: A Review.* Unpublished manuscript prepared for World Wildlife Fund, Washington, D.C.

Chapman, C. A., and L. J. Chapman. 1995. Survival without dispersers: Seedling recruitment under parents. *Conservation Biology* 9, no. 3 (June):675–678.

Coomes, Oliver. 1996. Book review of risky rivers: The economics of floodplain farming in Amazonia. *The Geographical Review* 86, no. 1:136.

Denevan, W. M. 1992. The Pristine Myth: The landscape of the Americas in 1492. *The Annals of the Association of American Geographers* 82, no. 3 (September): 369.

FAO (Oficina Regional para América Latina y el Caribe). 1994. Taller internacional sobre políticas de los sistemas de áreas protegidas en la conservación y uso sostenible de la biodiversidad en América Latina, Parque Nacional Iguazú, Argentina, 27 de septiembre al 1 de octubre de 1993. Informe. Proyecto FAO/PNUMA FP/6105-90-38. Santiago, Chile: FAO/RLC. 75 pp. (FOR45).

Forster, N. 1992. Protecting fragile lands: New reasons to tackle old problems. *World Development* 20, no. 4:571–85.

Freemuth, J. C. 1991. *Islands under Siege: National Parks and the Politics of External Threats*. Lawrence: University of Kansas Press.

Gradwohl, J., and R. Greenberg. 1988. *Saving the Tropical Forests*. Washington, D.C.: Island Press.

Grumbine, E. R. 1992. *Ghost Bears: Exploring the Biodiversity Crisis*. Washington, D.C.: Island Press.

Heiner, R. A. 1983. The origin of predictable behavior. *American Economic Review* 73, no. 4 (September):560–595.

Holling, C. S., and S. Sanderson. 1996. "The Dynamics of (Dis)harmony in Ecological and Social Systems." In *Rights to Nature*, S. Hannah, ed. Washington, D.C.: Island Press.

Hopkins, J. W. 1995. *Policymaking for Conservation in Latin America: National Parks, Reserves, and the Environment*. Westport, Conn.: Praeger Publishers.

IUCN/UNEP/WWF. 1991. *Caring for the Earth: A Strategy for Sustainable Living*. Gland, Switzerland.

Naughton-Treves, L., and S. Sanderson. 1995. Property, politics, and wildlife conservation. *World Development* 23, no. 8 (August):1265.

Norton, B. G. 1992. Epistemology and environmental values (The intrinsic value of nature). *The Monist* 75, no. 2 (April):208.

Oltremari, J. V. 1996. "Lineamientos de políticas para el desarrollo institucional y el manejo de los sistemas de áreas protegidas en la Amazonia." *Flora, Fauna, y Areas Silvestres* 10, no. 23, pp. 3–9.

Ostrom, Eleanor. 1988. "Institutional Arrangements and the Commons Dilemma." In *Rethinking Institutional Analysis and Development: Issues, Alternatives, and Choices*, V. Ostrum, D. Fenny, and H. Picht, eds. San Francisco: Institute for Contemporary Studies Press.

Padoch, C. and W. deJong. 1987. "Traditional Agroforestry Practices of Native and Ribereño Farmers in the Lowland Peruvian Amazon." In *Agroforestry: Reality, Possibilites and Potentials*, H. Gholz, ed., pp. 179–94. Netherlands: Martinus Nijhoff.

Pagden, A. 1987. *European Encounters with the New World: From Renaissance to Romanticism*. New Haven: Yale University Press.

Ponce, C. F. 1996. *Políticas, estrategias y acciones para la conservación de la diversidad biológica en los sistemas andinos de áreas protegidas*. Santiago, Chile: FAO/Programa de las Naciones Unidas para el Medio Ambiente.

Redford, K. 1992. The empty forest. *Bioscience* 42:412–22.

Redford, K., R. Godschalk, and K. Ascher. 1995. *What About the Wild Animals? Community Forestry in the Tropics*. Rome, Italy: FAO Community Forestry, Note 13.

Redford, K. H. and J. A. Mansour (eds.). 1996. *Traditional Peoples and Biodiversity Conservation in Large Tropical Landscapes*. Washington, D.C.: The Nature Conservancy.

Redford, K. H., and S. E. Sanderson. 1992. "The Brief Barren Marriage of Biodiversity and Sustainability." *Bulletin of the Ecological Society of America* 73:36–39.

Sanderson, S. E., and K. H. Redford. 1997. "Biodiversity Politics and the Contest for Ownership of the World's Biota." In *Last Stand: Protected Areas and the Defense of Tropical Biodiversity*, R. Kramer, C. von Schaik, and J. Johnson, eds., pp. 115–132. Oxford: Oxford University Press.

Seidman, H. 1986. *Politics, Position, and Power: From the Positive to the Regulatory State*. New York: Oxford University Press.

Southgate, D., and M. D. Whitaker. 1994. *Economic Progress and the Environment: One Developing Country's Policy Crisis*. New York: Oxford University Press.

Spooner, B. 1987. "Insiders and Outsiders in Baluchistan: Western and Indigenous Perspectives on Ecology and Development." In *Lands at Risk in the Third World*, P. D. Little and M. M. Horowitz, eds., pp. 58–68. Boulder, Colo.: Westview Press.

Stone, C. D. 1988. *Should Trees Have Standing? Toward Legal Rights for Natural Objects*. Palo Alto, Calif.: Tioga Publishing.

Vitousek, P. M. 1994. Beyond global warming: Ecology and global change. *Ecology* 75, no. 7 (October):1861.

Wells, M., and K. Brandon. 1992. *People and Parks: Linking Protected Area Management with Local Communities*. Washington, D.C.: World Bank.

Worster, D. 1993. *The Wealth of Nature: Environmental History and the Ecological Imagination*. New York: Oxford University Press.

Chapter 16

Agarwal, A. In press. *Community in Conservation: Beyond Enchantment and Disenchantment*. Conservation and Development Forum Working Paper. University of Florida, Gainesville.

Oryx. 1997. 31(2) pp. 94, summarizes a World Conservation Monitoring Center finding as reported in *Plant Talk*, January 1997, pp. 34.

Barzetti, V., ed. 1993. *Parks and Progress: Protected Areas and Economic Development in Latin America and the Caribbean*. Washington, D.C.: IUCN (The World Conservation Union).

Borrini-Feyerabend, G. 1996. *Collaborative Management of Protected Areas: Tailoring the Approach to the Concept*. Gland, Switzerland: IUCN.

Brandon, K. 1997. "Policy and Practical Considerations in Land-Use Strategies for Biodiversity Conservation." In *Last Stand: Protected Areas and the Defense of Tropical Biodiversity*, R. Kramer, C. von Schaik, and J. Johnson, eds., pp. 90–114. Oxford: Oxford University Press.

Bruggemann, J. 1997. "National Parks and Protected Area Management in Costa Rica and Germany: A Comparative Analysis."

FAO (Oficina Regional para América Latina y el Caribe). 1994. Taller internacional sobre políticas de los sistemas de áreas protegidas en la conservación y uso sostenible de la biodiversidad en América Latina, Parque Nacional Iguazú, Argentina, 27 de septiembre al 1 de octubre de 1993. Informe. Proyecto FAO/PNUMA FP/6105-90-38. Santiago, Chile: FAO/RLC. 75 pp. (FOR45).

Gardner, G. 1997. "Preserving Global Cropland." In *State of the World 1997*, Linda Starke, ed., pp. 42–59. New York: W.W. Norton.

Ghimire, K. B., and M. P. Pimbert, eds. 1997. *Social Change and Conservation: Environmental Politics and Impacts of National Parks and Protected Areas*. London: Earthscan Press.

Heathcote, T. 1994. *A Wild Heritage: The History and Nature of the New Forest*. Exeter, England: Ensign Publications.

IUCN (World Conservation Union). 1997. IUCN World Commission on Protected Areas. Newsletter no. 70.

Kramer, R., C. van Schaik, and J. Johnson. 1997. *Last Stand: Protected Areas and the Defense of Tropical Biodiversity*. New York: Oxford University Press.

Kuran, T. 1993. The Unthinkable and the Unthought. *Rationality and Society* 5: 473–505.

MacKinnon, K. 1997. "The Ecological Foundations of Biodiversity Protection." In *Last Stand: Protected Areas and the Defense of Tropical Biodiversity*, R. Kramer, C. von Schaik, and J. Johnson, eds. Oxford: Oxford University Press.

McCloskey, M. 1996. Conservation biologists challenge traditional nature protection organizations. *Wild Earth Winter* (1996–97):67–70.

McNeely, J. A. 1995. "Partnerships for Conservation: An Introduction." In *Expanding Partnerships in Conservation*, J. A. McNeely, ed., pp. 1–10. Washington, D.C.: Island Press.

McNeely, J. A., J. Harrison, and P. Dingwall. 1994. "Introduction: Protected Areas in the Modern World." In *Protecting Nature: Regional Reviews of Protected Areas*, J. A. McNeely, J. Harrison, and P. Dingwall, eds., pp. 1–28. Gland, Switzerland: IUCN.

National Parks and Conservation Association. 1994. *Our Endangered Parks*. San Francisco: Foghorn Press.

Redford, K., R. Godschalk, and K. Ascher. 1995. *What about the Wild Animals? Community Forestry in the Tropics*. Rome, Italy: FAO Community Forestry, Note 13.

Redford, K., and J. Mansour. 1996. *Traditional Peoples and Biodiversity Conservation in Large Tropical Landscapes*. Rosslyn, Va.: The Nature Conservancy.

Redford, K., and B. Richter. 1998. *Conservation of Biodiversity in a World of Use*. Submitted for publication.

Reid, W. V. 1996. "Beyond Protected Areas: Changing Perceptions of Ecological Management Objectives." In *Biodiversity in Managed Landscapes: Theory and Practice*, R. C. Szaro and D. W. Johnston, eds., pp. 442–453. New York: Oxford University Press.

Sanderson, S. E., and K. H. Redford. 1997. "Biodiversity Politics and the Contest for Ownership of the World's Biota." In *Last Stand: Protected Areas and the Defense of Tropical Biodiversity*, R. Kramer, C. von Schaik, and J. Johnson, eds., pp. 115–132. Oxford: Oxford University Press.

Southgate, D., and H. L. Clark. 1993. Can biodiversity projects save biodiversity in South America? *Ambio* 22: 163–66.

Stepanitsky, V. B. 1996. "Zapovedniks of Russia and the Modern State." *Parks* 6:4–7.

Weeks, W. W. 1997. *Beyond the Ark*. Washington, D.C.: Island Press.

Wells, M., and K. Brandon. 1992. *People and Parks: Linking Protected Area Management with Local Communities*. Washington, D.C.: World Bank.

The Editors

Katrina Brandon is a senior fellow with Parks in Peril Program (PiP) of The Nature Conservancy, adjunct professor in the Conservation Biology and Sustainable Development Program at the University of Maryland, and a consultant on conservation and development issues. From 1987 to 1991 she worked as staff at the World Wildlife Fund (WWF) on a range of projects at both policy and field levels that jointly promoted conservation and local socioeconomic development and coauthored a study entitled *People and Parks: Linking Protected Area Management with Local Communities.* She holds an M.A. in inter-American studies from the University of Miami and M.S. and Ph.D. degrees in development sociology and planning from Cornell University. She is currently writing a textbook on incorporating social and policy concerns into biodiversity conservation management.

Kent H. Redford was the director of the Conservation Science and Stewardship Department, Latin America and Caribbean Division, The Nature Conservancy, while this book was written. For several years at the Conservancy, Dr. Redford also directed the USAID PiP, served on the Conservancy's Conservation Committee, and was chair of the Conservancy's Ecoregion Working Group. Prior to joining the Conservancy in 1993, Dr. Redford was an associate professor in the Center for Latin American Studies and Department of Wildlife and Range Sciences at the University of Florida, where he also directed the Program for Studies in Tropical Conservation. He has published extensively in the fields of tropical ecology, neotropical mammalogy, and resource use and conservation by traditional forest-dwelling peoples, including editing or authoring eight books. He holds a Ph.D. in biology from Harvard University. He is currently director of Biodiversity Coordination at the Wildlife Conservation Society in New York.

Steven E. Sanderson is vice president of Arts and Sciences and dean of Emory College at Emory University. Professor Sanderson received his

Ph.D. in political science from Stanford University in 1978. As Ford Foundation program officer for Rural Poverty and Resources in Brazil, he led the design and implementation of the foundation's Amazon program in the mid-1980s. From 1994–1997, he was chair of the Social Science Research Council Committee for Research on Global Environmental Change, and served as a member of the International Scientific Steering Committee of the IGBP-IHDP Land Use and Cover Change Project. Dr. Sanderson has held fellowships from the Fulbright-Hays program, the Rockefeller Foundation, the Smithsonian Institution, and the Council on Foreign Relations, as well as grants from the Ford, Heinz, MacArthur, and Tinker Foundations. He has written three books about Latin American politics, most recently *The Politics of Trade In Latin American Development* (Stanford 1992). He is currently working on a book entitled *The Political Economy of Ecological Resilience.*

Chapter Contributors

Joann M. Andrews, a resident of the Yucatán since 1963, is the founder and former president of Pronatura Península de Yucatán, A.C. As the honorary president of this nongovernment organization (NGO), she remains engaged in many aspects of biodiversity conservation on the Peninsula and continues her long-time interest in the identification of peninsular orchids and their natural habitats.

Shawn Bird is a Ph.D. candidate in political science at the University of Florida undertaking dissertation research on institutional reform and democratization at the municipal level in El Salvador. He has spent extended periods of time in El Salvador and Guatemala. Mr. Bird earned a B.A. from the University of New Hampshire and an M.A. from the University of Colorado.

Timothy Boucher works in the Landscape Ecology Program at the Conservancy as the landscape ecology assistant. A South African, Mr. Boucher holds a bachelor of commerce from the University of Witwatersrand, South Africa. He previously worked at the Conservation and Research Center (CRC) of the Smithsonian Institution.

César Cuello was the director of programs at Fundación Neotrópica in Costa Rica, where he supervised the foundation's field activities. Dr. Cuello received his Ph.D. from the University of Delaware in 1997, where his dissertation focused on sustainable development in the Osa Peninsula. He is the founder and executive director of Fundación Tecnología y Naturaleza, an NGO based in Costa Rica.

Barbara Dugelby was the Local Peoples Specialist in the Latin America and Caribbean Division at the Conservancy. Dr. Dugelby holds a Ph.D. in tropical ecology and conservation from Duke University and conducted extensive research on the Maya Biosphere Reserve. She now works as an ecologist at The Wildlands Project.

Elba Alissie Fiallo was the technical coordinator of the Biodiversity and Protected Areas Program of Fundación Natura in Ecuador. She holds a master's in Latin American studies from the Tropical Conservation and Development Program at the University of Florida, Gainesville. She now resides in Greece.

Brian L. Houseal is vice president and regional director for Mexico at The Nature Conservancy. Mr. Houseal has spent over twenty years working on protected area management projects throughout the Latin American and Caribbean region, and is the architect and former director of The Nature Conservancy's PiP program.

Kelvin A. Guerrero is the executive director of EcoParque, an NGO that supports Del Este National Park in the Dominican Republic. Prior to that, he was the project director for the PiP program there. He has worked as a biologist on numerous projects and is the curator for insects at the National Museum of Natural History in Santo Domingo. He holds a licenciature in biology from the Universidad Autónoma of Santo Domingo.

Andreas Lehnhoff is currently the director of the Guatemala Program and regional policy advisor at The Nature Conservancy, after serving for four years as executive director of Defensores de la Naturaleza. Prior to this he was the first executive secretary of the Guatemalan National Protected Areas Council (CONAP). He wrote the Sierra de las Minas case study during his stay as a Fulbright fellow, completing his masters at Duke University.

Xiaojun Li is the senior landscape ecologist at the Latin American and Caribbean Division of the Conservancy. Prior to that he was a postdoctoral research associate at the U.S. Fish and Wildlife Service. He holds a Ph.D. in plant ecology from Michigan Technical University and is the author of over 20 articles.

Michelle Libby, a Costa Rican citizen, is the Community Conservation Program associate in the Latin America and Caribbean Division of the Conservancy. She holds a master's degree in Latin American studies/social and economic development from Tulane University.

Jane A. Mansour is a conservation science specialist in the Conservation Science and Stewardship Department in the Latin America and Caribbean Division of the Conservancy and is coeditor with Kent Redford of *Traditional Peoples and Biodiversity Conservation in Large Tropical Landscapes.* She holds a master of science degree in conservation biology from the University of Maryland.

Richard Margoluis is the director of the Biodiversity Support Program's (BSP) Analysis and Adaptive Management Program, where he manages BSP's portfolio of applied conservation and development research projects and its strategic planning, monitoring, and evaluation. Dr. Margoluis has a Ph.D. from Tulane University, where his academic training combined Latin American studies, international public health, and epidemiology.

Rodrigo Migoya Von Bertrab was the coordinator of the PiP program at Rías Lagartos and Celestún from May 1994 until September 1996 for Pronatura Península de Yucatán, A.C. He is now the reserve coordinator for the National Institute of Ecology for those sites.

Adolfo Moreno was the director of Amboró National Park in 1992 and from 1993 until 1997 he was head of the Conservation Department of Fundación Amigos de la Naturaleza (FAN), supervising activities in Amboró National Park, the surrounding area, and Noel Kempff National Park, where he also heads the Climate Action Project. His degree is in agronomy from the Federal University of Santa Maria, Brazil.

Lisa Naughton-Treves is an assistant professor of geography at the University of Wisconsin, Madison. She holds a Ph.D. in wildlife ecology from the University of Florida, Gainesville. Dr. Naughton-Treves formerly worked for the Wildlife Conservation Society, where she designed field research projects in conservation and consulted with international development agencies regarding local community participation in biological resource management.

Oscar Manuel Núñez Saravia is the executive director of Defensores de la Naturaleza in Guatemala. Prior to that he was Defensores's director of the

Sierra de las Minas Biosphere Reserve and a researcher and faculty member at the School of Agronomy of the University of San Carlos. Mr. Núñez holds a master of science degree in forestry from the University of Viçosa, Brazil, and an engineering degree in natural resource management from the University of San Carlos, Guatemala.

Mónica Ostria is director of operations for the PiP program, where she has worked with the Conservancy's Latin America and Caribbean Division since 1987 as part of the Stewardship Department. Ms. Ostria has been involved in the PiP program since the design phase and since 1991 she has been the administrative director, preparing technical proposals, reports, conducting financial analyses, and coordinating Conservancy relations with partner organizations.

Susana Rojas has been the general director of Pronatura Península de Yucatán, A.C. since 1994, where she is responsible for program supervision, fundraising, and general supervision. A biologist, she has experience in protected area conservation and environmental education.

Debra A. Rose is currently program officer for the Conservation and Development Forum, a joint program of the University of Florida and the Ford Foundation. Dr. Rose received her Ph.D. in political science at the University of Florida. Her dissertation was entitled *The Politics of Mexican Wildlife: Conservation, Development, and the International System.*

Armando Sastré Méndez received a licenciature in administration from the Universidad Nacional Autónoma in Mexico. He has worked at Pronatura Península de Yucatán, A.C. since 1992 and is currently the general coordinator for projects at the Calakmul Biosphere Reserve.

Roger Sayre is the director of biodiversity conservation in the Conservation Science and Stewardship Department of the Latin America and Caribbean Division of the Conservancy, where he is completing a manual for conservation practitioners on rapid ecological assessment. Dr. Sayre holds a Ph.D. in natural resources from Cornell University.

Stuart Sheppard is a geographic information system (GIS) specialist in Conservation Science and Stewardship in the Latin America and Caribbean Division of The Nature Conservancy. His degree is in geography from George Mason University.

Bolívar Tello is in charge of environmental education and community development for the Fundación Ecológica Arcoiris in Loja, Ecuador, where he works on issues related to Podocarpus National Park.

Jerome Touval is the Colombia program director and regional protected areas specialist at The Nature Conservancy. Prior to joining the Conservancy, he worked as an international affairs specialist for the U.S. Fish and Wildlife Service, where he was responsible for courses on conservation and environmental education for Latin American and Caribbean biologists and wildlife administrators. Mr. Touval holds a master of science degree in conservation biology from the University of Maryland.

Audrey E. Wallace is the senior project coordinator in administration and planning at Programme for Belize (PfB), where she is responsible for the administration and planning, research coordination, and marketing of PfB's nature tourism program. Ms. Wallace holds a master's degree in marketing from St. John's University, New York, where she also worked for Bankers Trust.

Luis Angel Yallico Madge is the technical director of Pronaturaleza, a Peruvian NGO, where he is responsible for protected area planning and conservation and development activities. A forester by training, he also received a master's in resource management from the University of Edinburgh, Scotland. Prior to that he was the director of Manu National Park in Peru.

Index